GW01551001

B.A.W. COLLER I.R. McKINNON I.R. WILSON

PRINCIPLES OF PHYSICAL CHEMISTRY

Edward Arnold

First published 1976 by
Holt Rinehart and Winston
a division of Holt Saunders Pty Ltd
9 Waltham St., Artarmon, NSW, Australia, 2064
First published in Great Britain 1978 by
Edward Arnold (Publishers) Ltd
41 Bedford Square, London WC1B 3DP

British Library Cataloguing in Publication Data

Coller, Bruce Arthur William
Principles of physical chemistry.
1. Chemistry
I. Title II. McKinnon, I R III. Wilson, I R
541'.3 QD453.2

ISBN 0-7131-2699-X

Preface

In this account of physical chemistry we have chosen to emphasise the behaviour of real materials, the patterns which may be discerned in this behaviour and their interpretation in terms of simple pictures (models) of the experiment which are able to "reproduce", "account for" or "explain" the observed general patterns of behaviour.

This approach differs radically from that used in another group of texts covering the same topics. Such texts follow this sequence in reverse. They begin with mental models constructed from atoms, molecules, nuclei and electrons, in such imaginary materials as "ideal gases", "ideal solutions" and "perfect solids". Actual experiments on real materials are discussed in terms of the way in which the models have to be modified to represent experimental behaviour. Properly handled, both approaches give the same answers to an actual experimental problem. The two approaches differ partly in priorities of teaching and partly in matters of philosophy. One approach emphasises the world around us and the other emphasises scientific theories as an approximation to ultimate truth. Both approaches are appropriate at the right time. Our belief is that our students are in more need of the approach we have taken here.

In this book we will seek to develop and weave together four major themes:

1. the general methods which are used to describe and classify chemical composition and its change;
2. the ways physical laws are used to summarise experimental situations;
3. the ways in which scientific concepts develop from everyday experiences and primitive ideas; and
4. the ways in which structural models are invented and used to interpret and correlate physical and chemical behaviour.

The experiments around which these themes are developed are shared by many elementary courses in physical chemistry. We believe that the approach we are using is new in many respects and that the changes we have made are worth the effort involved in the change.

People who come to the study of basic physical chemistry are often confused by parts of the texts (whether ours or those written by others). We have found that their confusion is often due to inadequate *ideas* in the texts and other resources used by the students and is not caused by the inadequacy of the writer or the inadequancy of the reader. Usually such learning difficulties are not taken seriously as symptoms of fundamental error in the ideas being presented. We believe they should be. We have found that careful attention to the precise points at which learners find inconsistencies and conflicts leads us to increase our own understanding of the principles of the subject and of their ranges of usefulness. We trust that others will be stimulated to join us on this fascinating journey.

Acknowledgments

We express our indebtedness to our many students, friends, associates and colleagues who have read and criticised the manuscript.

The words were made legible by Mrs Edna Peebles, Mrs Sharann Lampkin and Mrs Jean Parker.

The illustrations were designed by Mr Ben Baxter and his assistants.

Contents

1. Describing Chemical Systems

1.1 The Viewpoint of Physical Chemistry

1.1.1 Different viewpoints

We find it useful to think of each science as having a "view", looking out over everything from its own "lookout tower". In chemistry we classify materials according to the nature of the chemical substances they are made from, and we study the transformations which those substances undergo. So in chemistry we look at a forest and see not only trees but wood, cellulose, chlorophyll, water and air, undergoing changes by photosynthesis, transpiration and decay. In a limestone cave we see the beauty of the stalactites and stalagmites and also detect the removal of calcium salts from rocks and their decomposition again as calcium carbonate.

Where does physical chemistry fit in? In physical chemistry we develop and employ precise methods for describing and classifying materials. We use quantitative measurements as our basis for developing wide-ranging relationships between the properties of materials and their compositions. We develop interpretations of these patterns of behaviour by inventing structural models based on the modern physics of atoms and molecules, ions and electrons.

Query What is the use of physical chemistry?

Reply Dispelling fear of the unknown is in large measure the justification for any science. In addition there is the pleasure of finding new knowledge and the satisfaction of understanding. And there is the practical use by people—the technology. The quantitative chemical study of materials gives rise to new materials and new uses for existing materials.

Comment Then we use physical chemistry in most chemical study?

Reply Yes, and in other sciences such as botany, zoology, genetics, geology, physiology, materials engineering, electrical engineering and chemical engineering.

1.1.2 The analysis of real situations

Real situations present us with real problems —not with "chemical problems", "biological problems", "engineering problems" or "sociological problems". Real problems—like land use in an area of low rainfall—very often have "chemical aspects", "biological aspects", etc. There is insight to be gained by looking at such problems from the viewpoint of chemistry, from the viewpoint of biology, etc. Often we look at these aspects of a particular problem separately and combine the results.

The chemical study of a situation is often split into separate examination of the chemical transformations involving particular elements or substances. Examples include study of the "nitrogen cycle", the "carbon cycle", the "water cycle", or processes involving molybdenum or chlorine. The total chemical behaviour is then taken to be the sum of the separate behaviours, together with interactions between them. Study of a motor car is likely to involve separate treatment of the functional parts—steel frame, fuel intake system, gearbox, lubrication, tyres, etc., an analysis of what each part is made of, of what changes take place in the parts and of how change in each part is coupled with changes in others, under the kinds of conditions which exist during use.

1.1.3 Stages in using physical chemistry

Physical chemistry turns up at a number of stages in the study of a material. It usually comes in first in describing and cataloguing zones which have different states of aggregation (the different *phase types*). Figure 1.1 shows some examples.

The next step is the description of chemical compositions both of the whole body of material and of the individual parts (Fig. 1.2). Real situations are seldom static, and so the study usually involves the measurement of amounts and rates of chemical changes. These changes may be transformations of substances due to chemical reaction (Fig. 1.3), transformation of one phase type into another (Fig. 1.4) or transfer of substances from one phase to another (Fig. 1.5).

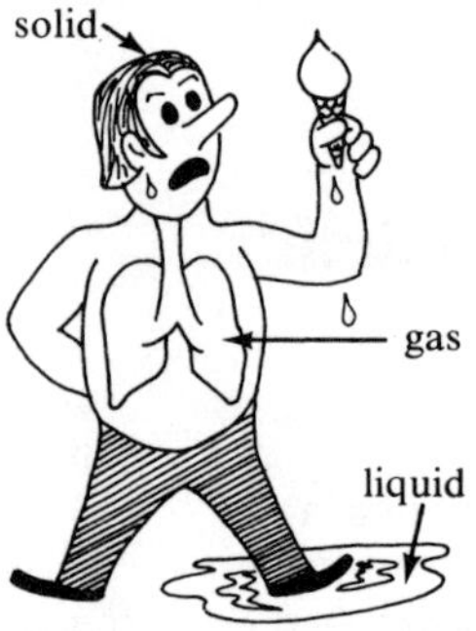

Fig. 1.1 Phases.

Fig. 1.2 Compositions of phases.

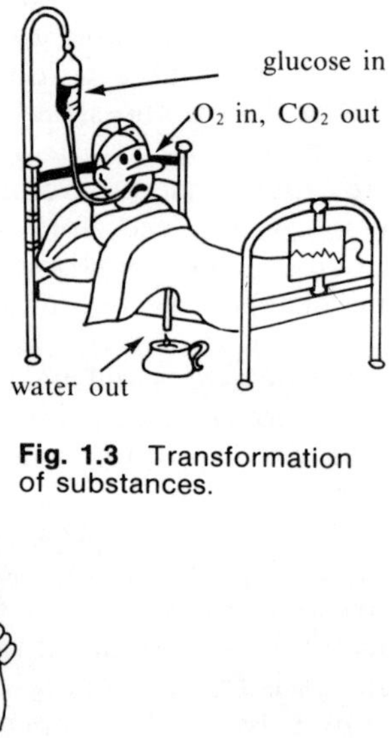

Fig. 1.3 Transformation of substances.

Fig. 1.4 Transformation of phases.

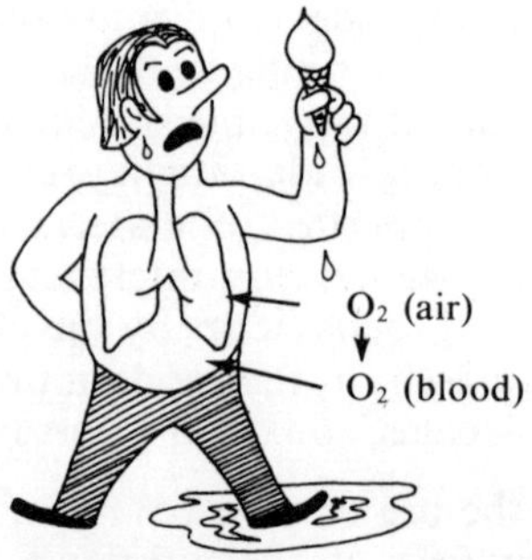

Fig. 1.5 Transfer of substances.

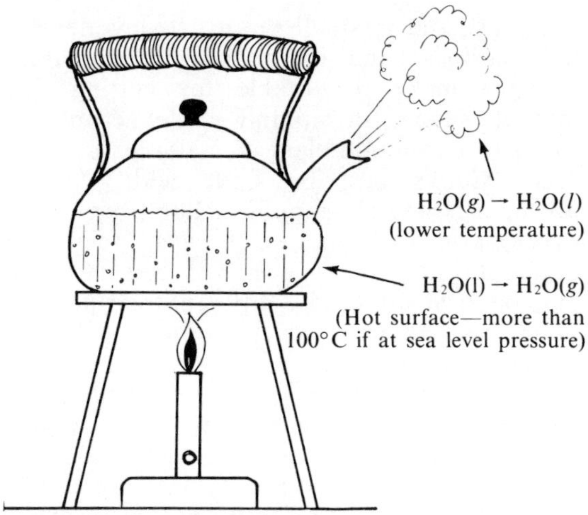

Fig. 1.6 Temperature and pressure often determine which phase is present.

Fig. 1.7 Phases present may control temperature — $H_2O(s) = H_2O(l)$.

A third step in the examination of a material may be description of other properties such as volume, density, temperature, pressure, energy and the rate at which changes occur. These both affect (Fig. 1.6), and are affected by (Fig. 1.7), the nature of the phases present and their chemical compositions.

The fourth step involves description of the interactions of the different parts of the sample or system with each other and with their environment (Fig. 1.8). These interactions are an important part of any examination from the physicochemical viewpoint. By controlling the environment (e.g. by adjusting the temperature and pressure) we can often control the contents. Hence we may be able to control the nature and chemical compositions of the phases.

Careful description is always essential to understanding. Most of physical chemistry, and hence most of this text, is concerned with describing and correlating the nature and behaviour of common materials in common situations. The more quantitative the description the more precise the language must be. The most precise language is that of mathematics. That is why quantitative ideas are often expressed in a mathematical form.

Some words which are used in physical chemistry have been altered in their meanings as the study has developed. This is true of any area of science. Failure to appreciate the differences between the scientific meaning and the common usage is one of the usual problems in gaining an understanding. So we will use the next section to illustrate this aspect of the development of scientific ideas.

Fig. 1.8 Parts of a system often interact.

1.2 The Development of Scientific Language

All sciences share a pattern of development of the language in which they express their ideas. When an idea is first expressed, it is described if possible in words which have common use outside the science. This enables the idea to be fitted into earlier experience. As the idea develops, the words used acquire precise meanings which may be very different from the ordinary ones. In fact the "common use" may become a hindrance. New, strange words are often introduced in the later development of an idea. By this stage it is as though the idea has developed independent existence; it has become a concept. In this section we take up, as examples, the ideas that have led to the important scientific concepts of *temperature*, *phase* and *material*.

1.2.1 The development of the idea of temperature

The concept of temperature has grown from the ordinary ideas of hot, warm, cool and cold. These need no explanation because they are defined by our senses, and this direct experience is the reason we start with this particular concept rather than a more "chemical" one.

The degree of hotness or coldness has been given the name *temperature*. Our feelings do not define it adequately, and so we need an instrument to measure temperature. Thus the definition changes, and "temperature" becomes "the reading of a thermometer". What kind of thermometer? Should it be mercury in glass, gas in constant volume, a platinum resistance, a thermocouple, a thermistor or a bimetallic strip? In this sense there can be as many definitions of temperature as there are thermometers. A great deal of effort goes into construction of thermometers which give compatible readings. For the most highly precise work, the measurement of temperature is defined in terms of standards set up by international agreement. The present agreement is "the International practical temperature scale of 1968". It defines a set of standard thermometers, how they are to be used, and over what ranges. It replaces a previous scale which was defined in 1948.

1.2.2 The development of the ideas of materials

We use the concept of materials to represent the kinds of matter in the situation with which we are concerned. The primitive idea of "a material" stems from our everyday experience. We know, without need for definition, that a given body of material is meat or milk, sand or sea water, air or wood.

A material can often be described by its origin and by its properties. For example, *balsa wood* is the kind of timber which originates from the particular kind of tree, *Ochroma lagopus*, and has characteristic properties which are similar in large logs and in small scraps. Its low density, low hardness and reasonably high breaking strength make it suitable for cutting into intricate shapes for building model aeroplanes. *Flour* is a common class of materials used for food. Cooks describe flour with care to distinguish between supplies made from wheat, rice, maize or potatoes, because they give different results. It may even be necessary, in the making of bread, to give different treatments to flours from different varieties of wheat. Among the physical properties of materials which may be of interest to the user are such attributes as the density and hardness or viscosity and elasticity, the electrical and thermal conductivities, the colour and texture, etc. In Chapter 2 we give particular attention to the properties which relate to the filling of space by particular materials.

Example 1.1
Give a brief description of the material milk, in terms of its origin and physical properties.

Answer
Milk is the liquid material secreted by the females of some species of animals (mammals) to feed their young.

Exercise 1.1
Describe the material cream, in similar terms to the above.

In chemistry we seek also to describe materials by analysing their compositions. The analysis can be carried to several successive levels. The first level of analysis is the differentiation of the material into parts or types of particles. For example, if we inspect closely a piece of good steak we see particles of protein and particles of fat. Thus, steak is a *composite material* made up of two different materials: protein and fat. Fibreglass, concrete and reinforced concrete are composite materials which are deliberately manufactured for their special mechanical properties and low cost. The collections of particles of each type of material in a given situation are what we refer to in this book as the *phases of material*. Steak has a phase of protein and a phase of fat. Fibreglass has a phase of glass wool and a phase of plastic; concrete has a phase of cement, a phase of sand and phases of crushed rock; reinforced concrete also has a phase of steel rods.

The phases of material in a given situation

may separate, under gravity, into layers. An example is the ice, water, salt bath (Fig. 1.9) which is often used for cooling in chemistry laboratories. Solid salt particles collect at the bottom. Among and above them will be the phase of liquid salty water. The remaining particles of solid ice rise to the top. Above these is the gaseous phase of air.

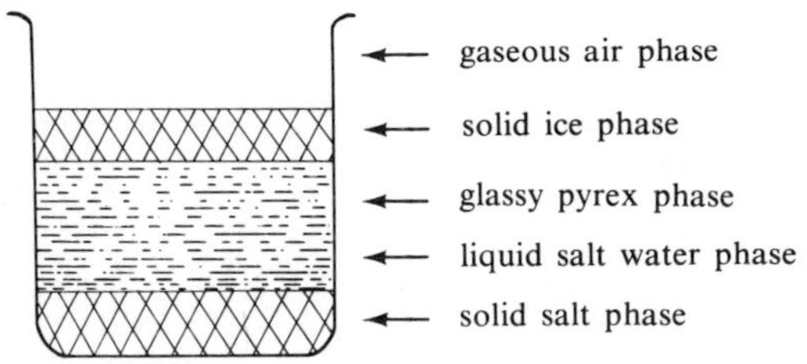

Fig. 1.9 Phases of material in a layered composite body of matter.

1.2.3 The development of the idea of phases

One of the simplest ways we have of classifying materials depends on their appearance and their flow properties. For example, a material which appears "free from cavities or empty spaces" and which retains its shape when pushed (such as wood or granite), we call a *solid.* If the substance appears to be "free from cavities" but flows when pushed (say water, orange juice, tomato soup or wet concrete), we call it a *liquid.* On the other hand, if it is distributed throughout that part of the container not occupied by "solid" or "liquid", flows when pushed, and is easily compressed (like steam, air, or natural gas), we classify it as *gas* or *vapor.*

Whether a given material exhibits the properties of a solid, liquid or gas phase type depends on the environment of the material and on its previous history. For example, at atmospheric pressure, water behaves as a solid, a liquid or a gas depending on whether the temperature is below 273K, between 273K and 373K, or above 373K. These three different forms of water are known respectively as its solid phase type, liquid phase type and gaseous phase type.

Query What does the word *phase* mean?

Reply The word *phase* is used to denote different aspects of an object or differences in its appearance. Two examples of this use are "the phases of the moon" and "the phases of material in a given situation".

Query Does this mean that separate objects are different phases?

Reply No. It is not *objects* which are being classified but the *material* from which they are constructed. For example, the crystals of sucrose in a bowl of sugar have the same form, or appearance, and hence are of the same phase. The water drops in a fog belong to the same phase (the water in each drop has the same properties as that in all the other drops). An ice-cube in a cocktail, an iceberg, and the snow in a snowfield, all consist of water in its solid phase.

In order to describe materials more accurately it is necessary to increase the number of phase types. As this happens, the names used for them become more descriptive and more closely defined. A simple example is a mixture of petrol and water at room temperature and pressure, as illustrated in Fig. 1.10. It is clear that there are two distinct liquid phases present. Experiment indicates that the lower phase has the higher density and consists mainly of water with a small amount of petrol dissolved in it. The upper phase has the lower density and is petrol containing a small quantity of water. The names given to the phases ("water-rich phase" and "petrol-rich phase") describe their chemical composition. Solids also may be of different types. There may be different crystalline forms of a single substance: solid carbon can be diamond or graphite; and calcium carbonate can be calcite or aragonite. Solids may also be distinguished as crystalline, glassy or plastic. Sometimes it is useful to talk of the fraction of a

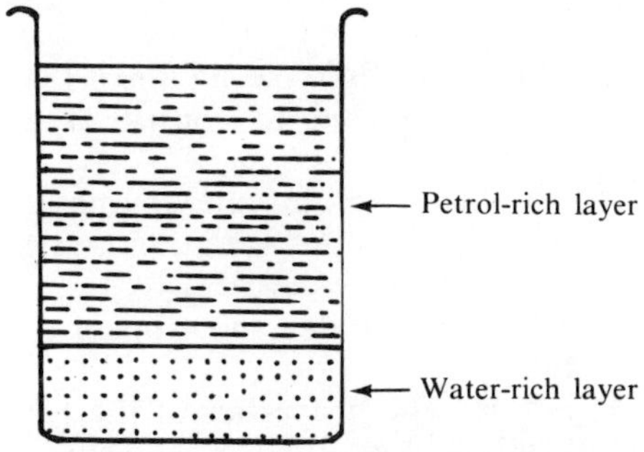

Fig. 1.10 Mixtures of petrol and water usually have two phases.

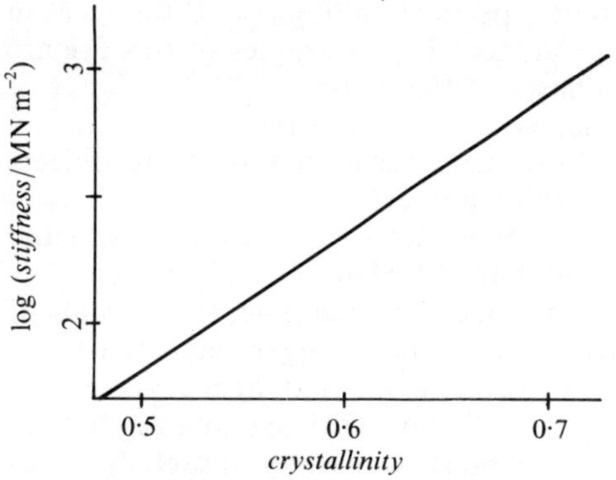

Fig. 1.11 Effect of degree of crystallinity on stiffness of polyethylene.

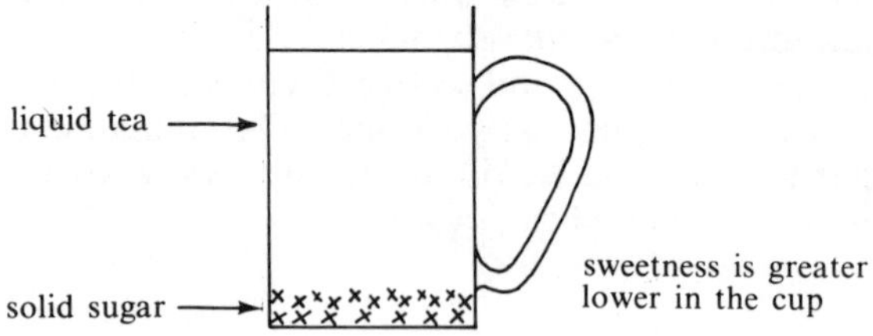

Fig. 1.12 Unstirred cup of tea with sugar added (solid sugar and tea!)

solid material which is crystalline. For example, an analysis of the effect of this "degree of crystallinity" on the stiffness of a sample of polyethylene is illustrated in Fig. 1.11.

While differences in physical form or in crystal structure give clear criteria for deciding that two samples belong to different phases, there are many situations which are not so obvious. Take, for example, an unstirred cup of well-strained tea to which sugar has been added (Fig. 1.12). There are two obvious regions with different properties (the solid sugar and the tea). However samples drawn from the top and bottom of the tea will have quite different properties (such as sweetness) and so it could make sense to classify them as belonging to different phases. The logical conclusion to this line of argument is to state that a sample consists of a single phase if, and only if, all the properties of its material are uniform throughout.

Query Does this mean that the liquid in the cup of tea might be said to consist of a number of phases?

Reply Yes. In fact, an infinite number, because there is a continuous variation in the concentration of sugar.

Comment That definition does not appear to be very useful.

Reply That is true. The properties of the tea change uniformly and gradually as the sugar concentration changes. Usually it is not profitable to call each region of different temperature and composition a separate phase.

Comment That reply appears to contradict the answer to the previous question.

Reply The key word is "profitable". Sometimes we need one definition of *phase*, sometimes another.

We have seen three stages in the development of the idea of *phase* as a means of classifying samples:

1. according to gross difference in appearance and mechanical behaviour of the material (solid, liquid or gas);
2. according to more closely specified differences in the form of the material (such as various crystalline forms, glasses, different liquid phases, etc.) and
3. according to differences in any measurable property of the material.

At each stage of the development, the descriptions of the phases and the criteria for classifying samples as belonging to various phases are made more precise. All three degrees of classification into phases are commonly used. The most appropriate one depends on the situation being examined and the reason for making the classification.

1.3 Describing Chemical Composition

Among the phases discussed in the previous section are several which chemists normally recognise as pure chemical substances. The main criterion used to identify a pure chemical substance is that its properties are definite and characteristic and that at least some of them do not merge gradually with the properties of other uniform materials. The major tool in identifying such pure substances, and in specifying other materials as well, is an examination of chemical composition. In the example illustrated in Fig. 1.12, the chemical composition of the solid sugar is essential to its recognition as a pure compound. Description of the tea in quantitative terms can only be made if we know things like its sugar content.

In order to describe materials and chemical substances, we seek their qualities or *intrinsic properties*. Other properties depend directly on the amounts of the materials in a particular system and we call them the *extensive properties* of the system. Masses and volumes of samples of material are extensive properties. Twice as much blood, measured by volume, is twice as much mass of blood. Densities of materials are "intrinsic properties". Similar blood samples of different sizes have the same density.

In this section we deal with the methods which are used to describe the compositions of materials and of bodies of materials in terms of concentrations (intrinsic properties) and amounts (extensive properties) of substances. When we are interested in the quality of a given material, it is likely to be more convenient to describe the composition in terms of the concentrations of the component substances. We begin by describing several alternative ways which are used for specifying concentrations.

1.3.1 Mass fractions of substances in a material

The most basic physical description of the composition of a body of material is a list of the mass fractions of its substances. That is, we list the masses, m_A, m_B. . ., of the substances present in a given total mass, m. To describe the quality of the material we list the mass fractions, w_A, w_B. . ., obtained by dividing the mass of each substance by the total mass m, of the body:

$$w_B = m_B/m$$

The complete list of mass fractions of the component substances specifies the composition of the particular material. This is often enough to identify the material itself.

1.3.2 Formulae for chemical substances

The basis for the growth of chemistry is the demonstration by Antoine Lavoisier, during the 1770s, that there is conservation of mass during chemical change and his development from this of the principle that the amounts of the chemical elements are conserved in chemical reactions. Thus chemical substances can be partly identified by identifying their component chemical elements and by determining the mass fractions or mole fractions of these elements. Thus, the formulae NaCl, H_2O and CO_2 are commonly used without further description of the corresponding substances. One formula may represent more than one substance, as $C_6H_{12}O_6$ applies to glucose, fructose, maltose, galactose, mannose and other sugars. It is then necessary to give further information to assist identification. When discussing general principles we use the symbol "B" to indicate the formulae and the identities of substances in a generalised way.

The chemical formula for a substance includes a coded statement of the mass fractions of its constitutent elements. Decoding of a formula involves multiplying the coefficient of each element by the appropriate "relative atomic mass" (A_r, Appendix A5) to obtain the relative formula mass, and the mass fraction of each constituent.

Example 1.2
The formula of ferric alum is given as $K_2SO_4{\cdot}Fe_2(SO_4)_3{\cdot}24H_2O$. What is the mass fraction of the element iron in ferric alum?

Answer
For making calculations the formula may be written more concisely as $K_2S_4O_{40}Fe_2H_{48}$. We identify the elements by their symbols and find the atomic weights in Appendix 5.

symbol	*element*	*coefficient*	× *"atomic weight"*
K	potassium	2	× 39.098
S	sulfur	4	× 32.06
O	oxygen	40	× 15.994
Fe	iron	2	× 55.847
H	hydrogen	48	× 1.0079

relative formula mass = 1006.27

Thus, the mass fraction of iron in ferric alum is

$$w_{Fe}\,(\text{ferric alum}) = \frac{2 \times 55.847}{1006.27}$$

$$= 0.110$$

Exercise 1.2
Calculate the mass fraction of the element carbon in the substance glucose ($C_6H_{12}O_6$).

Answer 72.066/180.126 = 0.4009

1.3.3 Molar masses and amounts of chemical substances

In the above example we have used the known formula of a substance to calculate one of its properties. Another property of a substance, which is implied by its formula, is the molar mass, $\bar{M}_B$. This quantity is equal to the mass of one mole for the given formula, B. The mole of a substance is defined (see Section 1.5.2) in such a way that the molar mass is simply the relative formula mass multiplied by 10^{-3}kg mol^{-1}. Thus the molar mass of ferric alum is

$$\bar{M}_{K_2SO_4.Fe_2(SO_4)_3.24H_2O} = 1006.316 \times 10^{-3} \text{kg mol}^{-1}$$

The molar masses of substances are used as the scale for calculating chemical amounts or mole numbers, n_B. Thus, for each substance in a body of material the mole number is

$$n_B \equiv m_B / \bar{M}_B$$

Exercise 1.3
Use the table of relative atomic masses to find the molar mass of cane sugar ($C_{12}O_{11}H_{22}$) and calculate the chemical amount of cane sugar in a 1 kg packet.

Answer
1 kg/342.299 g mol^{-1} = 2.92 mol

1.3.4 Mole fractions of chemical substances

When we know the amounts of chemical substances (their mole numbers), the compositions of materials may be described by listing the mole fractions x_B, of their components. To obtain the mole fractions, we divide the chemical amount (mole number) of each substance by the total of all the chemical amounts:

$$x_B \equiv n_B / n$$

with

$$n = n_A + n_B + \ldots$$

Example 1.3
A sample of formalin contains 15.0 g formaldehyde ($\bar{M}_{CH_2O}$ = 30.0 g mol^{-1}) and 3.6 kg water ($\bar{M}_{H_2O}$ = 18.0 g mol^{-1}). What are the mole fractions of formaldehyde and water in the formalin?

Answer
We require $x_{CH_2O} = n_{CH_2O}/n$.

The mole number of formaldehyde is

$$n_{CH_2O} = m_{CH_2O}/\bar{M}_{CH_2O} = 15.0 \text{ g}/30.0 \text{ g mol}^{-1} = 0.50 \text{ mol}$$

The mole number of water is

$$n_{H_2O} = m_{H_2O}/\bar{M}_{H_2O} = 3.6 \times 10^3 \text{ g}/18.0 \text{ g mol}^{-1} = 200 \text{ mol}$$

The total mole number of the sample is 200.5 mol.
The mole fraction of formaldehyde is thus

$$x_{CH_2O} = 0.50 \text{ mol}/200.5 \text{ mol} = 2.5 \times 10^{-3}$$

The mole fraction of water is

$$x_{H_2O} = 200/200.5 = (200.5 - 0.5)/200.5 = 1 - 2.5 \times 10^{-3} = 0.9975$$

The mass fractions, w_{CH_2O} and w_{H_2O}, can be calculated much more easily than the mole fractions but the mole fractions are more useful in explaining chemical change and the behaviour of solutions and of gas mixtures than the mass fractions.

1.3.5 Molalities of chemical substances

When dealing with solutions we often have one substance at high mole fraction (the solvent, A) and one or more substances at low mole fraction (the solutes, B, B′, B″...). The compositions of such solutions can be conveniently specified on another scale of concentration, the *molality*, m_B. The molality of a solute is its mole number, n_B, divided by the mass of solvent, m_A, in which it is present,

$$m_B \equiv n_B / m_A$$

Example 1.4
A 10 g sample of formalin (w_{CH_2O} = 0.37) is diluted by adding it to 1 kg of water. What is the molality of formaldehyde in the diluted solution?

Answer

$$n_{CH_2O} = 3.7\ \text{g}/30.027\ \text{g mol}^{-1} = 0.123\ \text{mol}$$

$$m_{CH_2O} = 0.123\ \text{mol}/1\ \text{kg water} = 0.123\ \text{mol (kg water)}^{-1}$$

When solutes are present at low concentrations, simple relationships between the scales of mass fraction, mole fraction and molality become increasingly accurate. Thus

$$w_B \equiv m_B/m = m_B/(m_A + m_B)$$
$$\simeq m_B/m_A$$
$$x_B \equiv n_B/n = n_B/(n_A + n_B)$$
$$\simeq n_B/n_A = m_B\bar{M}_A/m_A\bar{M}_B$$
$$\simeq w_B\bar{M}_A/\bar{M}_B$$
$$m_B \equiv n_B/m_A = n_B/\bar{M}_A n_A$$
$$\simeq x_B/\bar{M}_A$$
$$\simeq w_B/\bar{M}_B$$

These are examples of "limiting relationships". Here such relationships follow from the definitions of the quantities involved. In later chapters we meet examples of limiting laws which are found by experiment.

1.3.6 Volume concentrations of chemical substances

The most commonly used method for describing the compositions of dilute liquid solutions is by the volume concentrations (or mole densities) c_B of each of the dissolved substances (solutes). These are simply the mole numbers of the substances divided by the volume of material in which they are contained,

$$c_B \equiv n_B/V$$

Example 1.5

The suppliers of formalin describe their product as 37% formaldehyde (mass fraction of CH_2O $= 0.37$) with density $\rho = 1.09\ \text{g cm}^{-3}$ at 20°C. What is the volume concentration of formaldehyde in their product?

Answer

We require

$$c_{CH_2O} = n_{CH_2O}/V$$
$$= m_{CH_2O}/\bar{M}_{CH_2O}V$$
$$= w_{CH_2O}m/\bar{M}_{CH_2O}V$$
$$= w_{CH_2O}\,\rho_{\text{formalin}}/\bar{M}_{CH_2O}$$

Thus

$$c_{CH_2O} = 0.37 \times 1.09\ \text{g cm}^{-3}/30.0\ \text{g mol}^{-1}$$
$$= 0.0134\ \text{mol cm}^{-3}$$
$$= 13.4\ \text{mol dm}^{-3}\ \text{(at 20°C)}.$$

The density of a material of given composition alters with change in the temperature or pressure. Thus, the volume concentration of the solutes in a solution usually vary when these conditions are changed.

It has been accepted practice to express concentrations of chemical solutions in terms of moles per litre (that is, per 1 dm^3), rather than in moles per m^3. The special name *molarity*, symbol M, has been used for this scale. Thus, a 1 M solution of urea contains 1 mole NH_2CONH_2 per 1 dm^3 of solution. The name and symbol are not favoured by those who are developing the SI units (Section 1.5.2) and may not survive. We will not use it.

1.3.7 Two examples of composition

The ideas of composition presented above are very important. We therefore take two examples, soap and cement, to show how composition is used with common materials.

Originally the term "soap" was used for the material obtained by treating animal fats or vegetable oils with a solution of alkali, but in modern usage the term "soap" is applied to any material which consists of the salts of long-chain fatty acids. A typical specification of such a soap, in terms of its chemical composition, is given in Table 1.1.

Table 1.1
Composition of 100 g of a Soap Sample

water	19.26
fatty acid anions (RCO_2^-)	54.52
OH^-	0.09
CO_3^{2-}	0.11
Na^+	6.57
resin anions	19.45
Total	100.00

In this case, as often happens, the methods used for chemical analysis make it possible to determine the total mass of all the fatty acid anions. They do not give the separate amounts of the individual anions (stearate, oleate, palmitate).

A contract for the supply of "cement" normally lists the chemical composition as a major part of the specification of its properties. Table 1.2 is an example of such a statement of composition. Here the method of analysis used gives the amounts of each of the elements present. It is traditional to state the composition as though each element were present in its anhydrous oxide, that is, in the form given in Table 1.3. This does not state the chemical substances present in cement. If that is needed, a better representation describes the calcium, aluminium and iron as being present in the form of the compounds $(CaO)_3\,SiO_2$, $(CaO)_2\,SiO_2$, $(CaO)_4\,Al_2O_3\,Fe_2O_3$, $CaSO_4$ and CaO. Based on this assumption, the composition of the sample would be stated as in Table 1.4. The exact procedures by which the analysis is to be carried out and the methods to be used to present the results of the analysis are laid down by the Standards Association of each country. The Australian Standard (AS 2) carries the warning, "The compound percentages based on the hypothetical combinations, as calculated, do not necessarily mean that the oxides are actually, or entirely, present as such compounds". That is a warning that tests of other kinds may be necessary to establish the suitability of a cement for a particular purpose.

In all the tables describing cement composition the term *ignition loss* occurs. Before analysis the sample is strongly heated and the weight changes. This is due to loss of H_2O and, probably, of CO_2, both compounds which are absorbed again as the cement "sets". Reliable analysis does not need to take these compounds into account, but does need to remove uncertainty about their presence in the sample.

Table 1.2
Elemental Composition of 100 g of a Sample of Cement

	grams
Si	9.69
Al	2.84
Fe	2.08
Ti	0.16
Ca (as CaO)	0.39
Ca (remainder)	45.58
Mg	0.62
S (as sulfate)	0.90
K	0.33
Na	0.16
O	36.12
Ignition loss	1.31
Total	100.18

Table 1.3
Composition of 100 g of the Sample of Cement
(if all elements present as anhydrous oxides).

			grams
Silica	(SiO_2)	(includes 0.17 g insoluble)	22.07
Alumina	(Al_2O_3)		5.36
Ferric oxide	(Fe_2O_3)		2.97
Titanium	(TiO_2)		0.26
Lime	(CaO)	(includes 0.55 g free CaO)	64.32
Magnesia	(MgO)		1.03
Sulf. Anhyd.	(SO_3)		2.24
Ignition loss			1.31
Potash	(K_2O)		0.40
Soda	(Na_2O)		0.22
Total			100.18

Table 1.4
Composition of 100 g of the Sample of Cement
(calculated to give a closer representation of the amount of the actual substances present)

	grams
$(CaO)_3\cdot SiO_2$	46.5
$(CaO)_2\cdot SiO_2$	27.7
$(CaO)_3\cdot Al_2O_3$	9.2
$(CaO)_4\cdot Al_2O_3\cdot Fe_2O_3$	9.0
$CaSO_4$	3.81
CaO (free)	0.55
MgO (as silicate)	1.03
TiO_2	0.26
K_2O	0.40
Na_2O	0.22
Ignition loss	1.31
Insoluble SiO_2	0.17
Total	100.15

1.4 Describing Chemical Change

Chemical changes are changes in the amounts of substances in the phases of a body of material. When the system is fully enclosed by a boundary, the total amount of each element is found to remain unchanged whatever the changes in amounts of substances may be. This means that only those chemical changes which transfer elements from one set of substances to another are allowed.

1.4.1 Formal equations for chemical change

The changes in the amounts of substances in a chemical change are related in a simple and precise way. The formulae of the substances are used to write chemical equations to describe the processes which lead to change.

Example 1.6

A gaseous system contains the three substances, hydrogen (H_2), nitrogen (N_2) and ammonia (NH_3).

(a) What must be the changes in amounts of H_2 and N_2 when the amount of NH_3 increases by 1 mol?

(b) What is the equation for the process of chemical change in this system?

Answer

(a) The change in amount of NH_3 is $\Delta n_{NH_3} = +1$ mol. One mol NH_3 contains 1 mol of the element nitrogen and 3 mol of the element hydrogen. The only source of nitrogen is the substance, N_2, which contains 2 mol N per mol N_2. Thus the change in amount of N_2 must be $\Delta n_{N_2} = -0.5$ mol. The only source of hydrogen is the substance, H_2. The change in amount of hydrogen must be $\Delta n_{H_2} = -1.5$ mol.

(b) The list of changes, in amounts of the substances, is $\Delta n_{NH_3} = +1$ mol, $\Delta n_{N_2} = -0.5$ mol, $\Delta n_{H_2} = -1.5$ mol. The process of change may be described by the equation, $0 = +1\ NH_3 - 0.5\ N_2 - 1.5\ H_2$, which implies that the elements N and H are conserved when the process occurs.

Query This is not the usual way people write chemical equations. Why have you done it?

Reply Because it is useful in discussing many aspects of reactions. We call it the formal equation for reaction and will use it frequently in this text. For example, we may say that the change involved in the problem corresponds to 1 mol of the process described by the equation in this form. That is, the statement 1 mol ($0 = 1\ NH_3 - 0.5\ N_2 - 1.5\ H_2$) implies $\Delta n_{NH_3} = +1$ mol, $\Delta n_{N_2} = -0.5$ mol and $\Delta n_{H_2} = -1.5$ mol.

The coefficients of the formulae in the formal equation for reaction are known as stoichiometric coefficients. The substances whose formulae are multiplied by positive stoichiometric coefficients are the *products* when the process advances and those whose formulae have negative stoichiometric coefficients are the *reactants* when the process advances.

Query Are you suggesting that the equation for reaction does not necessarily correspond to the direction in which a chemical process occurs?

Reply Yes. Equations can be written before we know which substances accumulate and which are depleted. If the substances which are given positive coefficients are found to decrease in amount, we can simply say that there is a negative amount of the chemical process which is represented by our equation.

Exercise 1.4

(a) Rearrange the formal equation,

$$0 = NH_3 - 0.5\ N_2 - 1.5\ H_2$$

into the familiar form which has the formulae of reactants (N_2 and H_2) at one side and the formulae of products (NH_3) at the other.

(b) Multiply the equation given above by the factor necessary to obtain a new equation with integer values of the stoichiometric coefficients. What will be the changes in amounts of the substances when there is 1 mol of the process described by this equation?

Answer

(a) $0.5\ N_2 + 1.5\ H_2 = NH_3$

(b) $0 = 2\ NH_3 - 1\ N_2 - 3H_3$ or $N_2 + 3H_2 = 2NH_3$. Now +1 mol of this process leads to $\Delta n_{NH_3} = +2$ mol, $\Delta n_{N_2} = -1$ mol, $\Delta n_{H_2} = -3$ mol.

The formal equation for a reaction has the advantage that it can be written in a general algebraic form. Thus we use the general symbol B to represent the formulae of the chemical substances and the general symbol ν_B to represent their stoichiometric coefficients. For the equation $N_2 + 3H_2 = 2NH_3$ we may put

$$B = N_2 \quad , \nu_B = -1$$

$$B' = H_2 \quad , \nu_{B'} = -3$$

$$B'' = NH_3 , \nu_{B''} = +2$$

Thus the products are recognised by positive stoichiometric coefficients and the reactants by negative stoichiometric coefficients. Then the formal equation for the reaction is simply

$$0 = \nu_B B + \nu_{B'} B' + \nu_{B''} B''$$

Now we introduce the symbol $\sum_B$ to mean "the sum for all kinds of B" and reduce our equation to the form

$$0 = \sum_B \nu_B B.$$

Query What is gained by doing this?
Reply It makes the accountancy of other changes which accompany chemical processes much easier to carry out. We have already mentioned molar masses, $\bar{M}_B$. The change in mass of a system per mole of reaction can be expressed as

$$\Delta \bar{M} (0 = \sum_B \nu_B B) = \sum_B \nu_B \bar{M}_B$$

Of course this property of a chemical reaction is zero, because there is conservation of mass. In Chapter 2 we meet molar volumes of substances, $\bar{V}_B$. The change in volume per mole of a given reaction can be expressed as

$$\Delta \bar{V} (0 = \sum_B \nu_B B) = \sum_B \nu_B \bar{V}_B$$

and calculated using the known values of the molar volumes of the individual substances.

1.4.2 Extent of reaction

We have seen that when chemical reaction occurs, there are changes in the amounts of each of the substances which appear in the appropriate chemical equation, and these changes are all connected through that equation. The connection is such that the change in amount of every substance in the formal equation divided by its stoichiometric coefficient has a common value. This value we call "the change in the extent of reaction", $\Delta\xi$. ξ is "the extent of reaction" (Greek letter Xi, which is equivalent to x.)

Thus, for reaction following equation (1), consider consumption

$$Zn(s) + 2HCl(aq) = ZnCl_2(aq) + H_2(g) \quad (1)$$

of 0.1 mol of Zn(*s*). This requires reaction of 0.2 mol of HCl(*aq*). The change in extent of reaction is

$$\begin{aligned} \Delta\xi &= \Delta n_{Zn(s)} / \nu_{Zn(s)} \\ &= \Delta n_{Zn(s)} / (-1) \\ &= (-0.1 \text{ mol}) / (-1) \\ &= 0.1 \text{ mol} \end{aligned}$$

Exercise 1.5
Check that the change in extent of reaction does not depend on the substance used to measure it.

The unit for extent of reaction is mole. A different quantity, the "degree of reaction" has sometimes been used. It is of more restricted use, and is a simple number between zero and one. Note that we can write the extent of reaction only after the chemical equation has been written. Simply multiplying equation (1) throughout by a factor z will reduce all changes in extent of reaction by the same factor, z.

1.5 Quantities and Units in Physical Chemistry

Before we can proceed confidently to use measurements in physical chemistry, it is desirable to get clear what we are doing when we measure anything—that is, when we "determine a physical quantity".

1.5.1 Physical quantities

The result of any effective measurement is a physical quantity. Such a quantity has two parts which are treated separately—a *numerical value* and a *unit*. Thus, quite generally,

(*physical quantity*) ≡ (*numerical value*) × (*unit*)

Thus, for examples,

(fruit bought) = (15) × (orange)

(length of race) = (500) × (metre)

(car speed) = (100) × (kilometre hour^{-1})

(NaCl concentration) = (1.25) × (mol dm^{-3})

We usually assign symbols to the *physical quantities* (for example, $c_{NaCl} \equiv$ NaCl concentration). When we wish to manipulate these quantities, we separate the numerical values from the units. The numerical values can be handled by the ordinary rules of arithmetic. When we draw a graph, it displays analogues of the numbers. That is why our graphs show, on their axes, the *number* = (physical quantity)/(unit), for example:

length/(metre)
speed/(kilometre hour^{-1})
concentration/(mol dm^{-3})

and not "length", "speed" or "concentration".

1.5.2 SI units

It would be convenient to have a system of units which is fully "consistent". That would mean that any quantity which can be derived by arithmetic from other quantities has units which are described the same way, with no numerical constants involved in the system of units.

Query What does that mean?
Reply If *force* = *mass* × *acceleration*
then
(Unit of force) = (unit of mass) × (unit of acceleration).
There may be need for special names for some of these units, but that is only a matter of keeping names short enough for use.

There have been limited sets of such consistent units (for example, MKS, c.g.s.) for many years. The nearest approach so far to a complete set is the Système Internationale (SI), which is being introduced progressively throughout the world. It is used in this text.

The initial problem in achieving such a system of units is the selection of suitable basic quantities. The SI, as agreed in 1969 (*Pure and Applied Chemistry*, 1970, *21*, 1), treats seven basic quantities as independent. These quantities, and their symbols, are listed in Table 1.5.

Table 1.5
Basic Physical Quantities

Basic physical quantities	*Symbol for quantity*
length	l
mass	m
time	t
electric current	I
thermodynamic temperature	T
luminous intensity	I_v
amount of substance	n

We shall have no need to use "luminous intensity". The remaining six are sufficient to generate all the quantities we require, and so the units for these six are sufficient, in principle, to generate the units for all our other quantities.

The definitions of the SI basic units are:

(a) *length*—metre. The metre is the length equal to 1 650 763.73 wavelengths in vacuum of the radiation corresponding to the transition between the levels $2p_{10}$ and $5d_5$ of the krypton-86 atom.

(b) *mass*—kilogramme. The kilogramme is the unit of mass and is equal to the mass of the international prototype of the kilogramme.

(c) *time*—second. The second is the duration of 9 192 631 770 periods of the radiation corresponding to the transition between the two hyperfine levels of the ground state of the caesium-133 atom.

(d) *electric current*—ampere. The ampere is that constant current which, if maintained in two straight parallel conductors of infinite length, of negligible cross-section, and placed 1 metre apart in vacuum, would produce between these conductors a force equal to 2×10^{-7} newton per metre of length.

(e) *thermodynamic temperature*—kelvin. The kelvin is the unit of thermodynamic temperature and is the fraction 1/273.16 of the thermodynamic temperature of the triple point of water.

(f) *luminous intensity*—candela. The candela is the luminous intensity, in the perpendicular direction, of a surface of 1/600 000 square metre of a black body at the temperature of freezing platinum under a pressure of 101 325 newton per square metre.

(g) *amount of substance*—mole. The mole is the amount of substance of a system which contains as many elementary entities as there are carbon atoms in 0.012 kilogrammes of carbon-12.

Query What about the kelvin and the international practical scale of temperature?

Reply The way we define *thermodynamic temperature* includes recipes for its measurement. Those recipes are not as precise for use at the present time as other recipes which have less theoretical background. For the present, really precise measurements of temperature have to be made with the empirical scales. Many other quantities were previously in the same somewhat untidy situation—so there were "international ohms" for electric resistance and "thermochemical calories" for energy transfer as heat, for example, which are not needed any more.

A problem which arises with any set of consistent units is that the range of values which arise in experiments, given the set of units, is not always in the range of small numbers. Thus many experiments involve magnetic flux densities of a few gauss (an old unit). The corresponding SI unit, the tesla, is 10^4 gauss, so the same fields are now to be expressed as (a few) $\times 10^{-4}$ tesla. To avoid continual use of such exponential notation, a set of SI prefixes (Table 1.6) is used to change unit size.

So a length of 1.00×10^{-10}m (1.00 Ångstrom) may be written either as 0.100 nanometres (nm) or as 100 picometres (pm) and similarly with other quantities.

A list of the quantities, symbols, units and symbols for units used in this book is given as Appendix A for convenient reference.

Table 1.6
SI *Prefixes*

fraction	*prefix*	*symbol*	*multiple*	*prefix*	*symbol*
10^{-1}	deci	d	10	deca	da
10^{-2}	centi	c	10^{2}	hecto	h
10^{-3}	milli	m	10^{3}	kilo	k
10^{-6}	micro	μ	10^{6}	mega	M
10^{-9}	nano	n	10^{9}	giga	G
10^{-12}	pico	p	10^{12}	tera	T
10^{-15}	femto	f			
10^{-18}	atto	a			

PROBLEMS

1.1 Give examples to illustrate the following concepts:
   (a) materials—uniform and composite
   (b) phases—pure and solution or mixture
   (c) substances—elementary and compound
   (d) elements.

1.2 (a) What is implied about the composition of methane by use of the chemical formula, CH_4,
   (i) from the percentage composition point of view?
   (ii) in terms of structural models?
   (b) What is the molar mass of methane?
   (c) What information would you need in order to calculate the chemical amount (mole number) of methane in a sample of given volume?

1.3 Formalin contains 37% formaldehyde, CH_2O, by weight, in solution with water.
   What further information would you need in order to calculate the volume concentration ($c_B = n_B/V$) of formaldehyde in a given sample of formalin?

1.4 (a) State what is meant by the term "chemical change". List the properties of a chemical system which remain unaltered by chemical change.
   (b) Consider the statement. "The reaction
   $$C(s) + O_2(g) = CO_2(g)$$
   occurred to the extent of 1.5 mol". Write the formal equation for the chemical process. In what respects does conservation of mass apply? What is the displacement of composition?

1.5 The following set of independent equations is used as a basis for describing the combustion of methane

$$CH_4 + 2O_2 = CO_2 + 2H_2O \qquad (1)$$
$$CH_4 + 1\tfrac{1}{2}O_2 = CO + 2H_2O \qquad (2)$$

The amounts of water and carbon dioxide formed during a given period of time were found to be

$$\Delta m_{H_2O} = 36\text{g and } \Delta m_{CO_2} = 33\text{g}.$$

   (a) What were the increases in the extents of the reactions represented by equations (1) and (2)?
   (b) What were the changes in the amounts of the substances present?

1.6 In a study of the operation of an internal combustion engine, significant changes were found in the amounts of the following substances:
   iso-octane (C_8H_{18}), oxygen (O_2), nitrogen (N_2), carbon monoxide (CO),

carbon dioxide (CO_2), nitrogen dioxide (NO_2), water (H_2O).

(a) Write a set of independent equations which may be used as a basis for describing chemical displacements in this system. (The set will have three members).

(b) Express the change in amount of each substance in terms of the amounts of the reactions you have included in your set.

1.7 In the Welsh process for extraction of metallic copper, cuprous sulfide is obtained by roasting the ore and separating compounds of iron. When the cuprous sulfide is heated in a current of air, the chemical changes can be correlated on the basis of the two equations

$$2Cu_2S + 3O_2 = 2Cu_2O + 2SO_2 \quad (1)$$
$$Cu_2S + 2Cu_2O = 6Cu + SO_2 \quad (2)$$

(a) When there is 3 mol of reaction (1) and 1 mol of reaction (2), what amount of Cu_2O accumulates?

(b) In a given time interval, the accumulation of SO_2 is 6 mol and the accumulation of Cu is 12 mol. What are the amounts of reaction (1) and reaction (2) which have taken place?

1.8 A chemical reactor contains the substances $S(s)$, $O_2(g)$, $N_2(g)$, $SO_2(g)$, $SO_3(g)$, $NO(g)$, $NO_2(g)$.
How many independent chemical equations must be constructed to form a basis for describing changes in the composition of the contents of the reactor?

2.

Space-Filling Properties of Materials

2.1 Goal in this Chapter

Materials occupy space, and the space they occupy is an important property both for their use and for the examination of their behaviour. For the great majority of materials which are transported, it is the space they occupy rather than their mass which determines the cost of transportation. The efficient transport of fragile articles depends on proper use of packing materials which function precisely by occupying the right spaces in relation to the fragile articles and the containers in which they are to be carried. When we come to a quantitative study of the properties of materials we usually measure the quantity of material through the mass of material. The most convenient second quantity to measure (and thus the first "property" of the material) is often the space-filling property. For this reason, commercial and scientific specifications of uniform materials normally include a value for a space-filling property, (e.g. .880 ammonia).

In introducing space-filling properties at this early stage we have two goals in mind. As we have just argued, the quantities are very important in themselves, and will be familiar to most readers. This importance provides our first goal—to achieve familiarity with the space-filling properties of materials. Our second goal is to show how properties are manipulated in physical chemistry. We shall carry out similar treatments for other properties which we introduce, and it is desirable to start with one which is not abstract.

This chapter contains a treatment of the ways in which measurements of the space-filling properties are converted to useful forms for filing and for future uses, and an example of the ways in which measurements in physical chemistry are handled and transformed.

Table 2.1
Properties of Materials which Describe Space Filling

Name	*Symbol*	*Definition*	*SI Units*
specific volume	v	volume occupied per unit mass ($v = V/m$)	$m^3 kg^{-1}$
molar volume	$\bar{V}$	volume occupied per 1 mol of material ($\bar{V} = V/n$)	$m^3 mol^{-1}$
density or mass density	ρ	mass per unit volume of material ($\rho = m/V$)	$kg\ m^{-3}$
concentration or mole density	$c = 1/\bar{V}$	amount of substance per unit volume ($c = n/V$)	$mol\ m^{-3}$

2.2 Properties Used to Describe Space Filling

Four properties will concern us closely in the discussion of volumes. They are listed with their definitions in Table 2.1. All are useful. There are, of course, only two independent ones. If we are concerned with masses of material, we shall be using either specific volume or its reciprocal, the density. If the material happens to have a known chemical constituent (and if its *chemistry* comes into the problem) then we shall be wanting to use either the molar volume or its reciprocal, the concentration.

Query Why this concern for volume? Isn't mass very much more important?

Reply Yes, it is. If you compare tea sweetened with granulated sugar with tea sweetened with icing sugar, first using equal *volumes* of sugar and then equal *masses*, you will notice a difference.

Nevertheless, volume is important, if only because we can often find the volume much more easily than the mass. We might want to know the amount of rock in a pile, the quantity of nickel in an ore body, the amount of protein in a glass of milk or the number of teaspoons of fertiliser required per square yard of garden. In each case we have to deal with a volume.

Query Why not tabulate the volumes occupied by such things?

Reply Do you want separate entries for one brick and one hundred bricks?

Comment Of course not!

Reply Then we must look at properties of materials—properties of a specified amount. We can always use multiples of the property for other amounts.

One very common use of space-filling properties is to assist in characterising substances and materials. For this purpose we are likely to use the density, or perhaps the molar volume. In Table 2.2 we have listed values of the molar mass, density and molar volume of a number of substances. From the table it is

Table 2.2
Molar Masses, Densities and Molar Volumes of Some Substances at 25°C, P = 101.3 kPa

Substance	*Symbol*	$\bar{M}$ / g mol^{-1}	ρ / g cm^{-3}	$\bar{V}$ / cm^3 mol^{-1}
Solids				
Iron	Fe	55.85	7.86	7.10
Copper	Cu	63.55	8.96	7.09
Sodium	Na	22.99	0.97	23.7
Tantalum	Ta	180.95	16.6	10.90
Titanium	Ti	47.90	4.50	10.64
Asbestos	-	-	2.4	-
Quartz	SiO_2	60.09	2.6	23.1
Bone	-	-	1.9	-
Balsa wood	-	-	0.2	-
Marble	$CaCO_3$	100.09	2.7	37.1
Liquids				
Mercury	Hg	200.59	13.53	14.8
Bromine	Br_2	159.81	3.12	51.2
Water	H_2O	18.02	0.997	18.1
Acetone	$(CH_3)_2CO$	58.08	0.785	74.0
Ethanol	C_2H_5OH	46.07	0.785	58.7
Benzene	C_6H_6	78.11	0.874	89.4
Olive oil		-	0.9	
Gases				
Hydrogen	H_2	2.02	0.08234	24.48×10^3
Nitrogen	N_2	28.01	1.1453	24.46×10^3
Oxygen	O_2	32.00	1.3087	24.45×10^3
Fluorine	F_2	38.00	1.553	24.46×10^3
Argon	Ar	39.95	1.6333	24.45×10^3
Nitric oxide	NO	30.01	1.2272	24.45×10^3
Methane	CH_4	16.04	0.6568	24.42×10^3
Difluoromethane	CH_2F_2	52.02	2.1616	24.07×10^3
Silicon tetrafluoride	SiF_4	104.08	4.277	24.33×10^3
Sulfur dioxide	SO_2	64.06	2.663	24.05×10^3

immediately clear that the numbers are different enough to be used to distinguish between many pure materials. The values of molar volume for gases are so similar that it is clear that they can, alternatively, be used as a convenient and precise measure of the amount of substance in a sample at known temperature and pressure.

2.3 Simple Manipulation of Space-filling Properties

The kinds of things we most often do with space-filling properties may be expressed very easily in terms of a few examples.

Example 2.1
What mass of sand (density $1.6\ g\ cm^{-3}$) is delivered by a front-end-loader of $0.4\ m^3$ capacity?

Answer
We require m_{sand}, the mass of the sand.
We know

$$\rho_{sand} = 1.6\ g\ cm^{-3} = \frac{m_{sand}}{V}$$

so

$$\begin{aligned} m &= \rho V \\ &= 1.6\ g\ cm^{-3} \times 0.4\ m^3 \\ &= 1.6 \times 10^{-3} kg \times 10^6 m^{-3} \times 0.4 m^3 \\ &= 640\ kg \qquad (0.64\ \text{tonne}) \end{aligned}$$

We often measure properties, like density, on a small sample of a particular material and use them in practical operations.

Example 2.2
Use the data from Table 2.2 to estimate the amount of hydrogen in a plastic bag which occupies $1\ m^3$ in the atmosphere at 25°C.

Answer

$$\bar{V} = V/n \simeq 24.5 \times 10^{-3} m^3\ mol^{-1}$$

for a gaseous substance at $P = 101$ kPa.

$$\begin{aligned} n_{H_2} &= 1\ m^3/24.5 \times 10^{-3}\ mol^{-1} \\ &= 40.8\ mol \end{aligned}$$

Amounts of gases are often measured by measuring volumes under controlled conditions.

Example 2.3
0.4 mol of aqueous Cl^- is to be added to a solution of silver nitrate. What volume of aqueous HCl ($2\ mol\ dm^{-3}$) is required?

Answer $c = n/V$
We require $n_{Cl^-} = 0.4\ mol = n_{HCl}$ to be added.
We know

$$c_{HCl} = 2\ mol\ dm^{-3} = \frac{n_{HCl}}{V}$$

So

$$\begin{aligned} V = n/c &= \frac{0.4\ mol}{2\ mol\ dm^{-3}} \\ &= 0.2\ dm^3 \end{aligned}$$

Volumes of solutions are often used to measure chemical amount of their constituents.

Example 2.4
A charge of $4.5\ dm^3$ of toluene is added to a reaction vessel. What quantity (number of moles) of toluene does the vessel contain? Density of toluene = $0.87\ g\ cm^{-3}$, molar mass $\bar{M} = 92\ g\ mol^{-1}$.

Answer
We require $n_{toluene}$
We know

$$n_{toluene} = \frac{m_{toluene}}{\bar{M}_{toluene}}$$

and that

$$\rho_{toluene} = \frac{m_{toluene}}{V}$$

hence $m = \rho V$

and

$$\begin{aligned} n &= \rho V/\bar{M} \\ &= \frac{0.87\ g\ cm^{-3} \times 4.5\ dm^3}{92\ g\ mol^{-1}} \\ &= \frac{0.87\ g\ cm^{-3} \times 4.5 \times 10^3 cm^3}{92\ g\ mol^{-1}} \\ &= 43\ mol \end{aligned}$$

The density provides a convenient means to reinterpret volumes of pure liquids in terms of chemical amount.

PROBLEMS

2.1 What size of tanker is required to transport 1.00×10^3 kg milk (density = $1.03\ g\ cm^{-3}$).
Answer $10\ m^3$
(*Note*: exactly $9.7\ m^3$ would not be enough to allow an air space for expansion of the contents.)

2.2 Water is added slowly to a measuring cylinder containing 50 cm^3 of concentrated, aqueous ammonia (14 mol dm^{-3}). To what volume should the cooled solution be made up in order to obtain aqueous ammonia at 2 mol dm^{-3}?

Answer 350 cm^3

2.3 What volume of concentrated sulphuric acid (18 mol dm^{-3}) must be added to water so that, when the hot solution is allowed to cool and is made up to a volume of 2 dm^3, the final concentration will be 2 mol dm^{-3}.

Answer 222 cm^3

(What is the significance of the "cooling" in questions 2.2 and 2.3?)

2.4 An instant coffee manufacturer used a jar size of 425 cm^3. By law it must not contain less than the amount stated on the label (113 g). However if the volume occupied by the coffee is less than 380 cm^3 the product is avoided by housewives as the jar does not appear to be full. Within what limits must the plant chemist or engineer control the specific volume of the coffee powder for the product to be saleable?

Answer $v_{min} = 3.4\ cm^3\ g^{-1}$
$v_{max} = 3.8\ cm^3\ g^{-1}$

2.4 The Effects of Experimental Conditions on the Space-filling Properties

From our general knowledge we know most of the things which alter properties like the specific volume, molar volume, density or concentration. A summary is given in Table 2.3 of the effects of changing some parameters, one at a

Table 2.3
Changes in Specific Volume and Molar Volume with Condition or Situation of the Material

Parameter	*"Usual" effect*
Composition	change in composition may increase or decrease the volume
Temperature	increase in temperature normally increases the volume—except for water between 0° and 4°C
Pressure	increase in pressure decreases the volume
Phase type (solid, liquid, gas, etc.)	often large changes in volume occur when a material changes from one phase to another
Height above earth	no direct effect
Voltage above "earth"	negligible effect
Speed	effect negligible unless it approaches that of light
Light	negligible effect apart from (a) temperature changes caused by absorption of the radiation (b) chemical reactions initiated by the radiation
Size of particles	altered size alters packing properties of the particles. This gets more important as the surface area/volume ratio becomes larger
History	earlier treatment of the material may be important

time. From this it is clear that the things we need to examine are the composition, temperature, pressure, phase type and history.

The history is sometimes very important in fixing the properties of materials. One of the causes of failure in metal objects is "metal fatigue", a picturesque term to describe the build-up of distortions to crystal structure which can lead to abrupt fracture. The "tempering" of metals and annealing of metals or glass is also a matter of controlled exposure to conditions of temperature (and perhaps pressure) which lead to desired changes in properties. In a different sense, we control the behaviour of canned foods and medical supplies by subjecting them to conditions of temperature or radiation which are sufficient to prevent the growth of undesired organisms in storage. When we can, we like to work with materials which are free from these "history" effects. When we have to, it is simplest if the history effects can be treated by themselves. For example, a carpenter is doing that when he chooses to work only with "dry" (and not "green") timber. The future behaviour of his products will be less dependent on the weather and the loadings which they suffer.

2.5 Composition and Specific or Molar Volume

The chemical composition affects both the specific volume, v, and the molar volume, $\bar{V}$, of a uniform mixture. To examine these effects we take a particular example, the liquid in "lead-acid" batteries.

This liquid is a mixture of sulfuric acid and water. Its composition and specific volume both change progressively as the battery is "charged"

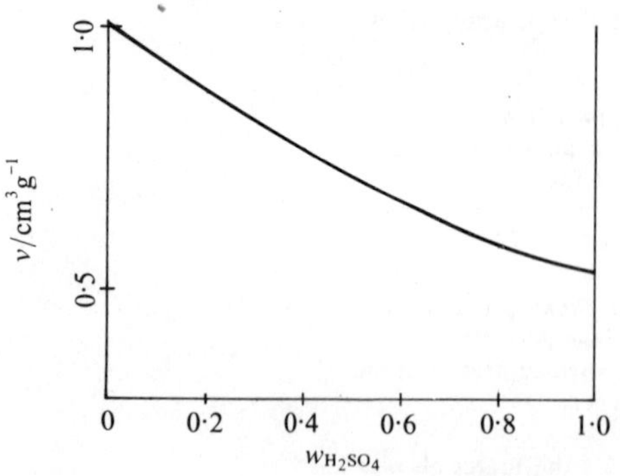

Fig. 2.1 Specific volume v of sulfuric acid–water mixtures of varying mass fraction $w_{H_2SO_4}$ of sulfuric acid.

or "discharged". Figure 2.1 illustrates the way in which the specific volume changes in this mixture as the composition changes.

Question
What is the concentration scale used in Fig. 2.1?

Answer
The *specific volume* is plotted against the *mass fraction* $w_{H_2SO_4}$ of sulfuric acid in the solution. The mass fraction is defined, in terms of the masses m of the substances, by

$$w_{H_2SO_4} \equiv \frac{m_{H_2SO_4}}{m_{H_2SO_4} + m_{H_2O}}$$

and so

$$w_{H_2O} \equiv \frac{m_{H_2O}}{m_{H_2SO_4} + m_{H_2O}} = 1 - w_{H_2SO_4}$$

Query What is the use of a graph like this?

Reply It can be used for analysis or for characterisation. A lead-acid battery is known to be "fully charged" when the specific volume is 0.8 $cm^3\ g^{-1}$ —a test which is made with a hydrometer. The battery needs to be recharged when the specific volume rises to 0.9 $cm^3\ g^{-1}$ whether due to use or to standing idle.

Query Density is used rather than specific volume, isn't it?

Reply Yes, it is. We decided to put it the other way up so that we could comment that the *mass* of the battery stays constant, but the *volume* of the liquid doesn't. So one always has to design equipment to allow for volume changes! (See the answer to problem 2.1)

Query What about another example?

Reply The urine excreted by a healthy adult has a density between 1.01 and 1.03 $g\ cm^{-3}$. Outside that range you have a health problem.

An alternative way of putting the same information as in Fig. 2.1, transforms it to values of molar volume and mole fraction (Fig. 2.2). The mole fraction of sulfuric acid is defined as

$$x_{H_2SO_4} \equiv \frac{n_{H_2SO_4}}{n_{H_2SO_4} + n_{H_2O}}$$

and so

$$x_{H_2O} \equiv \frac{n_{H_2O}}{n_{H_2SO_4} + n_{H_2O}} = 1 - x_{H_2SO_4}$$

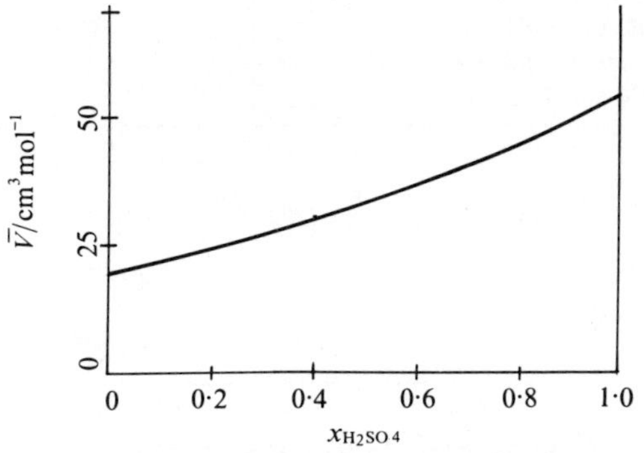

Fig. 2.2 Molar volume $\bar{V}$ of sulfuric acid–water mixtures of varying mole fraction $x_{H_2SO_4}$ of sulfuric acid.

Query Could the specific volume of the solution be expressed in terms of its mole fraction or the molar volume in terms of its mass fraction?

Reply Yes. Any measure of composition may be used. Usually we express the composition dependence of the molar volume of a mixture in terms of its mole fraction. Both the mole fraction and mass fraction scales are in common use to express the composition dependence of the specific volume and its reciprocal, the density. (For a review of composition scales, and the relationships between them see Chapter 1.) To determine the molar volume of a material we normally need to know its composition first so molar volumes are not much use for characterising materials. The effects of pressure on chemical equilibria and on equilibria between phases (Section 3.4) are related to the molar volumes of the materials involved. This provides the largest single use of molar volume.

Query What do you mean by "molar volume" for a mixture?

Reply What we meant in Fig. 2.2 is the volume V of a sample of the mixture divided by the amount of substance ($n_{H_2SO_4}+n_{H_2O}$) in it. Later we will meet a different, more useful quantity, the "partial molar volume" of a substance in a mixture.

Example 2.5
What is the composition of a mixture of sulfuric acid and water with $v=0.85$ cm^3 g^{-1}.

Answer
From Fig. 2.1,

$$w_{H_2SO_4}=0.25$$

i.e., 1 g of solution contains 0.25 g sulfuric acid and 0.75 g water.

Example 2.6
What is the concentration of sulfuric acid in battery acid with density 1.25 g cm^{-3}?

Answer
We require concentration of sulfuric acid

$$c_{H_2SO_4}=\frac{n_{H_2SO_4}}{V}$$

We know the density of the solution

$$\rho=\frac{m}{V}=1.25\ \text{g cm}^{-3}$$

hence the specific volume

$$v=1/\rho$$
$$=0.8\ \text{cm}^3\ \text{g}^{-1}$$

from Fig. 2.1,'$w_{H_2SO_4}=0.34$

hence 0.8 cm^3 of solution contain 0.34 g of H_2SO_4. We know that the molar mass of sulfuric acid $\bar{M}_{H_2SO_4}=98.1$ g mol^{-1} hence 0.8 cm^3 of solution contains

$$\frac{0.34\ \text{g}}{98.1\ \text{g mol}^{-1}}=3.46\times10^{-3}\ \text{mol } H_2SO_4$$

i.e.

$$c_{H_2SO_4}=\frac{n}{V}$$
$$=\frac{3.46\times10^{-3}\ \text{mol}}{0.8\ \text{cm}^3}$$
$$=4.3\times10^{-3}\ \text{mol cm}^{-3}$$

PROBLEMS

2.5. Table 2.4 lists values reported for the specific volume, v, of a number of ethanol/water mixtures as a function of the weight fraction, w (ethanol).
   (a) Plot the graph of specific volume of ethanol/water mixtures at 30°C as a function of the weight fraction of ethanol.
   (b) Estimate the specific volume of a mixture which contains 75% ethanol by mass.
   Answer 1.18 cm^3 g^{-1}
   (c) A sample of vodka had a density of 0.952 g cm^{-3}. Estimate
   (i) the mass fraction of water, and
   (ii) the mole fraction of ethanol in it.
   Answer (i) $w_{H_2O}=0.724$
   (ii) $x_{ethanol}=0.13$

2.6. It is found that the molar volumes of most gases and gaseous mixtures at 25°C and 101 kPa lie within the range

Table 2.4
Specific Volume of Ethanol/water Mixtures at 30° C, P = 101 kPa

$w_{ethanol}$	0	0.1	0.2	0.3	0.4	
$v/\mathrm{cm^3 g^{-1}}$	1.00434	1.02171	1.03740	1.05551	1.07794	
$w_{ethanol}$	0.5	0.6	0.7	0.8	0.9	1.0
$v/\mathrm{cm^3 g^{-1}}$	1.10400	1.13317	1.16404	1.19799	1.23576	1.28082

$V(25°\mathrm{C}, 101\ \mathrm{kPa}) = 0.0245 \pm .0005\ \mathrm{m^3\ mol^{-1}}$ (see Table 2.2 and Section 2.2).

(a) The densities of pure nitrogen and pure oxygen at 25° C and 101.3 kPa are 1.145 g dm^{-3} and 1.309 g dm^{-3} respectively. Use the data above to estimate the density of a mixture of oxygen and nitrogen at
$x_{O_2} = 0.50$, $T = 25°$ C, $P = 101.3$ kPa.
Answer 1.227 g dm^{-3}

(b) Use the three points available to plot the graph of density against mole fraction of oxygen for mixtures of oxygen and nitrogen at 25°C and 101.3 kPa.

(c) Samples of air were taken from ground level during the passage of a fire through scrubland. After drying and removal of the carbon dioxide, the lowest density found at 25°C and 101.3 kPa was 1.165 kg m^{-3}.
  (i) Use the graph to determine the minimum value of the ratio n_{O_2}/n_{N_2} found.
  Answer $n_{O_2}/n_{N_2} = 0.14$
  (ii) Normal dry air has a composition 78.08 mole % N_2, 20.95 mol % O_2, 0.93 mol % Ar, 0.05 mol % other gases. If most of the oxygen consumed by the fire is converted to CO_2, estimate the minimum mole fraction of oxygen at ground level during such a fire.
  Answer $x_{O_2}(\mathrm{min}) = 0.10$
  (This is sufficient for survival if a person is protected from the flames and radiation.)

2.6 General Patterns of Behaviour

Now that we have looked at the experimental behaviour of a few mixtures we can ask whether it fits any general pattern. This is the way a scientific enquiry usually proceeds—observe behaviour, then identify patterns of behaviour and formulate laws to describe them.

The simplest statement which summarises the main features of the pattern is the *ideal law*. It provides a standard with which actual behaviour can be compared. An ideal law is useful provided it enables us to make an intelligent guess (an "estimate") about the properties of systems whose behaviour is not known, or for which we need only a quick estimate of approximate behaviour. When we proceed to develop a theory with which to describe, and so to understand, behaviour in terms of models of the molecular structure of mixtures, the ideal law is also of great value. Basically we begin by devising simple theories to account for the ideal laws. Then we introduce extensions to the theory to improve the extent to which it may be used to simulate observed behaviour. The ideal law often corresponds to a very simple model.

Query Surely a "law" should be exact?

Reply No! Nature is too varied. We use ideal laws to classify experimental behaviour. If laws had to be exact there would be too many of them and the classes would be too narrow to be useful. By the time enough experiments were done to establish an "exact law", there would be no new experimental results to predict!

Query But a good "law" should fit?

Reply Yes, but only well enough to satisfy our needs. We need to know the range in which a given law is likely to be sufficiently accurate for forecasting the properties which we wish to know. The ideal law often becomes a precise description of behaviour in some extreme conditions—such as very low pressures or concentrations.

2.6.1 Forecasting molar volumes of mixtures

As a general statement (ideal law) about the volume of mixtures of two substances, let us try: "when two substances are mixed, the volume occupied by the mixture at some temperature and pressure is roughly equal to the sum of the volumes of the substances (at the same temperature and pressure) before they were mixed."

Example 2.7
To get clear what that statement means, let us use it to estimate the final volume when 2 kg of cane sugar (sucrose, $v = 0.63\ \text{cm}^3\text{g}^{-1}$) is dissolved in 1 dm^3 of water at 20°C.

Answer
We require V(solution) at 20°C. We know that roughly

$$V\,(\text{solution}) = V(\text{water}) + V(\text{sucrose})$$
$$V(\text{water, } 20°\text{C}) = 1\ \text{dm}^3$$
$$V(\text{sucrose, } 20°\text{C}) = vm\,(\text{sucrose}) = 0.63 \times 2 \times 10^3\ \text{cm}^3 = 1.26\ \text{dm}^3$$

∴ we estimate

$$V(\text{solution}) = 1.26\ \text{dm}^3 + 1\ \text{dm}^3 = 2.26\ \text{dm}^3$$

Note: in fact this estimate is very good indeed. Such a solution has a measured volume of 2.25 dm^3.

Example 2.8
Estimate the molar volume at 25°C of a mixture of ethanol and benzene in which the mole fraction of benzene, x_{benzene}, is 0.4. The molar volumes of pure ethanol and benzene at 25°C are

$$V_{\text{ethanol}} = 58.7\ \text{cm}^3\ \text{mol}^{-1}$$
$$\bar{V}_{\text{benzene}} = 89.4\ \text{cm}^3\ \text{mol}^{-1}$$

Answer
We want the volume occupied by 1 mol of the mixture, that is

$$n_{\text{benzene}} + n_{\text{ethanol}} = 1\ \text{mol}$$
$$n_{\text{benzene}} = x_{\text{benzene}} \times 1\ \text{mol} = 0.4\ \text{mol}$$

the volume occupied by 0.4 mol benzene $= 0.4 \times 89.4\ \text{cm}^3$.
Similarly,

$$n_{\text{ethanol}} = x_{\text{ethanol}} \times 1\ \text{mol} = (1 - x_{\text{benzene}}) \times 1\ \text{mol} = 0.6\ \text{mol}$$

the volume occupied by 0.6 mol ethanol $= 0.6 \times 58.7\ \text{cm}^3$.
From our general statement we estimate that

$$V(1\ \text{mol mixture}) \simeq 0.4 \times 89.4 + 0.6 \times 58.7\ \text{cm}^3 \simeq 71\ \text{cm}^3$$

(The experimental result is 71.3 cm^3.)

Comment It looks as though the general statement could be put as

$$\bar{V}(\text{mixture}) \approx \bar{V}_{\text{A(pure)}}x_{\text{A}} + \bar{V}_{\text{B(pure)}}x_{\text{B}}$$

for mixtures of A and B?

Reply Yes.

Query Is there a similar equation for specific volumes?

Reply Yes. $v\,(\text{mixture}) = v_{\text{A}}w_{\text{A}} + v_{\text{B}}w_{\text{B}}$ where v(mixture) is the volume occupied by 1 g of the mixture, v_{A}, v_{B} are the specific volumes of A and B and $w_{\text{A}}, w_{\text{B}}$ are the weight fractions of A and B in the mixture.

Example 2.9
Estimate the density at 25°C of a mixture of 50g carbon tetrachloride ($\rho_{\text{CCl}_4,298\text{K}} = 1.584\ \text{g cm}^{-3}$) and 20g of diphenyl ether ($\rho_{\text{ether},298\text{K}} = 0.714\ \text{g cm}^{-3}$)

Answer
We want the density of the mixture, ρ_{mixture} and know that this is $1/v_{\text{mixture}}$. We estimate that

$$v_{\text{mixture}} = v_{\text{CCl}_4}w_{\text{CCl}_4} + v_{\text{ether}}w_{\text{ether}}$$
$$v_{\text{CCl}_4} = 1/\rho_{\text{CCl}_4} = 0.6313\ \text{cm}^3\ \text{g}^{-1}$$

Similarly,

$$v_{ether} = 1.401 \text{ cm}^3 \text{ g}^{-1}$$

$$w_{CCl_4} = \frac{m_{CCl_4}}{m_{CCl_4} + m_{ether}} = \frac{50}{50+20} = 0.714$$

Similarly

$$w_{ether} = 0.286$$

Hence,

$$v_{mixture} \simeq 0.631 \times 0.714 + 1.401 \times 0.286 \text{ cm}^3 \text{g}^{-1}$$
$$\simeq 0.851 \text{ cm}^3 \text{ g}^{-1}$$

Thus we estimate the density of the mixture as $1/0.851 = 1.17 \text{ g cm}^{-3}$.
(The experimental value is 1.19 g cm^{-3}.)

Question
How useful are these simple laws for estimating the compositions of solutions?

Answer
In practice (except for gases at low pressures), the answers are seldom accurate enough to be useful. Measurements of density are often used for chemical analysis but the experimental behaviour of the mixture under consideration often needs to be known to a higher degree of accuracy than is given by such simple statements. The following examples show this up very clearly.

Example 2.10
Estimate the composition (w_{CH_3COOH}) of a sample of white vinegar (a solution of acetic acid in water) which has a density at 15°C of 1.0213 g cm^{-3}.

At 15°C, $\rho_{CH_3COOH} = 1.0545 \text{ g cm}^{-3}$, and $\rho_{H_2O} = 0.9991 \text{ g cm}^{-3}$, for the pure substances.

Answer
We know that roughly,

$$v \text{ (mixture)} = v_{CH_3COOH} w_{CH_3COOH} + v_{H_2O} w_{H_2O}$$

and that

$$w_{H_2O} + w_{CH_3COOH} = 1$$

or

$$w_{H_2O} = 1 - w_{CH_3COOH}$$

hence

$$v \text{ (mixture)} = v_{CH_3COOH} w_{CH_3COOH} + v_{H_2O} (1 - w_{CH_3COOH})$$

We also know that

$$v = 1/\rho$$

hence

$$v \text{ (mixture)} = 0.9791 \text{ cm}^3 \text{ g}^{-1}$$
$$v_{CH_3COOH} = 0.9483 \text{ cm}^3 \text{ g}^{-1}$$
$$v_{H_2O} = 1.0009 \text{ cm}^3 \text{ g}^{-1}$$

The predicted value of w_{CH_3COOH} from this treatment is 0.414.
(The actual composition of such a mixture is $w_{CH_3COOH} = 0.15$.)

Example 2.11
Estimate the composition ($x_{C_6H_5Br}$) of a mixture of bromobenzene and chlorobenzene such that, at 40°C, 0.2 mol of mixture occupies a volume of 20.84 cm^3. For the pure components at 40°C

$$\bar{V}_{C_6H_5Cl} = 103.88 \text{ cm}^3 \text{ mol}^{-1}$$

and

$$\bar{V}_{C_6H_5Br} = 107.25 \text{ cm}^3 \text{ mol}^{-1}$$

Answer
We require $x_{C_6H_5Br}$. The law states that

$$\bar{V} \text{(mixture)} \simeq \bar{V}_{C_6H_5Cl} x_{C_6H_5Cl} + V_{C_6H_5Br} x_{C_6H_5Br}$$

and we know that

$$x_{C_6H_5Cl} + x_{C_6H_5Br} = 1$$

or

$$x_{C_6H_5Cl} = 1 - x_{C_6H_5Br}$$

$$\bar{V}\text{(mixture)} = V \text{(mixture)} / n \text{(mixture)}$$
$$= 104.2 \text{ cm}^3 \text{ mol}^{-1}$$

Hence the value of $x_{C_6H_5Br}$ is approximately

$$\frac{104.2 - 103.88}{107.25 - 103.88} = 0.10$$

(The actual composition of such a mixture is $x_{C_6H_5Br} = 0.20$)

Example 2.12
Estimate the composition (x_{NO_2}) of a gaseous mixture of NO_2 and N_2O_4 at 300 K and $P = 100$ kPa if under these conditions

$$\rho\text{(mixture)} = 3.130 \text{ kg m}^{-3}$$
$$\rho_{NO_2} \text{ (pure)} = 1.845 \text{ kg m}^{-3}$$
$$\rho_{N_2O_4} \text{ (pure)} = 3.690 \text{ kg m}^{-3}$$

Answer
We require x_{NO_2}. The available data are the densities. The simplest method is to estimate w_{NO_2} and hence x_{NO_2}. We know that roughly,

$$v\,(\text{mixture}) = v_{NO_2}w_{NO_2} + v_{N_2O_4}w_{N_2O_4}$$

and

$$w_{N_2O_4} = 1 - w_{NO_2}$$

also

$$v = 1/\rho$$

That is

$$v\,(\text{mixture}) = 1/3.130\ \text{m}^3\ \text{kg}^{-1} = 0.3195\ \text{m}^3\ \text{kg}^{-1}$$

and

$$v_{NO_2} = 0.5420\ \text{m}^3\ \text{kg}^{-1}$$

$$v_{N_2O_4} = 0.2710\ \text{m}^3\ \text{kg}^{-1}$$

hence

$$w_{NO_2} = 0.1790$$

We estimate that 1 g of mixture contains 0.1790 g NO_2 and 0.8210 g N_2O_4.

We know that

$$n = \frac{m}{\bar{M}}$$

and

$$\bar{M}_{NO_2} = 46.01\ \text{g mol}^{-1},\quad \bar{M}_{N_2O_4} = 92.02\ \text{g mol}^{-1}$$

hence in 1 g of mixture:

$$n_{NO_2} = \frac{0.1790}{46.01}\ \text{mol} = 0.00389\ \text{mol}$$

and

$$n_{N_2O_4} = 0.00892\ \text{mol}$$

Now

$$x_{NO_2} = \frac{n_{NO_2}}{n_{NO_2} + n_{N_2O_4}}$$

i.e. we estimate

$$x_{NO_2} = 0.304$$

(In such a mixture x_{NO_2} is 0.303 — very good agreement.)

Those space-filling properties which may be estimated with useful accuracy (to within a few per cent of the experimental value) by use of the ideal law are as follows:

(a) the *volume* of liquid mixtures of given composition;
(b) the *volume* of gaseous mixtures of given composition;
(c) the *composition* of gases of given density, and
(d) in a few selected cases, the volume of solutions of solids in liquids.

There are many situations when the behaviour of a solution cannot be estimated usefully using the ideal law. In particular:

(a) when the two pure components are in different phases (think of the change in total volume when gaseous ammonia is dissolved in water!);
(b) for estimation of the composition from a knowledge of the density (illustrated by Examples 2.10 and 2.11); and
(c) when mixing is accompanied by chemical reaction—such as mixtures of water and acetylchloride.

2.6.2 Partial molar volumes of substances in solutions

We often need to be able to estimate molar volumes accurately for solutions and, as we have just stated, the ideal law is not always adequate for our needs. We solve this problem by assigning a "partial molar volume", $\bar{V}_A$, to the solvent, A, and "partial molar volumes", $\bar{V}_B$, $\bar{V}_{B'}$. . . to each solute species, B, B′, . . . We assign the values of the partial molar volumes in such a way that the volume of a solution is given by the simple relation

$$V = \bar{V}_A n_A + \bar{V}_B n_B + \bar{V}_{B'} n_{B'} + \ldots$$

This is done by adopting as our definition of partial molar volumes the equation:

$$dV \equiv \bar{V}_A dn_A + \bar{V}_B dn_B + \bar{V}_{B'} dn_{B'} + \ldots$$

in which dV is the small increase of volume which occurs when the mole numbers of the substances are increased by the small amounts dn_A, dn_B, $dn_{B'}$. . ., the temperature and pressure being maintained at their initial values.

Query How can several quantities be defined by one equation?
Reply The use of small changes makes this possible. The small increases in n_A, n_B, $n_{B'}$ can

be carried out independently. When only n_A is altered,

$$dV = \bar{V}_A \, dn_A$$

When only n_B is altered

$$dV = \bar{V}_B \, dn_B$$

By such a method, each partial molar volume can be measured separately. Table 2.5 lists molar volumes and partial molar volumes for mixtures of benzene and cyclohexane.

Query How is a "partial molar volume" different from a "molar volume"?

Reply A molar volume is a property of a material, either a solution (or mixture) or a pure substance. A partial molar volume refers to a particular substance in a solution or mixture.

For a pure substance the partial molar volume and the molar volume are identical.

2.6.3 Standard molar volumes of solutes and solvents

Handbooks of physicochemical data cannot list molar volumes and partial molar volumes for every composition of solutions which we may wish to use. What we often find in such collections are the limiting values of partial molar volumes of solutes and solvents in dilute solutions. These are referred to as "standard molar volumes" and are indicated by adding a small superscript zero to the usual symbol for partial molar volume.

The standard molar volume of a solute B in a given solvent A is the value approached by the partial molar volume of B as the relative concentration of solutes $(\Sigma x_B/x^\circ)$ approaches zero. The measurements are made at standard pressure and the selected temperature. That is,

$$\bar{V}^\circ_{B(\text{in A},T)} \equiv \text{limiting value of } \bar{V}_B \text{ as } \Sigma x_B/x^\circ \to 0 \text{ in solvent A, with } P = P^\circ \text{ and } T \text{ as stated}$$

The data in Table 2.5 reveal that the standard molar volume of benzene in cyclohexane at 298K is

$$\bar{V}^\circ_{\text{benzene (cyclohexane, 298K)}} = 92.134 \text{ cm}^3 \text{ mol}^{-3}$$

The standard molar volume of a solvent A is the value approached by its partial molar volume as the concentration of *solutes* $(\Sigma x_B/x^\circ)$ approaches zero. The measurements are taken at standard pressure and the selected temperature. This quantity is a property of the pure substance in the given phase type (either solid or liquid). Thus, for a liquid solvent A, the standard molar volume is

$$\bar{V}^\circ_{A(l,T)} \equiv \text{limiting value of } \bar{V}_A \text{ as } x_B/x^\circ \to 0, \text{ with } P = P^\circ \text{ and } T \text{ as stated.}$$

$$\equiv \bar{V}_{A(l, \text{ pure}, P^\circ, T)}$$

Thus according to Table 2.5, the standard molar volume of liquid cyclohexane at 298 K is $V_{\text{cyclohexane (l, 298K)}} = 108.753 \text{ cm}^3 \text{ mol}^{-1}$.

Exercise 2.1
Use Table 2.5 to find the values of

(a) $\bar{V}^\circ_{\text{cyclohexane (in benzene, 298K)}}$

and

(b) $\bar{V}^\circ_{\text{benzene (l, 298K)}}$.

The advantage gained by introducing standard molar volumes of solvents is a great improvement in the reliability of the ideal law for forecasting the molar volumes of dilute solutions.

We find that

$$\bar{V} \approx \bar{V}^\circ_{B(\text{in A})} x_B + \bar{V}^\circ_A x_A$$

is usually a good approximation for solutions with up to 20% solute.

Exercise 2.2
Compare the estimates of V for a solution with $x_{\text{benzene}} = 0.20$, $x_{\text{cyclohexane}} = 0.80$ using the following assumptions:

Table 2.5

Molar Volumes of Mixtures of Cyclohexane (A) *and Benzene* (B) *at* 298K, P°

x_B	$\bar{V}/\text{cm}^3 \text{ mol}^{-1}$	$\bar{V}_A/\text{cm}^3 \text{ mol}^{-1}$	$\bar{V}_B/\text{cm}^3 \text{ mol}^{-1}$
0	108.753	108.753	92.134
0.2	105.311	108.868	91.083
0.4	101.645	109.193	90.324
0.6	97.766	109.703	89.808
0.8	93.688	110.402	89.510
1.0	89.411	111.296	89.411

(a) $\bar{V}_{\text{benzene}} \approx \bar{V}^{\circ}_{\text{benzene(l)}} = 89.411 \text{ cm}^3 \text{ mol}^{-1}$

$\bar{V}_{\text{cyclohexane}} \approx \bar{V}^{\circ}_{\text{cyclohexane(l)}} = 108.753 \text{ cm}^3 \text{ mol}^{-1}$

(b) $\bar{V}_{\text{benzene}} \approx \bar{V}^{\circ}_{\text{benzene(cyclohexane)}} = 92.134 \text{ cm}^3 \text{ mol}^{-1}$

$\bar{V}_{\text{cyclohexane}} \approx \bar{V}^{\circ}_{\text{cyclohexane(l)}} = 108.753 \text{ cm}^3 \text{ mol}^{-1}$

Answer

(a) $\bar{V} \approx 104.885 \text{ cm}^3 \text{ mol}^{-1}$;

(b) $\bar{V} \approx 105.429 \text{ cm}^3 \text{ mol}^{-1}$;

which may be compared with experiment, $\bar{V} = 105.311 \text{ cm}^3 \text{ mol}^{-1}$.

Table 2.6 (a) lists values of standard molar volumes of a number of substances as solutes in water and as solvents.

Standard specific volumes of solutes and solvents may be derived simply from the experimental data and used, in an ideal law with mass fractions, to forecast values of specific volumes and densities of solutions. The standard specific volumes of a number of solutes in water are listed in Table 2.6 (b).

Example 2.13
The standard specific volume of ammonia in water at 20°C is $\nu^{\circ}_{NH_3(aq)} = 1.44 \text{ cm}^3 \text{ g}^{-1}$. Estimate the density at 20°C of a solution containing 4% by weight NH_3. At 20°C,

$$\nu^{\circ}_{H_2O} = 1.0018 \text{ cm}^3 \text{ g}^{-1}$$

Answer
We require ρ (solution).
We know

$$\rho = 1/\nu$$

and that

$$\nu(\text{solution}) \simeq \nu^{\circ}_{H_2O}\, w_{H_2O} + \nu^{\circ}_{NH_3(aq)}\, w_{NH_3}$$

with

$$w_{NH_3} = 0.04, \quad w_{H_2O} = 0.96.$$

Table 2.6 (a)
Standard Molar Volumes for a Number of Substances as Solutes in Water and as Solvents at 20° C

Substance B	$\bar{V}^{\circ}_{B(aq)}/\text{cm}^3 \text{ mol}^{-1}$	$\bar{V}^{\circ}_{B(l)}/\text{cm}^3 \text{ mol}^{-1}$
Acetic acid	49.1	59.8
Acetone	60.7	73.7
Ethanol	48.6	58.5
Ethylene glycol	54.5	55.7

Table 2.6 (b)
Standard Specific Volumes, $\nu^{\circ}_{B(aq)}$, of a Number of Solutes in water at 20°C.

Solute B		$\nu^{\circ}_{B(aq)}/\text{cm}^3\text{g}^{-1}$
Ammonia	NH_3	1.44
Sodium chloride	$NaCl$	0.305
Citric acid	$C_6H_8O_7$	0.595
Potassium oxalate	$C_2O_4K_2$	0.30
Creatinine	$C_4H_7N_3O$	0.75

Thus

$$\nu(\text{solution}) \approx (1.0018 \times 0.96 + 1.44 \times 0.04) \text{ cm}^3\text{g}^{-1} = 1.019 \text{ cm}^3 \text{ g}^{-1}$$

and

$$\rho(\text{solution}) = 0.981 \text{ g cm}^{-3}$$

The experimental value is 0.982 cm^{-3}.

Example 2.14
Estimate the density at 20°C of a 1 mol (kg water)$^{-1}$ solution of ethylene glycol.

Answer
We require $\rho = m/V$.
We know that the solution contains 1 mol (that is 62.07 g) glycol, for which $\bar{V} \approx 54.5 \text{ cm}^3 \text{ mol}^{-1}$, and 1 kg water, for which $\bar{V} \approx 1001.8 \text{ cm}^3$.

$$\text{Estimated total volume} \simeq 1001.8 + 54.5 = 1056.3 \text{ cm}^3$$

The total mass is 1062.1 g and thus $\rho \simeq 1.0055 \text{ g cm}^{-3}$.
(The density of such a solution is 1.0072 g cm^{-3}.)

PROBLEMS

2.7. (a) Estimate the final volume when 0.2 m^3 of benzene is added to 30.0 m^3 of petrol.

Answer 30.2 m^3

(b) Estimate the final volume when 20 cm^3 of benzene is added to 30 cm^3 of water. (They dissolve almost not at all.)

Answer 50 cm^3

2.8. A series of mixtures contain substances A and B, for which, when pure, $\nu_A = 7 \text{ cm}^3 \text{ g}^{-1}$, $\nu_B = 3 \text{ cm}^3 \text{ g}^{-1}$, $\bar{M}_A = 50 \text{ g mol}^{-1}$, $\bar{M}_B = 100 \text{ g mol}^{-1}$.

(a) Estimate the specific volume of a mixture with $w_B = 0.2$

Answer $v = 6.2\ \text{cm}^3\ \text{g}^{-1}$

(b) Estimate the density of a mixture with $x_A = 0.4$

Answer $\rho = 0.25\ \text{g cm}^{-3}$

(c) Estimate the molar volume of a mixture with $x_A = 0.4$

Answer $320\ \text{cm}^3\ \text{mol}^{-1}$

(d) Estimate the composition (mass fraction of B) in a mixture of density $\rho = 0.2\ \text{g cm}^{-3}$.

Answer $w_B = 0.5$

(e) Estimate the mole fraction of A in a mixture where $c_B = 1\ \text{mol dm}^{-3}$.

Answer $x_A = \frac{2}{3}$ or 0.67

2.9. A solute is to be extracted from aqueous solution by shaking the solution with a suitable solvent which is immiscible with water, so that on standing the mixture will separate into an aqueous layer and a solvent layer. The solvent is prepared by mixing 70 parts by volume of 1,2-dichlorethane ($C_2H_2Cl_2$) with 30 parts by volume of *n*-pentane (C_5H_{12}). Determine the density of the solvent layer if, at 25°C, the properties of the pure substances are $\rho_{C_2H_2Cl_2} = 1.246\ \text{g cm}^{-3}$, $\rho_{C_5H_{12}} = 0.621\ \text{g cm}^{-3}$ and $\rho_{H_2O} = 0.997\ \text{g cm}^{-3}$.

Answer $\rho_{\text{solvent}} = 1.058\ \text{g cm}^{-3}$

2.10 Removal of oxygen, water vapor and carbon dioxide from air leaves mainly nitrogen and argon. In one experiment such a residue had a density, at 300 K and 100 kPa, of $1.1288\ \text{g dm}^{-3}$. Estimate the ratio n_{Ar}/n_{N_2} in the original sample of air (for most gases at 300 K and 100 kPa, $\bar{V} \simeq 24.942\ \text{dm}^3\ \text{mol}^{-1}$).

Answer $n_{Ar}/n_{N_2} = 0.012$

2.11. The densities at 20°C of $0.1\ \text{mol dm}^{-3}$ solutions of hydrochloric acid, sodium chloride, cesium chloride and water are $1.000\ \text{g cm}^{-3}$, $1.0018\ \text{g cm}^{-3}$, $1.0127\ \text{g cm}^{-3}$, and $0.99823\ \text{g cm}^{-3}$ respectively. Estimate the effective molar volumes of these substances in aqueous solution. Assume that the effective molar volume of the chloride ion is the same for each salt and compare the effective molar volume of H_3O^+, Na^+, Cs^+.

Answer $\bar{V}\{H_3O^+ + Cl^-(aq)\} = 36.8\ \text{cm}^3\ \text{mol}^{-1}$

or $\bar{V}\{H^+ + Cl^-(aq)\} = 18.8\ \text{cm}^3\ \text{mol}^{-1}$

$\bar{V}\{Na^+ + Cl^-(aq)\} = 21.8\ \text{cm}^3\ \text{mol}^{-1}$

$\bar{V}\{Cs^+ + Cl^-(aq)\} = 23.7\ \text{cm}^3\ \text{mol}^{-1}$

2.12. A dispersion of two immiscible solvents is more stable (the solvents show less tendency to separate into layers) if they have the same density. An insecticide which is not soluble in water is soluble in mixtures of 1,2–dichloroethane and *n*–pentane. Estimate in what proportions these liquids must be mixed to make a solvent suitable for dispersing the insecticide in water. At 25°C —

$\rho_{C_2H_2Cl_2} = 1.246\ \text{g cm}^{-3}$,
$\rho_{C_5H_{12}} = 0.621\ \text{g cm}^{-3}$,
$\rho_{H_2O} = 0.997\ \text{g cm}^{-3}$.

Answer 4 parts C_5H_{12}:6 parts $C_2H_2Cl_2$ by volume

2.13. A soft drink manufacturer buys sugar as a concentrated syrup. One shipment had a density at 20°C of $1.287\ \text{g cm}^{-3}$. Estimate the volumes of bulk sugar and water required to make a $10\ \text{m}^3$ batch of syrup containing 5% by weight of sugar. At 20°C $\bar{V}^\circ_{\text{sugar (aq)}} = 0.627\ \text{cm}^3\ \text{g}^{-1}$

$\bar{V}_{H_2O} = 1.0018\ \text{cm}^3\ \text{g}^{-1}$

Answer $0.52\ \text{m}^3$ bulk syrup and $9.48\ \text{m}^3$ water

2.14. When using an analytical balance for the accurate determination of mass, it is necessary to know the density of the air. An analyst found that during one hot humid day (temperature 35°C, pressure 98 kPa) the relative humidity of the air in the balance room was 70%. That is, the partial pressure of water vapor (p_{H_2O}) was 3.92 kPa. From tables he found that the density of dry air at 35°C and 98 kPa is $1.108\ \text{g dm}^{-3}$, and that the molar volumes of gases under these conditions are approximately

$\bar{V}$(most gases) $\simeq 26.142\ \text{dm}^3\ \text{mol}^{-1}$ at 35°C, 98 kPa.

Estimate the density of the air in the room.

Answer $1.0912\ \text{g dm}^{-3}$

2.7 Pressure Change and the Space-filling Properties

Moderate changes in pressure have very little effect on the specific or molar volume of solid or liquid materials. What change there is, is approximately proportional to the change in pressure over a range of at least 100 P°. (P° is the standard atmospheric pressure, 101.3 kPa).

$$v_{P_2} - v_{P_1} = k(P_2 - P_1)$$

$$\bar{V}_{P_2} - \bar{V}_{P_1} = k'(P_2 - P_1)$$

The constants are expressed in terms of the *coefficient of compressibility*, β, measured at constant temperature.

$$\beta = -\frac{1}{v}\frac{dv}{dP}$$

The form of this coefficient is chosen so that its value is independent of the units of volume. So it describes the changes in the values of both the specific volume and the molar volume of a material and the change in volume of a body made of that material.

Example 2.15
At 298 K, β for water is 4.5×10^{-10} Pa^{-1}. How much does the molar volume at 10 P° differ from that at P°?

Answer
The change in molar volume is

$$\Delta\bar{V} = \bar{V}_{10P^\circ} - \bar{V}_{P^\circ}$$

$$= -\bar{V}_{P^\circ}\beta(10\,P^\circ - P^\circ)$$

$$\Delta\bar{V} = -(18.02)\times(4.5\times10^{-10})$$

$$\times(9\times101\ \times10^3)\ \text{cm}^3\ \text{mol}^{-1}$$

$$= -0.007\ \text{cm}^3\ \text{mol}^{-1}$$

Query This is very small. Is the change important?
Reply Not often. Unless much higher pressures than this are involved, we are unlikely to have to consider the change in molar or specific volume of a liquid or solid due to change in pressure.

The changes in volume of gases with change of pressure are very different from those of liquids and solids. Boyle found that, within the accuracy of his measurements, the volume of air at room temperature is inversely proportional to the pressure. This relation is valid within about 2% for most gases up to P°, although it may become very inaccurate at higher pressures. As an example of this more extended variation, consider the values of the product, $P\bar{V}$, for ethylene at 298 K (Fig. 2.3). As the pressure decreases toward zero, the value of the product increases toward a definite value (RT). Over the range to about 5 P°, the decrease in $P\bar{V}$ is linear (Fig. 2.3a).

Extension of the measurements to higher pressures shows that the curve becomes non-linear (Fig. 2.3b) and, if measurement is carried to really high pressures, the dependence can become complex (Fig. 2.3c).

The treatment of such behaviour usually involves fitting a polynomial (part of a power series) to it. Two polynomials are used, both

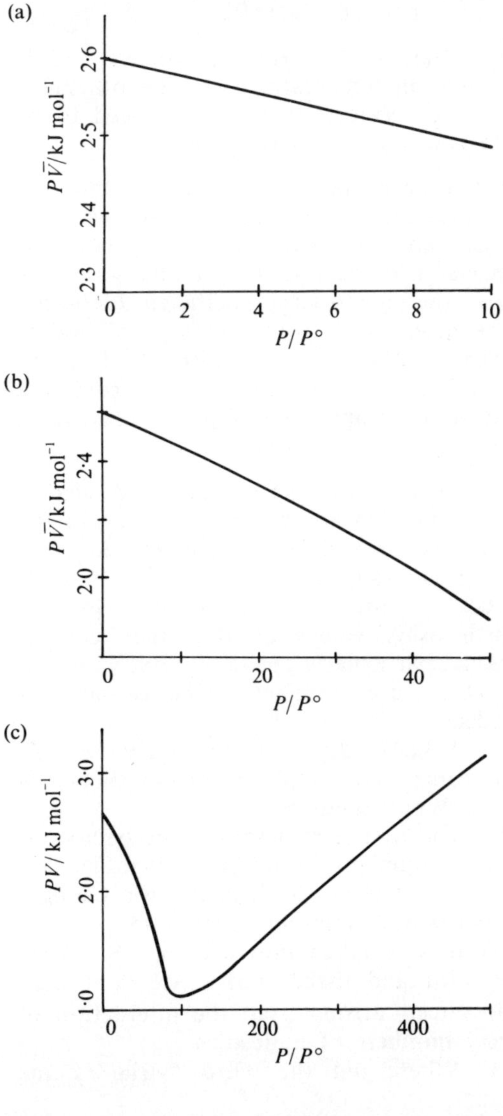

Fig. 2.3 The product PV for ethylene at 298 K.

called the *virial equation of state.* The "volume series" is

$$P\bar{V} = RT(1 + B\bar{V}^{-1} + C\bar{V}^{-2} + \ldots)$$

and the "pressure series" is

$$P\bar{V} = RT + B^*P + C^*P^2 + \ldots$$

The coefficients RT, B, C ... are called the first, second, third, ... virial coefficients.

Query Why use two series?
Reply They are useful for different purposes. The coefficients of one set can be converted to those of the other

$$B^* = B$$

$$C^* = (C - B^2)/RT$$

Query The first virial coefficient (RT) depends on temperature. Do the others?
Reply Yes. Values of B, C, ... need to be determined for each temperature.

As has been stated, the value of the first term, RT, is determined from the properties of the gas at low pressures. Its value is independent of the nature of the gas. The value of the second virial coefficient (B or B^*) can be determined from the slope of graphs like that in Fig. 2.3a. The third virial coefficient (C or C^*) is determined from the curvature in graphs like Fig. 2.3b (or from equivalent numerical procedures).

Query How many coefficients are needed?
Reply That depends on the accuracy of the data. There are a good many second virial coefficients available, and a few reliable values of third virial coefficients. So it is worth using values of B if they can be found,‡ particularly above $P°$. But values of C, D, ... are not likely to be available or needed.
Query I know the "first virial coefficient", RT, comes into simple molecular theory of gases. What about the others?
Reply The first term describes the behaviour of large numbers of independent molecules. The second term, B, describes the changes in behaviour when interaction of pairs of molecules is taken into account. Similarly the third and higher terms are connected with effects arising from the interaction of larger numbers of molecules.
Query Where did the word "virial" come from?
Reply Its meaning is related to force and it was meant to draw attention to the forces between molecules.

‡Dymond, J.H. and Smith, E.B., *The Virial Coefficients of Gases*, Oxford University Press, Oxford, 1969.

Example 2.16
For the refrigerant gas, dichlorodifluoromethane, CCl_2F_2, at 373.16 K

$$B = -282 \text{ cm}^3 \text{ mol}^{-1}$$

$$C = 23\,515 \text{ cm}^6 \text{ mol}^{-2}$$

What is the pressure for which $\bar{V}$ is 400 $\text{cm}^3 \text{ mol}^{-1}$?

Answer

$$P\bar{V} = RT(1 + B/\bar{V} + \bar{C}/\bar{V}^2)$$

$$= 8.315 \times 373.16\,(1 + (-282)/400 + 2.35 \times 10^4/(400)^2)$$

$$= 8.315 \times 373.16\,(1 - 0.705 + 0.147)$$

$$P = 3.43 \times 10^6 \text{ Pa}$$

$$= 33.8\ P°$$

Another useful way of representing gas behaviour involves the compressibility factor, Z.

$$Z \equiv \frac{P\bar{V}}{RT}$$

The deviation of Z from unity is a measure of the extent to which the behaviour of the gas at some pressure, P, differs from the simple low-pressure behaviour. If the third, and higher, virial coefficients can be neglected, Z is connected simply with the second coefficient, B.

$$Z = 1 + \frac{BP}{RT}$$

Example 2.17
At 250 K, dichlorodifluoromethane has $B = -586 \text{ cm}^3 \text{ mol}^{-1}$.
What is Z at $P°$?

Answer

$$Z = 1 - \frac{586 \times 10^{-6} \times P/\text{Pa}}{8.315 \times 250}$$

$$= 1 - 2.82 \times 10^{-7}\ P/\text{Pa}$$

$$= 0.972 \text{ (at } P°)$$

Comment So this gas shows 3% deviation from "perfect" gas behaviour under these conditions.

2.8 Temperature Change and Volume of Uniform Materials

The effect of temperature change on the volume of uniform solid or liquid materials is usually slight, though it is often important. For example, it is necessary to build expansion joints into buildings and bridges. This effect is also the basis of many thermometers.

When a correction needs to be made, the change in volume is usually proportional to the change in temperature.

$$V_{T_2} - V_{T_1} = k(T_2 - T_1)$$

The value of k is expressed, like the corresponding quantity β for pressure change, in a form which makes the coefficient independent of the volume units and hence equally applicable to changes in v, $\bar{V}$ and V.

$$k = \alpha V = (\mathrm{d}V/\mathrm{d}T)$$

or

$$\alpha = \frac{1}{V}(\mathrm{d}V/\mathrm{d}T)$$

The quantity α is the *coefficient of thermal expansion.*

Example 2.17
The value of α for liquid water is $4.8 \times 10^{-4}\ \mathrm{K}^{-1}$ at 25°C. How much does the molar volume at 50°C differ from that at 25°C?

$$\Delta\bar{V} = \bar{V}_{50\,\mathrm{C}} - \bar{V}_{25\,\mathrm{C}} = \bar{V}_{25\,\mathrm{C}}\,\alpha(50-25)$$
$$= (18.02)\times(4.8\times10^{-4})\times(25)\ \mathrm{cm^3\ mol^{-1}}$$
$$= 0.21\ \mathrm{cm^3\ mol^{-1}} \quad (1.1\%\ \text{change})$$

Comment The range of such an equation must be limited? Water has a maximum density!

Reply Yes. This equation is a first approximation for use close to the point where it is determined, and there are a few awkward cases like water.

Query Suppose we have changes in both temperature and pressure?

Reply The first, and usually sufficient, approximation is to sum separately the effects due to temperature and pressure. So, for example,

$$\Delta\bar{V} = (\mathrm{d}\bar{V}/\mathrm{d}T)\Delta T + (\mathrm{d}\bar{V}/\mathrm{d}P)\Delta P$$
$$= \bar{V}(\alpha\Delta T + \beta\Delta P)$$

Example 2.18
What is the difference in molar volume of water at 50°C and 10 P° from the value at 25°C and P°?

Answer
From previous examples,

$$\Delta\bar{V} = -0.007 + 0.21\ \mathrm{cm^3\ mol^{-1}}$$
$$= 0.20\ \mathrm{cm^3\ mol^{-1}}$$

PROBLEMS

2.15. Estimate the change in the pressure when the temperature of a silica tube completely filled with benzene is changed from 20°C to 30°C.

$$\alpha(\text{benzene}) = 1.2\times10^{-3}\ \mathrm{K}^{-1}$$

$$\beta(\text{benzene}) = 9\times10^{-10}\ \mathrm{Pa}^{-1}$$

(The volume of the tube does not alter significantly.)

Answer $1.3\times10^{7}\ \mathrm{Pa} = 130\ P^\circ$

For gases at low pressures the effect of change in temperature is given by

$$P\bar{V} \simeq RT$$

At higher pressures, it is necessary to add the changes which result from the variation of the virial coefficients with temperature. The next problem shows the extent to which departure from ideality alters at a single pressure, as reflected in the changes in the compressibility factor.

2.16. The second virial coefficients of acetone are quoted as

T/K	$B/\mathrm{cm^3\ mol^{-1}}$
300	−2000
320	−1520
340	−1200
360	−960

Calculate the values of the compressibility factor at 0.1 P° for acetone at these temperatures.

2.9 Changes of Phase and the Space-filling Properties

In the preceding sections we found it necessary to discuss the effect of composition, pressure and temperature for gases in different ways to those used for solids and liquids. When a change in conditions leads to transfer from the liquid or solid state to the gaseous state, there is a large change in the specific (and the molar) volume. There is one exception—we shall see (Section 3.3.6) that there is a region very close to the "critical point" where liquid and gas have very similar properties, including specific volume. The difference in space-filling properties between liquids and solids or between different solid forms of the same substance are smaller (a few per cent) but definite. Thus for sulfur at 18°C, the rhombic form of S_8 has $v = 0.485\ cm^3\ g^{-1}$ whereas the monoclinic form of S_8 has $v = 0.510\ cm^3\ g^{-1}$. The S_6 solid is reported to have $v = 0.468\ cm^3\ g^{-1}$. Differences like these are readily detected and so may be used to follow the transformation of one solid into the other. The specific volume of liquid sulfur is close to 10% larger than that of rhombic S_8. These differences are also large enough to form the basis for methods of separation of the solids from one another or from the liquid.

2.10 Models and Space-filling Properties

The simplest model with which we might try to account for the properties of liquids and solids pictures the molecules as spheres. If we assume that pure solid or liquid substances are composed of closely-packed rigid spheres, we can estimate the size of a molecule from the molar volume. On this model, the molar volumes of mixtures should be related simply to the fractions of each type of molecule present. Then we may expect that the molar volume of a mixture of A and B would obey the equation:

$$\bar{V}(mixture) = x_A \bar{V}_A + x_B \bar{V}_B$$

Thus the theory can explain the simple ideal law which we have tested.

When this sort of simple mixture law does not fit, the strategy may be to introduce forces between pairs of molecules. Such forces will be chosen to make the values of $\bar{V}(mixture)$ smaller or larger than those calculated on the simpler theory. In some cases, an alternative is to replace the spherical model by some less symmetrical or more flexible shape, based on our guesses from molecular models. The iodine molecule, for example, might be better regarded as a dumbell. Larger molecules may be like long rods or discs. Thus the molar volumes of sulfanes given in Table 2.7 increase in a very regular way as though extra segments were being added to a rod. For crystalline solids, there is evidence from diffraction experiments that the molecules are arranged in regular repeating units. For liquids, the most satisfactory models at the present time are erratic assemblies of molecules with sufficient forces of attraction to hold them together.

For gases, any such "packed sphere" model is inadequate. As a first approximation, the molecules are treated as being independent and as having a volume which is negligible by comparison with the total volume available to the sample. Higher approximations allow for interactions between molecules across small distances as well as the exclusion of each molecule from the volume contained *within* the others.

Table 2.7
Specific and Molar Volumes of Liquid Sulfanes (at 20°C)

	$v/cm^3\ g^{-1}$	$\bar{V}/cm^3\ mol^{-1}$
H_2S	1.007	32.2 (at −60°C)
H_2S_2	0.750	49.6
H_2S_3	0.671	65.9
H_2S_4	0.632	82.3
H_2S_5	0.608	98.7
H_2S_6	0.592	115.2
H_2S_7	0.581	131.6
H_2S_8	0.572	148.0

3.

Behaviour of Phases

3.1 The Importance of Phases

In this chapter we will consider the conditions under which solid, liquid and gaseous phases exist together in equilibrium, or transform into one another. In doing so we extend the analysis in Section 2.9 of the marked dependence of the values of molar volume and density of a substance on the phase which is present.

Such differences between the properties of phases are very important in practice. If we have to design a tank to store 1000 kg of oxygen, we had better ask whether it is to be liquid oxygen or gaseous oxygen before we begin. The design of pumps for transferring liquid oxygen into the tank will also be very different from the design of pumps for transferring gaseous oxygen. The changes which accompany transformation from one phase to another are also of great practical value—for example, we are very much accustomed to controlling our body temperature by controlling the rate of evaporation of water at its surface! Most of the methods we use in separating substances depend on the differences between the properties of phases. So, for example, we often clarify the water from muddy streams by adding some polymeric substance. This causes the mud particles to come together ("flocculate") and then the difference in density between the flocculated particles and the water causes the particles to sink, leaving the water clear. Briefly, we may say that differences in phase are important because they give rise to differences in physical properties and in extent of chemical reaction. To help in discussing situations like these we use "phase diagrams".

3.2 Phase Diagrams

Phase diagrams are maps which show the conditions under which the various phases are capable of stable existence and the conditions under which one phase can be transformed into another. As in other maps, we use contour lines or sections in phase diagrams whenever three or more dimensions would be required to form a complete map. Many properties of substances may be represented by these maps—not only pressure, temperature, and molar volume, but also the speed of sound or light, viscosity and many other properties.

3.3 Maps of Pressure and Temperature for a Simple System

A useful way to begin looking at phase diagrams is to study the kind of map represented by Fig. 3.1. This displays the values of pressure and temperature for which carbon dioxide exists as a solid, as a liquid or as a gas.

3.3.1 Conditions for having one phase

Points which lie within the area ABD of Fig. 3.1 involve values of pressure and temperature for which only the *solid* phase is present. Points within the area DBC involve values for which there is only *liquid* present. Points in the area under the lines AB and BC involve only *gas*.

3.3.2 Conditions for having two phases

If we want the conditions which *limit* the existence of the solid as the stable phase, we look at the boundaries of the area ABD. The map shows that transformation to liquid ("melting") begins when conditions reach those on the line BD. Transformation to gas ("sublimation") occurs for the range of conditions on the line AB. This sublimation transformation is used whenever solid carbon dioxide ("dry ice") is used to cool objects. Clearly the temperature attained at equilibrium can have a range of values depending on the pressure—if we pick a pressure there can only be one equilibrium temperature for the presence of both solid and gas.

In a similar way, the limits to the existence of the liquid as the stable phase are the line BD (where liquid forms solid—"freezes") and the line BC (where liquid forms gas—"boils"). Carbon dioxide is most commonly sold in cylinders, which contain liquid in equilibrium with gas. The pressure inside the cylinder will depend upon the storage temperature.

The idea of gas, in the sense of transformed liquid, was used long before any other idea of "gas" had been accepted. The Shorter Oxford Dictionary quotes a use of the word "vapor" as early as 1610 which has exactly the meaning we have in mind—"vapour is a moist kind of fume extracted chiefly out of the water". Most of us can recall seeing events which fit that quotation on warm mornings after rain. We continue to use the word "vapor" in terms like "vapor pressure". Vapors are simply substances present in a gas phase that is in equilibrium with a liquid or a solid phase.

3.3.3 Melting, boiling and sublimation points

People often talk of melting or boiling as though a single definite temperature were involved. When we look for these changes on Fig. 3.1 it is obvious that the temperature for each process depends on the experimental pressure. The values usually quoted are the "normal" melting and boiling points and refer to the standard atmospheric pressure, P° (101.325 kPa). For carbon dioxide the process at this pressure is sublimation and it takes place at 195 K. Melting or boiling can occur only at pressures for which there is a liquid phase, i.e. above 527 kPa.

Exercise 3.1
What are the boiling and melting temperatures of carbon dioxide at 6000 kPa? (See Fig. 3.1.)

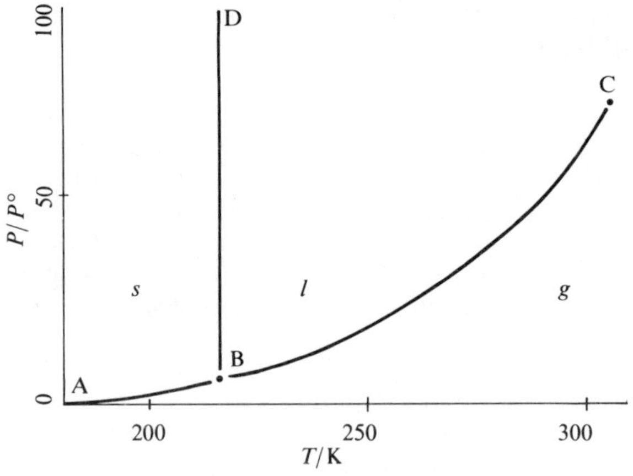

Fig. 3.1 A phase equilibrium map for carbon dioxide.

Most substances have a pressure less than P° at point B and so their melting and boiling temperatures are observed more frequently than their sublimation temperatures.

3.3.4 Triple points

The three lines in Fig. 3.1 intersect at B. Since this point represents the junction of lines for the equilibrium of:

solid and gas
solid and liquid
and liquid and gas,

all three phases are present at B. Such points are very significant in the treatment of systems containing one substance. Because *three* phases are in equilibrium, these points are known as *triple points*. The triple point for water has already been met—it defines the Kelvin temperature scale (see Section 1.2.1). The values for other compounds can be used in characterising them. The lines which correspond to BD have such steep slopes that the melting temperatures depend very little on the pressure. This is why melting points measured at any pressure close to P° can be used to characterise pure substances. Even though the triple point would give more certain information, it is less convenient to measure.

3.3.5 Slopes of the lines

When the theory of equilibrium is developed further than this book attempts, it can be used to predict the slope, $(\mathrm{d}P_{eq}/\mathrm{d}T)$, of any of the lines of a phase diagram like Fig. 3.1 (for example, see W.J. Moore, *Physical Chemistry*, 5th ed., Longman, London, 1972, p. 211). For equilibrium involving change from a phase α to any other phase β of *any* substance, the result obtained is known as the Clapeyron equation:

$$\frac{\mathrm{d}P_{eq}}{\mathrm{d}T} = \frac{\Delta \bar{H}(\alpha \rightarrow \beta)}{T \Delta \bar{V}(\alpha \rightarrow \beta)}$$

where $\Delta \bar{H}(\alpha \rightarrow \beta)$ is the difference in molar enthalpy of the phases, $\Delta \bar{V}(\alpha \rightarrow \beta)$ is the difference in the molar volume of the phases, and $\mathrm{d}P_{eq}$ is a change in pressure measured along the line which corresponds to the equilibrium of α and β. This very simple equation provides a way to measure the slopes.

Query What exactly is the meaning of $\Delta \bar{H}(\alpha \rightarrow \beta)$ and of $\Delta \bar{V}(\alpha \rightarrow \beta)$?

Reply There is a definite value of $\bar{V}_{\alpha}$, the molar volume in phase α, and a different value, $\bar{V}_{\beta}$ for phase β. $\Delta \bar{V}(\alpha \rightarrow \beta)$ is simply the difference between them, $(\bar{V}_{\beta} - \bar{V}_{\alpha})$. In exactly the same way, $\Delta \bar{H}(\alpha \rightarrow \beta)$ means the difference between the two values of the molar enthalpy, $\bar{H}_{\beta}$ and $\bar{H}_{\alpha}$. So the "Δ" means "difference between the properties of". Values of $\Delta \bar{V}$ and $\Delta \bar{H}$ can be obtained from measurements involving the two phases separately without direct study of the transformation of the phase α into the phase β.

Query What general conclusions about the slopes follow from this equation?

Reply Take melting first. For this, $\Delta \bar{H}$ (solid → liquid) is always positive. Almost all substances have positive values of $\Delta \bar{V}$ (solid → liquid). That is, substances almost always melt at higher temperatures if the pressure is higher. The exceptions, ($\Delta \bar{V}$(solid → liquid) negative), of which the best known case is water, have lower melting temperatures when the pressure is higher. This exception is highly significant. It is essential to the survival of fish below ice packs and to the "flow" of ice in glaciers.

The formation of gas phases from liquids or solids always has positive values of both $\Delta \bar{H}$ and $\Delta \bar{V}$. So the pressure of a pure gas in equilibrium with a pure liquid or pure solid always increases with rising temperature.

When solids change to other solids (like diamond forming graphite, or white phosphorus forming red), there is no simple way of predicting the direction of change apart from measurement either of the slope itself or, more likely, of $\Delta \bar{H}$ and $\Delta \bar{V}$.

At low pressures (below, say, 2 to 5 P°) the value of $\bar{V}$ for a gas (for example, of the order of 25 000 $\mathrm{cm^3\ mol^{-1}}$ at P°) is so much greater than that for a solid or liquid (for example, $10 - 100\ \mathrm{cm^3\ mol^{-1}}$) that $\Delta \bar{V}$ for formation of a gas from a liquid or a solid is very close to the molar volume of the gas. That means we simplify the general equation for such processes to get the equations:

$$\frac{\mathrm{d}P_{eq}}{\mathrm{d}T} \approx \frac{\Delta \bar{H}(l \rightarrow g)}{T\bar{V}(g)}$$

and

$$\frac{\mathrm{d}P_{eq}}{\mathrm{d}T} = \frac{\Delta \bar{H}(s \rightarrow g)}{T\bar{V}(g)}$$

If the pressure is not high, we can get useful accuracy by using the ideal gas approximation

$$\bar{V}(g) = RT/P_{eq}$$

leading to

$$\frac{\mathrm{d}P_{eq}}{\mathrm{d}T} \approx \frac{\Delta \bar{H}(l \rightarrow g)}{T(RT/P_{eq})}$$

and

$$\frac{\mathrm{d}P_{eq}}{\mathrm{d}T} \approx \frac{\Delta \bar{H}(s \rightarrow g)}{T(RT/P_{eq})}$$

which can be written in the further forms, known as the Clausius–Clapeyron equations:

$$\frac{\mathrm{d} \ln P_{\mathrm{eq}}}{\mathrm{d}T} \approx \frac{\Delta \bar{H}(l \rightarrow g)}{RT^2}$$

and

$$\frac{\mathrm{d} \ln P_{\mathrm{eq}}}{\mathrm{d}T} \approx \frac{\Delta \bar{H}(s \rightarrow g)}{RT^2}$$

since

$$\frac{\mathrm{d}P}{P\mathrm{d}T} = \frac{\mathrm{d} \ln P}{\mathrm{d}T}$$

These equations can be integrated to give the value P_2 of the gas pressure at one temperature T_2, given its value P_1 at another temperature T_1. This is especially simple if $\Delta \bar{H}$ is close enough to constant for the purpose in hand, as it often is even over a range of temperature of 100 K. Then:

$$\ln(P_2/P_1) \simeq -\frac{\Delta \bar{H}}{R}\left(\frac{1}{T_2} - \frac{1}{T_1}\right)$$

that is,

$$2.303 \log \frac{P_2}{P_1} \simeq \frac{\Delta \bar{H}}{R} \frac{(T_2 - T_1)}{T_1 T_2}$$

These relationships between equilibrium gas pressures and temperatures have three common uses. Frequently they are used to get values of $\Delta \bar{H}$ by measuring the pressures over a range of temperatures. They permit correction of boiling temperatures to the values at the standard pressure, P°, or the estimation of boiling temperatures and pressures for the low-pressure distillation of substances which decompose at their normal boiling point. They are also the basis of a convenient method (Fig. 3.2) of rescaling maps like Fig. 3.1. The lines in the new form are almost straight, which makes them easy to handle.

Query The sign "≃" ought to mean those equations are not accurate. Why does this last graph have such straight lines?

Reply That is a happy accident. $\Delta \bar{H}$ decreases as T increases, but so does $\Delta \bar{V}$. So the two things often cancel one another.

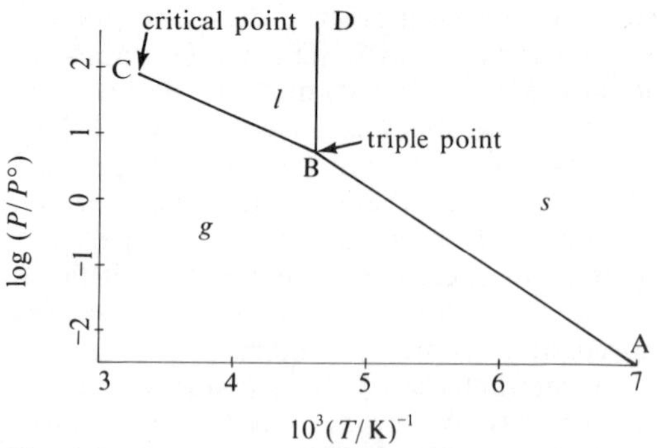

Fig. 3.2 A more convenient phase equilibrium diagram for carbon dioxide.

3.3.6 The critical point

Figures 3.1 and 3.2 show lines for the equilibrium between liquid and gas which have a definite upper end. At this end, the substance is in its *critical state* and the point is called the *critical point*. No matter what property we measure, the values for liquid and gas get closer together the closer we get to the critical state. For example, both $\Delta \bar{H}(l \rightarrow g)$ and $\Delta \bar{V}(l \rightarrow g)$ get smaller and smaller and both approach zero at this point. Beyond this point there is no sensible way of distinguishing between a liquid and a gas. There is one phase and not two.

Query That is a nuisance. What will we call it?

Reply The old name "fluid" is available and it is used whenever there is only one phase and we are not sure whether to say "liquid" or "gas". If you think of an experiment which takes the system near the critical point you will see that the map predicts some strange behaviour. If we start with liquid and gas, we can boil the liquid until it is all gas. If we then alter the pressure and temperature systematically, we can get back to the starting point without crossing the liquid–gas line. So you ought to be able to boil the "liquid" over and over without ever having it "condense".

Query Does it work?

Reply Yes.

Query How do we know whether we have liquid or gas?

Reply We don't. The assignment of the names can only be made unambiguously when both the liquid and the gas are present.

Query Figures 3.1 and 3.2 have a solid–liquid line which looks like the liquid–gas line. Does it end in a critical point, too?

Reply Not as far as we know. At the highest pressures tried there is no sign that $\Delta \bar{H}(s \rightarrow l)$, $\Delta \bar{V}(s \rightarrow l)$ or similar properties of the change of phase are approaching zero.

3.4 Maps of Pressure and Molar Volume

There are many other phase diagrams which can be drawn to show the properties of a substance like carbon dioxide. Such maps may look very different but they contain similar information. One of these displays the pressure as it varies with the volume per mole (molar volume). Figure 3.3 shows the properties of carbon dioxide plotted in this form. Two constant temperature lines (isotherms) are shown.

3.4.1 Tie lines

At low pressures the molar volumes of liquid and gas are very different when they are in equilibrium. So the map (Fig. 3.3) shows the properties of these phases as points on different lines, PQ and RQ. The point Q where the lines join is the critical point. Some name is needed for lines like XY which connect points corresponding to different phases in equilibrium with one another. They are called *tie lines* and often occur in phase diagrams.

A sample of carbon dioxide with values of pressure and of overall molar volume corresponding to a point M in the tie line XY will be part liquid and part gas. To approach the liquid extreme X, a greater proportion of the body of material must become liquid. Thus the overall molar volume will become smaller.

Query What happens in the space the tie lines traverse?

Reply No *single* phase has properties within that region. The tie lines connect the values of the property for the two phases which exist at the given temperature and pressure.

Query What meaning, if any, is attached to the distance along a tie line?

Reply The value of the property, somewhere along a tie line, must be an average over the two phases taking their relative amounts into account. A point near X corresponds to nearly all of the carbon dioxide sample being in the liquid phase. A point near Y implies that nearly all of the sample is gaseous.

It is easy to show that the amounts of the phases for a total value of molar volume corresponding to a point such as M must be related by a "lever rule":

(amount of phase Y) × (distance to Y)

= (amount of phase X) × (distance to X)

Exercise 3.2

A phase diagram for oxygen is shown in Fig. 3.4. What is the highest temperature at which liquid and gaseous oxygen can be in equilibrium? What other properties can be deduced for this state? What are the normal boiling and freezing points of liquid oxygen?

Exercise 3.3

The phase diagrams of carbon and of water are

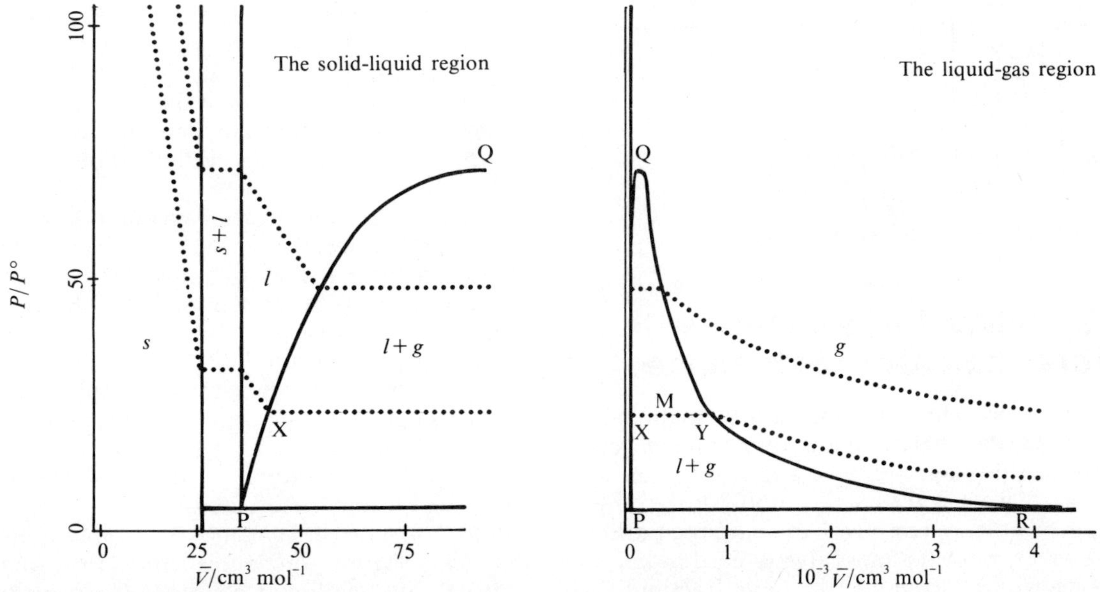

Fig. 3.3 Part of the pressure-molar volume map for carbon dioxide.

shown in Figures 3.5 and 3.6. What conditions are needed if diamonds are to be made? Under what conditions can the solid known as "ice V" be formed?

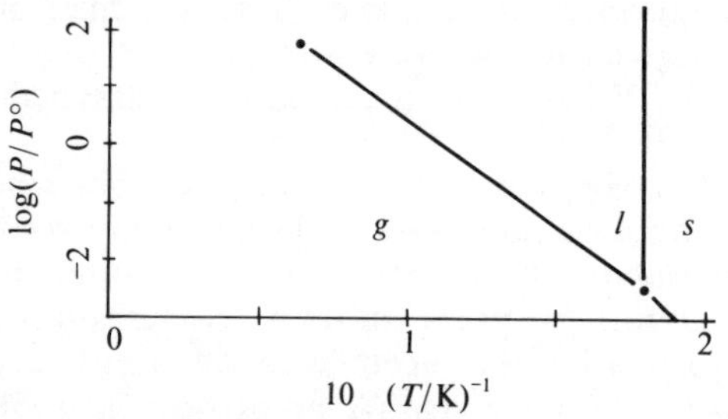

Fig. 3.4 Phase map for oxygen.

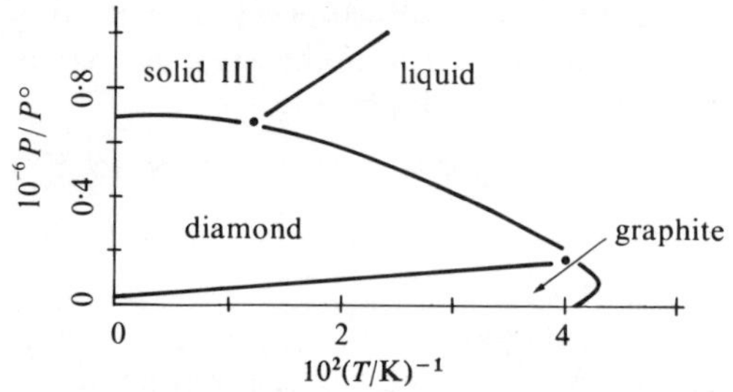

Fig. 3.5 Phase map for carbon.

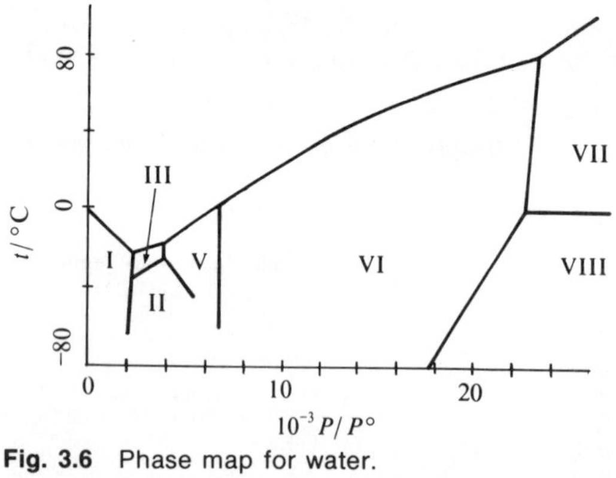

Fig. 3.6 Phase map for water.

3.5 Maps for Systems with More than One Substance

When a system contains more than one substance, the total behaviour of the system gets harder to display. We have a new question to answer—"what is the composition"? For example, when the system involves sodium chloride and water, we will almost always need to know the degree (for example, the mole fraction) to which there is sodium chloride in every solid, liquid and gas phase. If there is only one phase, three variables can be altered (for example, the pressure, the temperature, the mole fraction of NaCl) without necessarily shifting from, say, a liquid phase to a solid or gas phase.

Query What sort of map does this require?
Reply A three-dimensional map—made literally, as a sculpture; or made mathematically, by using a computer. Very often we decide to look at the behaviour in two dimensions. Then we draw maps on which some variable (temperature, composition, etc.) is kept constant. With composition variables there are more useful aspects to look at and some new types of behaviour. We will look mostly at just two substances at a time. Putting in more than two substances gives more scope for varying behaviour, but does not alter the patterns drastically.

3.6 Maps for Two Liquids in Equilibrium

Let us begin with a type of behaviour which is not observed when there is only one substance. We are often told that "oil and water don't mix". Anyone who has run into trouble with water in a petrol tank will know that problem only too well. If you carefully separate the "water" and the "petrol" layers, it is easy to show that the "water" does contain petrol—it tastes—and that the "petrol" does contain water—which you can get out with a "drying agent" like calcium chloride. This is a general result: oil and water *do* mix (but not very much). A more accurate description is: two liquid phases; one mostly water and one mostly oil.

Changing the temperature changes the extent to which the substances dissolve in one another. The map for a system like this, 2,4,6–trinitrotoluene (TNT) in α–naphthol, is shown in Fig. 3.7. In this case the solubility *increases* with rising temperature until 126°C. Above 126°C, the two substances can be mixed in any ratio to form a single liquid phase. (They are completely miscible!) Whether one is working above or below that temperature is rather important if you are using such mixtures and so the temperature merits a name—the *upper consolute temperature*. Above this temperature, there can be only one liquid phase, as is always the case for, say, ethyl alcohol and water. Below it,

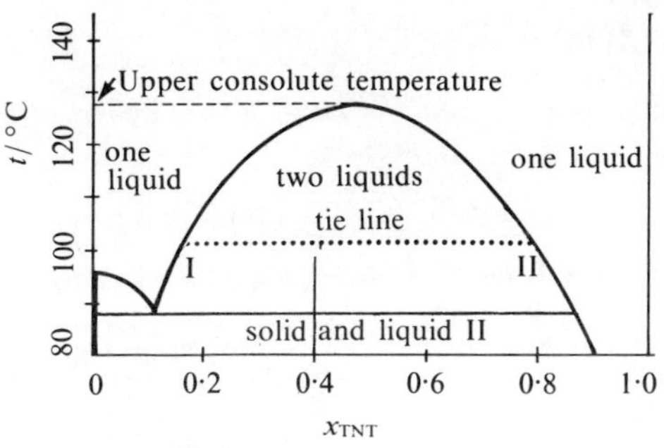

Fig. 3.7 Solubility of 2,4,6-trinitrotoluene (TNT) and -naphthol.

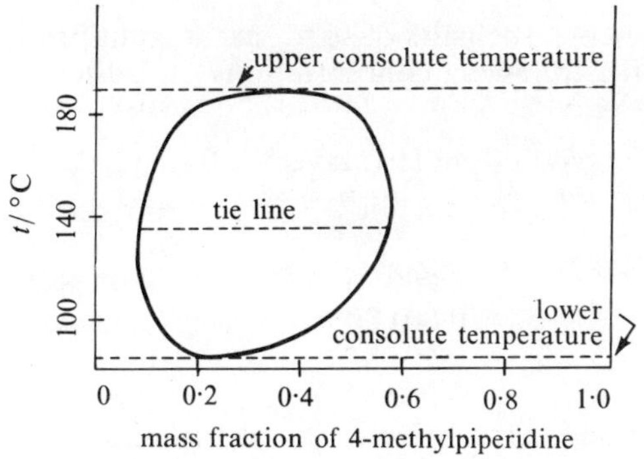

Fig. 3.8 Solubility of 4-methylpiperidine and water.

there may be two liquid phases, as with oil and water. Thus, at 100°C, mixing 2 mmol of TNT and 3 mmol α–naphthol gives two liquid phases. One (II) has

$$x_{TNT} = 0.78$$

and the other (I) has

$$x_{TNT} = 0.18$$

By the "lever rule", these are predicted to be present in the (mole) ratio

$$\frac{\text{Amount of II}}{\text{Amount of I}} = \frac{0.40-0.18 \text{ mol}}{0.78-0.40 \text{ mol}}$$

$$= 0.58.$$

Exercise 3.4

(a) How much TNT can be dissolved in α–naphthol at 120°C?

(b) What are the compositions of the liquids which are in equilibrium at 90°C?

(c) What relative amounts of the liquids are formed at 90°C, starting from the solution in (a) above?

(d) What fraction of the original TNT will be present in the richer liquid?

Answer

$x_{TNT} = 0.33$; $x_{TNT} = 0.14, 0.84$; 0.37; 0.69

If the solubilities *decrease* as the temperature rises, there can be a temperature below which there is unlimited mixing. Such mixtures are said to have a *lower consolute temperature*, as in Fig. 3.8. This happens less often than upper consolute temperatures, as one substance often solidifies before the stage of complete mixing is reached. That happens with α–naphthol in Fig. 3.7.

We may want two liquids to have low solubility in one another, as when skimming fat from soup, or we may want them to have high solubility, as in dispersing oil slicks on waterways. In either case, adding a third substance can help. A substance which dissolves in both the initial liquids increases their solubility in one another, often by enough to make them completely miscible with one another. That is part of the story with soaps and detergents used for washing greasy dishes. When the third substance has high solubility in one liquid and very low solubility in the other, its addition to the mixture lowers the mutual solubility. This principle lies behind the method of "salting out" which is used, for example, when a salt is added to a mixture of ether and water to improve the efficiency with which ether is recovered.

3.6.1 Distribution of a solute between two liquid phases

When a small quantity of a solute is added to a system which has two liquid phases, the solute usually distributes itself between them. If the amount of the added substance is small, the concentration of the substance in one phase is found to be proportional to the concentration in the other phase. That is,

$$c_{\text{solute in phase } \alpha} \,/\, c_{\text{solute in phase } \beta} \simeq K$$

The *distribution constant K* depends on what is the solute, what are the two liquid phases, and on the temperature.

Example 3.1

A series of solutions of iodine in chloroform, $CHCl_3$, were shaken with water at 298 K until equilibrium was established. In four experiments, the pairs of solutions at equilibrium had the following concentrations of iodine:

Experiment	$c\{I_2$ in H_2O layer$\}$ mol dm^{-3}	$c\{I_2$ in $CHCl_3$ layer$\}$ mol dm^{-3}
1	0.000 25	0.0338
2	0.001 20	0.1546
3	0.001 84	0.2318
4	0.002 42	0.3207

What is the value of the distribution constant?

Answer
For experiment 1,

$$c_{I2 \text{ in } H2O \text{ layer}} \ / \ c_{I2 \text{ in } CHCl2 \text{ layer}} \approx K$$

$$0.00025 \quad / \quad 0.0338 \quad = 0.0074$$

For the other experiments the estimates of K are

0.0076; 0.0079; 0.0075.

The average estimate of the distribution constant is 0.0076.

The distribution constant reflects the relative solubilities of the solute in the two separate pure liquid substances. The solubility of iodine in pure water at 298 K is 2.25×10^{-3} mol dm^{-3}, while its solubility in pure chloroform is 0.352 mol dm^{-3}. The ratio of these solubilities is 0.0064; similar to, but not equal to, the distribution coefficient measured in the above example. The difference arises because water and chloroform dissolve in one another to some degree. We could set up equilibrium between solid iodine, water and chloroform together in one container. The iodine concentrations in the layers are altered slightly by the presence of the small fraction of chloroform in the water and by the small fraction of water in the chloroform.

It will be clear that repeated use of a solvent like this can reduce the concentration in one phase to a low value. Let us take the example further to see how quickly it happens.

Example 3.2
How many times would it be necessary to equilibrate 1 dm^3 of a solution of iodine in water with 10 cm^3 samples of chloroform in order to reduce the iodine concentration to 10^{-2} of an initial value (X)?

Answer
At equilibrium

$$\frac{I_{2 \text{ in } H_2O}}{c_{I_2 \text{ in } CHCl_3}} = K \simeq 0.0076$$

$$\frac{c_{I_2 \text{ in } H_2O} \times V_{H_2O}}{c_{I_2 \text{ in } CHCl_3} \times V_{CHCl_3}} = \frac{0.0076\, V_{H_2O}}{V_{CHCl_3}}$$

$$\Rightarrow \quad \frac{n_{I_2 \text{ in } H_2O}}{n_{I_2 \text{ in } CHCl_3}} = 0.0076 \times 100$$

$$\frac{n_{I_2 \text{ in } H_2O}}{n_{I_2 \text{ in } H_2O} \ n_{I_2 \text{ in } CHCl_3}} = \frac{0.76}{1+0.76} = 0.432$$

In successive equilibration experiments, the value of $n_{I_2 \text{ in } H_2O}$ from one becomes $(n_{I_2 \text{ in } H_2O} + n_{I_2 \text{ in } CHCl_3})$ for the next.
So, for p experiments

$$\frac{(n_{I_2 \text{ in } H_2O}) \text{ after exp.} p}{(n_{I_2 \text{ in } H_2O} + n_{I_2 \text{ in } CHCl_3}) \text{ after exp.1}} = 0.432^p$$

$$10^{-2} = 0.432^p$$

Hence $$p = 5.5$$

Thus *six* extractions will be sufficient.

Exercise 3.5
Is trichloroethylene CCl_2CHCl a better choice for extracting iodine from water than chloroform? The following data are available:

$c_{I_2 \text{ in } H_2O}$	$c_{I_2 \text{ in trichlorethylene}}$
0.002 499	0.2749
0.004 498	0.5098
0.006 623	0.7748
0.007 873	1.087

Answer
Only slightly.

Exercise 3.6
The use of 1-pentanol is proposed for possible use in extracting hydrogen peroxide from aqueous solutions. Is this likely to be worth study?

$c_{H_2O_2 \text{ in } H_2O}$	$c_{H_2O_2 \text{ in 1-pentanol}}$
0.0940	0.0134
0.1935	0.0280

Answer
No.

3.7 The Maps showing a Solid and a Liquid in Equilibrium

There are many common examples in which a liquid and a solid are close to equilibrium, including surfaces in icebergs, glaciers and iced drinks. The maps which describe their behaviour often look like Fig. 3.9. Temperatures above the curved lines AZ and ZB correspond to conditions when there is only a liquid phase. These curved lines give the conditions for equilibrium of the liquid and one or other of the solids. The maps are drawn for a selected value of pressure, often the standard atmospheric pressure.

We have developed the habit of thinking of these curved lines in different ways, depending on the place we start map reading. If we start with the mole fraction, we will think of the curve as giving freezing temperatures (for example, a solution of *p*-dichlorobenzene (mole fraction 0.2) in cyclohexane freezes at 6°C, according to Fig. 3.9). If, on the other hand, we start with the temperature, we will think of the same curve as giving the solubility of one substance in another (for example, the solubility of *p*-dichlorobenzene in cyclohexane at 0°C is $x = 0.16$, according to Fig. 3.9).

Exercise 3.7
(a) What is the solubility of *p*-dichlorobenzene in cyclohexane at 40°C?
(b) What is the freezing temperature of *p*-dichlorobenzene from a solution which has a mole fraction 0.8?

When we get to the point Z, there are two solids and a liquid phase in equilibrium. This special sort of triple point (three phases in equilibrium) is known as the *eutectic point*—it involves the lowest freezing point for mixtures of the two given substances. If one starts with the solids in exactly the mass ratio which corresponds to the composition of the liquid at this point, they will melt at a single temperature, namely −13°C, to give that liquid. Such mixtures find many uses, for example in the eutectic alloys used in welding and soldering—because of their low melting point and because of their mechanical properties.

If we were to start with some other composition initially in the liquid phase (for example, X, (45°C, $x_B = 0.6$)) and cool it, behaviour is different from that for the eutectic composition. At 36°C, this particular mixture would begin to freeze, depositing solid *p*-dichlorobenzene. The effect of this crystallisation is to leave a liquid phase richer in cyclohexane. Thus at 20°C, the liquid has the composition corresponding to P($x_{\text{p-dichlorobenzene}} = 0.34$). Now, (the amount of the liquid) × (distance to the liquid end of the tie line) equals (the amount of the solid) × (distance to the solid end of the tie line). That is,

$$n_{\text{liquid}} \times (0.6 - 0.34) = n_{\text{solid}} \times (1.0 - 0.6)$$

When the temperature drops to −13°C, the liquid attains the composition of the eutectic, Z. At all temperatures below the eutectic point, the system contains the two solids in the ratio

$$\frac{0.6}{0.4} = 1.5$$

Exercise 3.8
Consider the changes which occur on cooling a mixture of mole fraction 0.5 from 40°C.

Often the solids which are formed are not pure substances but have a range of composition. One such case involves lead and antimony (Fig. 3.10). These solids of variable composition

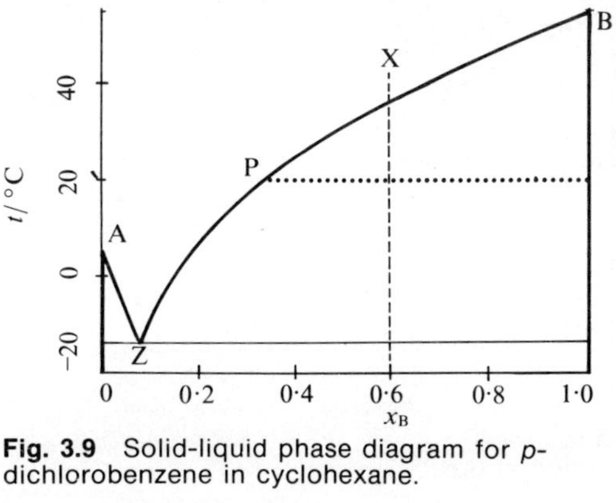

Fig. 3.9 Solid-liquid phase diagram for *p*-dichlorobenzene in cyclohexane.

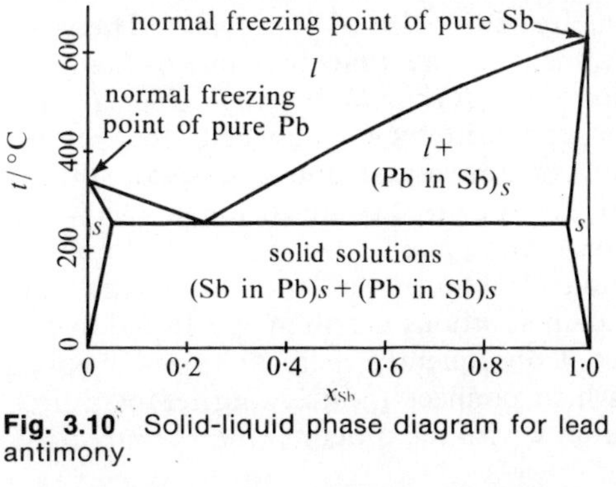

Fig. 3.10 Solid-liquid phase diagram for lead and antimony.

behave just like liquid solutions except that they are solid. They are called *solid solutions.* This particular example was chosen because lead–acid batteries normally use lead containing antimony, as a solid solution. The advantages in this case are greater ease in casting the complex shapes used for the electrodes, and more uniform corrosion than occurs with the pure metal. In short, solid solutions have their own chemical properties, sometimes desirable, sometimes not, like any liquid solution.

Exercise 3.9
How much antimony will dissolve in lead at the eutectic temperature to form

(a) the solid solutions,
(b) the liquid solution?

3.7.1 Nearly pure substances

The curves which relate freezing temperature to liquid composition in Fig. 3.9 are not strongly curved over a small range (say 0.1) in mole fraction. So, if we choose to look at liquids whose composition ranges from $x_B = 0$ to $x_B = 0.1$, or from $x_B = 0.9$ to $x_B = 1.0$, it looks as though a linear relationship between t and x would fit. This is true, not only in this case but in general.

In Fig. 3.11 we have sketched the freezing points of some dilute solutions of glucose and of urea in water as they vary with the mole fraction of the solute. It is clear that the experimental points for both substances fit on the same straight line. This line can be represented by the equation

$$T_f = T_f^\circ + k_f x_B$$

where x_B is the mole fraction of the solute and T_f° is the freezing point of pure water. Theoretical treatment reveals that the value of k_f in this equation is $-RT^2/\Delta\bar{H}$ where $\Delta\bar{H} = \bar{H}_l - \bar{H}_s$, namely the difference between the molar enthalpy of the liquid and the solid solvent (water). The region of composition in which this equation is adequate is also the region in which the mole fraction, the molality‡ and the molarity of solutions are all proportional to one another. Consequently there are similar equations which connect the freezing temperatures of solutions with the molality and the molarity.

Example 3.3
In Fig. 3.11, the experimental value of k_f is

$$\frac{-0.30}{3 \times 10^{-3}} = -100 \text{ K}$$

From the theoretical treatment, since

$$\Delta\bar{H}_{H_2O} = 6.02 \text{ kJ mol}^{-1}$$
$$k_f = -103 \text{ K}$$

Exercise 3.10
The freezing points of aqueous solutions of sucrose in solutions of varying molality are:

mol sucrose $(\text{kg } H_2O)^{-1}$	*freezing point*/ °C
0.10	−0.19
0.20	−0.37
0.30	−0.58
0.40	−0.78
0.50	−0.98

Find the value of k_f.

‡Molality is amount of solute per unit *mass* of *solvent*, $n_{solute}/m_{solvent}$.

3.8 Maps showing a Liquid and a Gas in Equilibrium

Systems in which there is equilibrium between a liquid phase and a gas phase are of very great theoretical and technical importance. They come up in problems as varied as the absorption of substances in the lungs, the separation of substances by distillation, and consideration of the inflammability of solvents. They also provide the basis for a useful measure of the

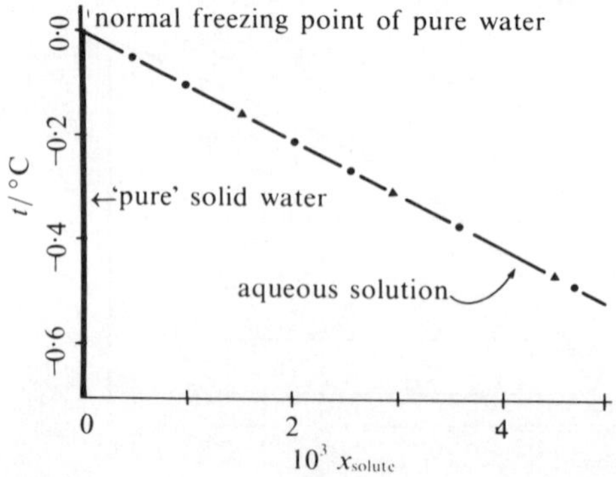

Fig. 3.11 Freezing points for aqueous solutions of glucose and of urea.

tendency for reactions to occur. In this section we will consider the relation of gas composition to liquid composition and take up other aspects of the same types of system later.

3.8.1 A simple case of liquid–gas equilibrium

The map in Fig. 3.12 shows a phase diagram for a liquid–gas system. Points in the area labelled "*g*" above the upper line involve values of temperature and mole fraction for which the material in the system is entirely in the gas phase at the pressure (namely P°) for which the diagram was drawn. Similarly, points below the lower line in the diagram correspond to conditions under which the material in this system is all in the liquid phase. So the two lines represent the limits to the existence of gas and liquid phases. If we take points (such as L,M) on these lines which are at equal temperature, they give the composition of a pair of gas and liquid phases which would be in equilibrium with one another. The line LM is a *tie line.* To be specific, if we started with gas at a composition and temperature represented by X and let it cool, it would begin to form liquid when the temperature reached that of L and M, and the first liquid formed would have the composition M.

3.8.2 Distillation

The success of distillation processes in separating chemical substances may be interpreted in terms of these maps. Consider a vertical tube (Fig. 3.13a) in which a region near the lower end is maintained at one temperature, T_2, and another region near the upper end is maintained at a lower temperature, T_1. These two temperatures are chosen so that they correspond to T_2, T_1 in the gas–liquid equilibrium diagram sketched in Fig. 3.13b. Intermediate regions of the tube are kept at temperatures between T_1 and T_2. At some region, F, the temperature will have the value, T_{LM}. At this point we choose to provide an entry hole, through which we can add a gas mixture. If the temperature and composition correspond to L (Fig. 3.13b), the gas will reach equilibrium with liquid phase of composition M, provided that we get the rate of addition right. The liquid is denser than the gas and will tend to flow down the tube. The gas will tend to ascend. As the liquid and gas travel, each encounters a variation of temperature which leads to a change of composition in both the gas and liquid phases in order to maintain equilibrium. So, with proper choice of conditions, the tube will eventually have gas emerging at the top with composition Q and liquid emerging at the bottom with composition J. The device has separated a "feed" of composition L into "outputs" each enriched in one of the constituents. If the device is large enough and appropriate conditions can be maintained, J and Q may approach as close to the pure substances as we choose.

In the account we have given, there are two main conditions to be met, namely:

(a) control of the temperatures to desired values;
(b) rate of addition controlled so that the flows in the vertical tube allow liquid–gas equilibrium to be closely approximated —the liquid should run out at a suitable rate, neither too fast nor too slow, for equilibrium to be reached efficiently.

In designing a practical apparatus, condition (a) gives only minor problems. In fact, minor

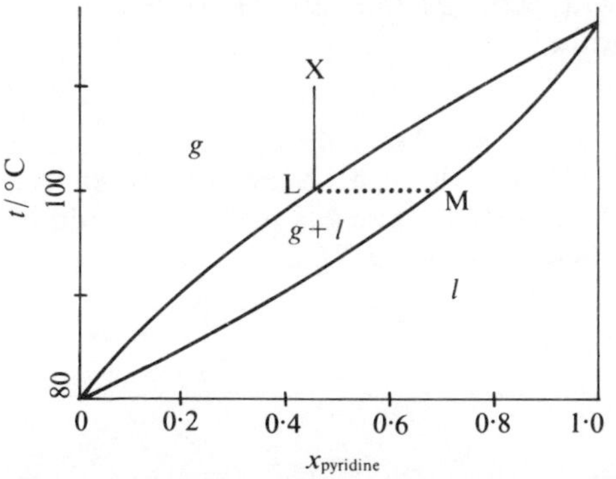

Fig. 3.12 Boiling temperatures of solutions of benzene and pyridine at 101 kPa.

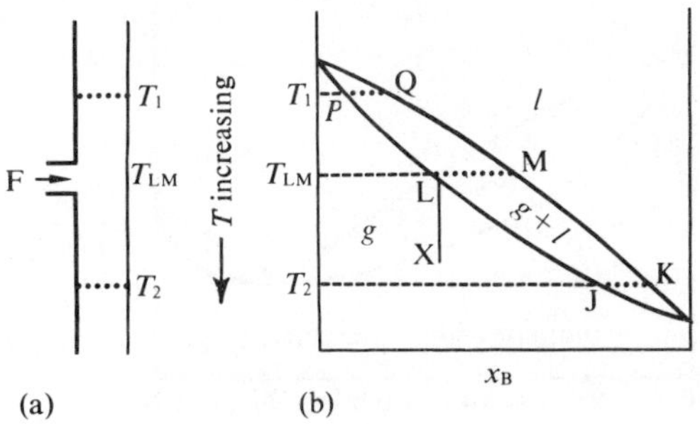

Fig. 3.13 Liquid-gas equilibrium in a fractionation column (a) in the column; (b) on the phase map.

variations from the best conditions tend to readjust themselves rather well. Condition (b) has two aspects. A large area of liquid surface is required. Usually this involves packing the tube loosely with some material on which the liquid forms a fairly uniform film and the packing is arranged so that the liquid flows steadily down the tube. (There is scope for technology here!) The rate of addition of material involves design and control problems which get more important the more efficient the equipment is required to be.

3.8.3 Azeotropic mixtures

The map in Fig. 3.12 depicts a particularly simple case of equilibrium between gas and liquid. It often happens that the curves for liquid composition and gas composition meet, either at a maximum (Fig. 3.14) or at a minimum (Fig. 3.15) temperature. Such points of meeting involve *equal* gas and liquid compositions (known as *azeotropic compositions* or *azeotropes*). An azeotropic mixture cannot be separated by distillation under the conditions to which the map refers. If further separation is required, then either a change in total pressure or addition of a third substance may be effective.

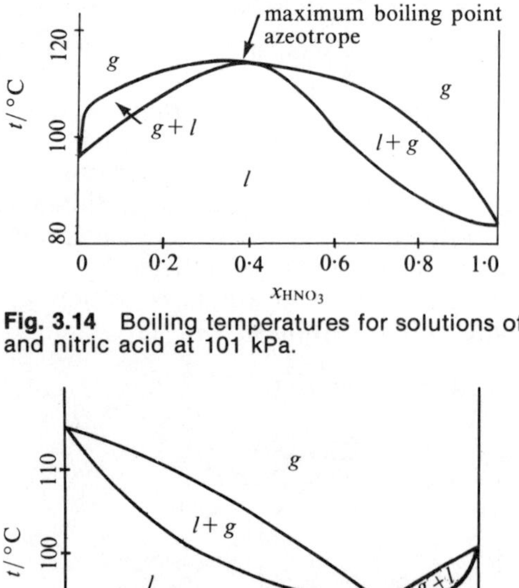

Fig. 3.14 Boiling temperatures for solutions of water and nitric acid at 101 kPa.

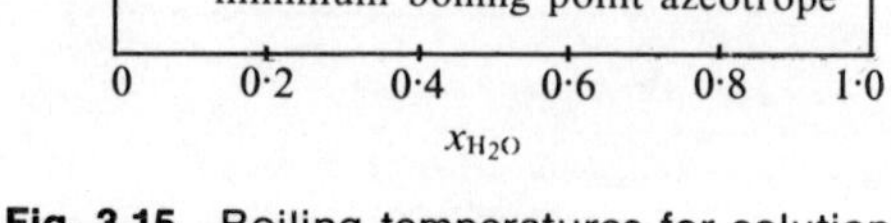

Fig. 3.15 Boiling temperatures for solutions of pyridine and water at 101 kPa.

Exercise 3.11

A fractionation column of length 1 m is to be used to separate methanol and ethanol. The temperature at the top of the column is 60°C and at the bottom is 80°C. The temperature decreases with height by 0.2 K cm^{-1} The equilibrium values of the mole fraction of methanol in the liquid phase, x_l, and in the gas phase, x_g, for this system at various temperatures, t, and at 101 kPa are:

t/°C	78.1	76.6	75.0	72.2	70.0	67.6	66.6	65.6	64.6
x_l	0	0.134	0.240	0.402	0.545	0.726	0.814	0.907	1.000
x_g	0	0.184	0.326	0.533	0.676	0.811	0.873	0.939	1.000

Draw a map to show the composition of the liquid and the composition of the gas at equilibrium at various distances along the column.

How far up the column should a gaseous mixture of methanol, mole fraction 0.3, be introduced to the column?

Answer

25 cm from the lower end.

Example 3.4

Consider a fractionation column designed to separate pyridine from benzene. It is to operate at atmospheric pressure, with the gases supplied at a pyridine mole fraction of 0.4.

(a) Use Fig. 3.12 to suggest suitable values for the temperatures at the ends of the column and for the gas mixture fed to the column.
(b) Consider such a column operating continuously and close to equilibrium with an input rate of 10 kg hr^{-1}. The products have the compositions:

pyridine, mole fraction 0.98
benzene, mole fraction 0.96

Ignoring other substances which might be present, what is the rate of production of pyridine?

Answer

(a) The base of the column should be at $T_2 \simeq 115$°C. The top of the column should be at $T_1 \simeq 80$°C. The gas mixture should be supplied at $T_{LM} \simeq 97$°C.

(b) The input is gas of pyridine mole fraction 0.4, that is,

$$\frac{m_{\text{pyridine}} / \bar{M}_{\text{pyridine}}}{m_{\text{pyridine}} / \bar{M}_{\text{pyridine}} + m_{\text{benzene}} / \bar{M}_{\text{benzene}}} = 0.4$$

$$\frac{m_{\text{pyridine}} / \bar{M}_{\text{pyridine}}}{m_{\text{benzene}} / \bar{M}_{\text{benzene}}} = \frac{0.4}{0.6}$$

$$\bar{M}_{\text{pyridine}} = 79.1 \text{ g mol}^{-1}$$

$$\bar{M}_{\text{benzene}} = 78.1 \text{ g mol}^{-1}$$

$$\frac{m_{\text{pyridine}}}{m_{\text{benzene}}} = 0.67$$

$$\text{gas input rate} = 10 \text{ kg hr}^{-1}$$

$$\text{pyridine input rate} = \frac{0.67 \times 10}{1 + 0.67} \times \text{kg hr}^{-1}$$

$$= 4.0 \text{ kg hr}^{-1}$$

$$= 51 \text{ mol hr}^{-1}$$

Similarly, benzene input rate $= 76 \text{ mol hr}^{-1}$

Pyridine is output in the liquid stream at 0.98 (r mol/hr) and in the gas stream at 0.04 (s mol/hr) while benzene is output at $(0.02r + 0.96s)$ mol/hr.

Thus $0.98r + 0.04s = 51$

$0.02r + 0.96s = 76$

$\Rightarrow \quad r = 48.9$

Pyridine recovery rate $= 48.9 \times 0.98$ mol/hr

$= 48$ mol/hr

Exercise 3.12
What input rate is necessary if methanol, mole fraction 0.99, is to be recovered at 0.5 kg hr^{-1} from a methanol ethanol mixture in which the methanol mole fraction is 0.3?

Exercise 3.13
Consider a fractionation column set up to separate the pair of substances depicted in Fig. 3.14.

What temperatures should be established in the column?

What degree of separation may be achieved for an input gas of mole fraction 0.3?

Exercise 3.14
Repeat the analysis used in Exercise 3.13 for the pair of substances depicted in Fig. 3.15.

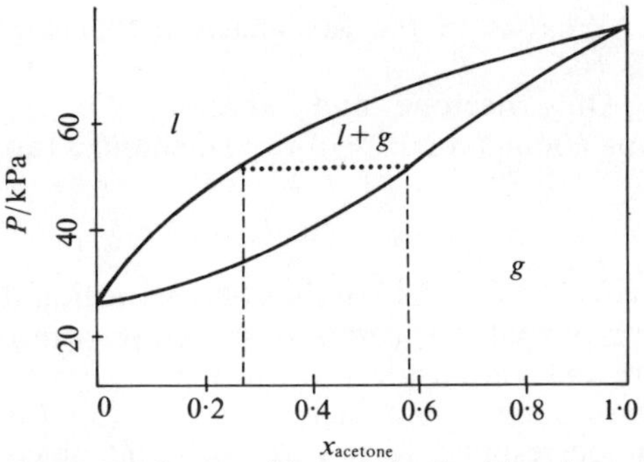

Fig. 3.16 Composition of the liquid and gas phases for acetone and ethanol at 48°C.

3.8.4 Pressure–composition maps

Instead of looking at behaviour at a constant pressure, as we have just done, we often look at a different section, namely behaviour at constant temperature. Figure 3.16 shows this map for mixtures of acetone and ethanol at 48°C. The upper curve gives values of gas pressure and liquid phase composition for which the liquid phase is in equilibrium with gas. The lower curve gives the corresponding points for gas pressure and gas phase composition.

Example 3.5
At 0.5 P° (50.5 kPa), equilibrium between liquid and gas phase involves (from Fig. 3.16) mole fractions of acetone equal to 0.27 in the liquid phase and 0.58 in the gas phase.

If the pressure of the gas is higher than the equilibrium value, liquid can deposit until equilibrium is reached. Likewise, a liquid phase in contact with a gas phase having a pressure less than the equilibrium value can form more gas. So the conditions of equilibrium of gas and liquid at the chosen temperature are displayed by the two lines drawn in the map. Tie lines can be drawn, such as the dotted line in Fig. 3.16, to connect phases in equilibrium. The lever rule will give the relative proportions of the phases in any composite body of material corresponding to a point on such a line.

Example 3.6
A mixture of *total* composition, $x_{\text{acetone}} = 0.5$, at $P = 0.5P^{\circ}$ will consist of liquid and gas in the mole ratio

$$\frac{\text{amount of liquid}}{\text{amount of gas}} = \frac{0.58 - 0.50}{0.50 - 0.27}$$

$$= 0.35$$

Query What is in the gas phase in the map Fig. 3.16?

Reply Only acetone and ethanol. We are talking about a mixture of two substances (no air).

In maps like Fig. 3.16, points *above* the liquid line correspond to conditions (temperature, pressure and composition) for which there is no gas phase present. Similarly, points *below* the gas line correspond to having no liquid phase present.

Another way of presenting the same information that is displayed in Fig. 3.16 involves a map of the partial pressures of the substances above solutions of different composition. The *partial pressure* of a substance is defined as the product of the mole fraction in the gas phase and the total pressure. Thus, for acetone

$$p_{\text{acetone}} \equiv x_{\text{acetone(g)}} P$$

In Fig. 3.17 the data used to construct Fig. 3.16 are represented in this alternative form. Comparing the two maps it will be seen that this later one has ceased to treat the two phases in identical ways. The compounds are presented in the gas phase as though they were independent. The total pressure has been submerged—it emerges as the sum of the partial pressures

$$P = p_{\text{acetone}} + p_{\text{ethanol}}$$

The attractiveness of partial pressures in analysing these equilibria is that they are very little altered by the presence of other gases which do not dissolve much in the liquid. So we can predict with confidence that a map like Fig. 3.17 will interpret the equilibria in the related system where sufficient nitrogen or air is present to give a constant total pressure of, say, P°. This is true although the mole fractions of both the acetone and the ethanol in the gas phase will have been altered very much by the addition of the air.

Query How could Fig. 3.16 be used in such a prediction?

Reply It can be used to calculate the mole fractions, and hence the partial pressures, of acetone and ethanol in the absence of other substances. These values of partial pressure can be used to calculate the changes in mole fraction which additions of the other substance give.

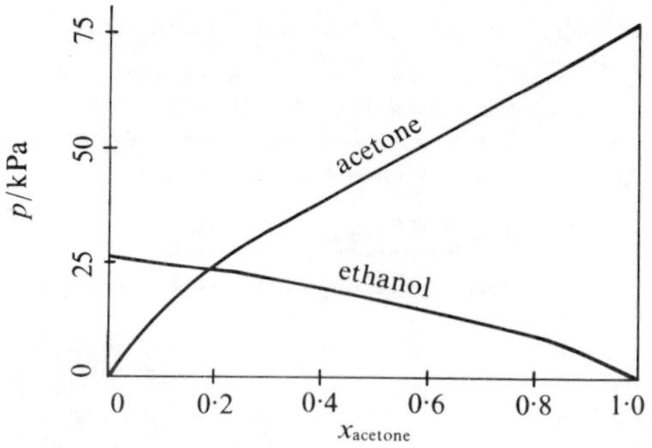

Fig. 3.17 Partial pressures above acetone-ethanol solutions at 48° C.

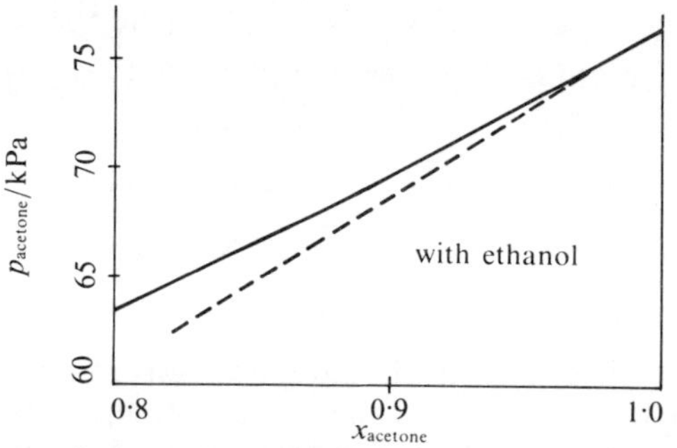

Fig. 3.18 Partial pressures of acetone above acetone-rich solutions with ethanol at 48° C.

3.8.5 Raoult's Law

The partial pressure of acetone for mole fractions greater than 0.8 is shown in Fig. 3.18. It is clear that the partial pressure of acetone is approximately proportional to the mole fraction of acetone in solution. So, in this region,

$$p_{\text{acetone}} \approx p^{\circ}_{\text{acetone}}\, x_{\text{acetone(l)}}$$

$$\approx 76.3\, x_{\text{acetone(l)}} \text{kPa}$$

The value of $p^{\circ}_{\text{acetone}}$ is the equilibrium vapor pressure at this temperature. The dashed line corresponds to values of p which are proportional to x. The approximation gets better as the mole fraction increases.

Behaviour like this is almost universal and may be put in the general form, for any solvent, A,

$$p_{\text{A(g)}} \simeq p^{\circ}_{\text{A}}\, x_{\text{A(l)}} \text{ when } x_{\text{A(l)}} \simeq 1.0$$

This, and any equivalent relationship, is commonly known as Raoult's law. First reported by François Marie Raoult (1887), it is a simple correlation, but is one of the most powerful means we have for the study of liquid solutions (see Section 4.4.7).

Example 3.7
Experiments to measure the useful life of paints, fabrics, etc. require control of the water content of the atmosphere in test enclosures. Suppose, for example, that a "relative humidity of 0.7" is called for at 35° C, that is,

$$p_{H_2O} = 0.7\, \mathring{p}_{H_2O}$$

We should get that in equilibrium with a liquid in which the mole fraction of water is about 0.7 and so will begin by thinking of adding substances like glycerol ($CH_2OH{\cdot}CHOH{\cdot}CH_2OH$) for which sufficiently concentrated solutions can be formed.

3.8.6 Henry's law

The vapor pressure of acetone is also almost proportional to the mole fraction in the liquid with ethanol when both vapor pressure and mole fraction are small. Figure 3.19 shows the region in which the mole fraction of acetone ranges from 0 to 0.2. The limiting behaviour at low mole fractions is represented by the dashed line. The equation for this line is

$$p_{\text{acetone}} = Kx_{\text{acetone(in ethanol)}}$$
$$= 142\ x_{\text{acetone(in ethanol)}}\ \text{kPa}$$

Behaviour like this was first reported (William Henry, 1803) in studies of the effect of pressure on the solubility of gases. It is observed for most substances in dilute solution. The general form of the equation, that is,

$$p_B \simeq Kx_{B\text{(as solute in liquid A)}}$$

(and other equivalent forms) is known as Henry's law. It is useful in analysis of the behaviour of substances in dilute solution (see Section 4.4.7).

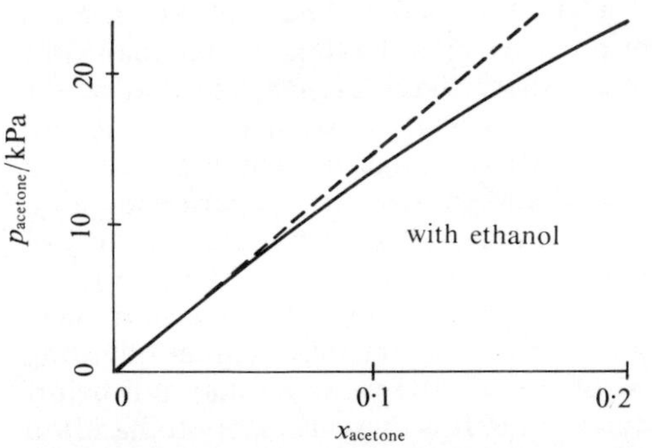

Fig. 3.19 Partial pressures of acetone above dilute solutions with ethanol at 48° C.

Example 3.8
The most common use of Henry's law has been connected with the solubility of gases in liquids. Thus it has been shown that argon is better than nitrogen for diluting the oxygen supplied to deep sea divers, because it has a smaller Henry's law constant, K. In the Apollo rocket system the fuel (dimethylhydrazine) was pumped with helium. The extent to which helium dissolves in the fuel was a factor to be taken into account in the design.

Sometimes Henry's K and Raoult's p° are nearly equal. Then graphs like Fig. 3.17 are almost straight lines over their whole range and are analysed easily. An example is provided by mixtures of benzene and dimethoxymethane (Fig. 3.20). More often there are intermediate regions between those in which Henry's law and Raoult's law apply. In such intermediate regions there is no simple proportional relationship between p and x_l. One way in which data in this region are recorded assigns values to the variable factor, f, using an equation related to Raoult's law

$$p_B = \mathring{p}_B f_B\, x_B$$

or another equation related to Henry's law

$$p_B = K f_B\, x_B$$

This extension is taken up in Section 4.4.7.

3.8.7 Boiling temperature and liquid composition

Some of the uses we make of the boiling temperatures of solutions have a connection with Raoult's law, or with Henry's law.

One of the simplest maps for these situations displays the boiling temperatures, at some constant pressure, of a range of dilute solutions

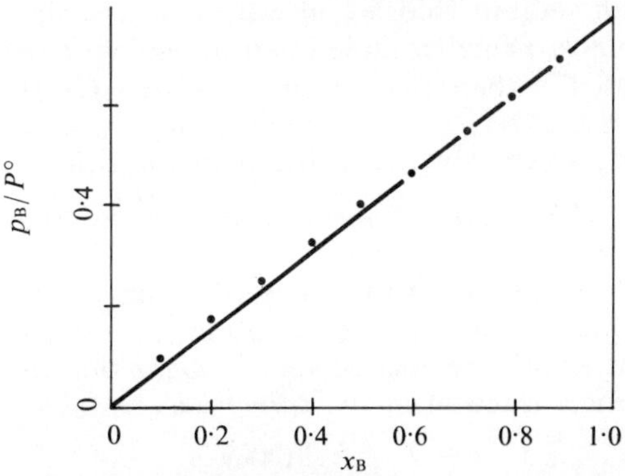

Fig. 3.20 Partial pressures of dimethoxymethane above benzene solutions at 35° C.

of solutes which are virtually absent from the gas phase. For example, Fig. 3.21 shows the results of measurements on aqueous solutions of glucose and urea. Clearly the same straight line represents the sets of solutions for both solutes. They obey an equation of the form

$$T_b = T_b^\circ + k_b\, x_{B(\text{in solvent A})}$$

in which x_B is the mole fraction of the solute (substance in small amount, here glucose or urea) in the solvent (water).

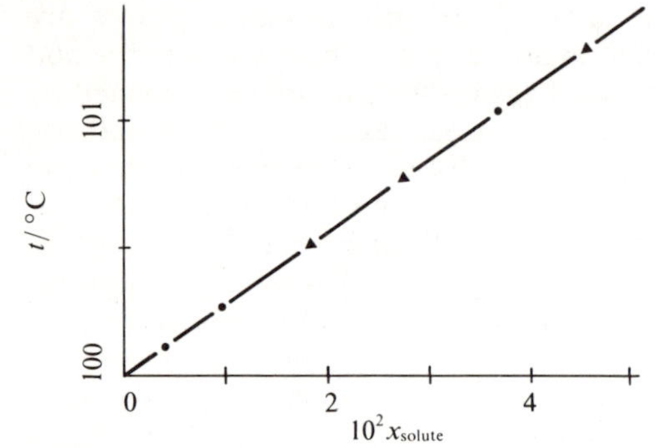

Fig. 3.21 Boiling temperatures of aqueous solutions of glucose and of urea at 101 kPa.

It appears that k_b is a property of the solvent, which does not depend on what solute is used. Some solutes appear, at first sight, to be exceptions. Sodium chloride in water, for example, gives close to twice as much elevation of the boiling point as the same amount of glucose. Benzoic acid in benzene leads to about half as much elevation of T as does an equal concentration of naphthalene. We have developed such confidence in the boiling temperature relationship that these cases are interpreted as involving dissociation of the solute to form more molecules than we allowed for (Na^+ and Cl^- with sodium chloride in water) or association to give a smaller mole fraction than we had calculated (benzoic acid, $2C_6H_5COOH = (C_6H_5COOH)_2$). This implies that k_b is a property of the solvent rather than the solute.

Query Why then does x_B appear in the relationship?

Reply An aspect of the theoretical analysis is the prediction that the elevation of boiling point should be related to the *logarithm* of the mole fraction of the *solvent*, as follows

$$T \simeq T^\circ - k_b \ln x_{A(l)}$$

We know that $x_{A(l)} = 1 - x_{B(l)}$ and that a good approximation to $\ln(1 - x_{B(l)})$ is $-x_{B(l)}$ when its value is small. This explains the presence of x_B in the equation.

Query Does the logarithmic relationship fit better? And, if so, why don't we use it?

Reply Yes, it does fit the results more closely and over a wider range of compositions. Perhaps it is less often used because logarithms are slightly more difficult to calculate and the simpler form is satisfactory when we don't use very high concentrations.

A theoretical analysis of the variation of boiling temperature with varying composition of solution can be carried out. It leads to the prediction that k_b is a particular function of properties of the solvent, namely,

$$k_b = RT_b^2/\Delta\bar{H}$$

with $\Delta\bar{H} = \bar{H}_{gas} - \bar{H}_{liquid}$ (the difference between the molar enthalpy of the gaseous and liquid solvent).

If we are dealing with highly volatile solutes, the boiling temperatures are also used. When we wish to remove air dissolved in water we often boil it. The Henry's law constants, K, for oxygen and nitrogen are large enough to ensure that the sum of the equilibrium partial pressures ($p_{O_2} + p_{N_2} + p_{H_2O}$) becomes equal to the external pressure, P, at a temperature well below that at which $p_{H_2O} = P$. At this lower temperature, bubbles of gas ($O_2 + N_2 + H_2O$) can form and rise through the liquid. We notice bubbles in a kettle of water some time before the water comes to a "steady boil". With the escape of such bubbles, they carry off most of the oxygen and nitrogen. The boiled water, when cooled, tastes "flat".

3.8.8 Critical behaviour and composition

When a liquid and a gas phase are in equilibrium, systems having more than one substance show behaviour just like that at the critical point of a single substance. In particular, there is a narrow range of temperatures over which liquid and gas lose their separate identity.

This situation is well known in the recovery of petroleum from some deposits. At the bottom of such wells, the pressure and temperature may be higher than the critical values. During removal of the oil, the pressure may fall below the critical value. It is then necessary to handle a mixture of liquid and gas—not an easy pumping problem. At lower pressures there will only be gas.

3.9 Isopiestic and Osmotic Equilibrium

3.9.1 Isopiestic equilibrium

Consider an experiment (sketched in Fig. 3.22) in which two sets of dishes containing liquid solutions are put side by side on a heavy metal plate which keeps the solutions very nearly at the same temperature. The lid encloses the arrangement and the pressure inside is reduced by pumping. Suppose that both sets of dishes contain the same solvent, A. In one set of dishes the solute is a well known substance of low vapor pressure, say, glucose. In the other set, the solute is a substance about which little need be known except that its vapor pressure is low.

This system can reach equilibrium. If it does, it will do so by adjusting the concentrations of the solutions by transfer of solvent so that they all have the same partial pressure of A in equilibrium with them. From Raoult's law:

$$p_A = p_A^\circ \, x_{A(l)}$$

we know that in *dilute* solutions, equal partial pressures involve equal mole fractions of A in the solutions. If we measure the mole fraction of glucose at equilibrium (and hence x_A for those solutions), we know x_A for the other solution, if it is dilute. The same method is also useful when the solutions are not so dilute; it then gives values of the relative activity of the solvent (compare Section 4.4.7). This method is called the *isopiestic method* for the study of solutions.

Query When is it used?
Reply It provides a quick method for getting approximate values of relative molar mass for solutes. It is also used to give accurate values for the activities of solvents in solutions which involve ionic solutes.

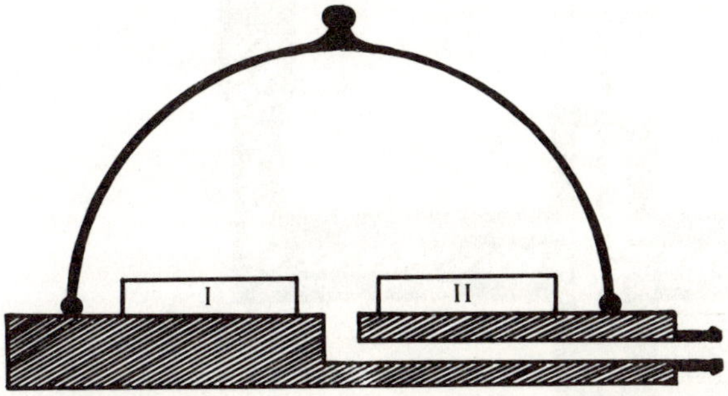

Fig. 3.22 Sketch of the apparatus for an isopiestic experiment.

Example 3.9
The following dilute aqueous solutions are reported to be in equilibrium with one another under isopiestic conditions:

Solute	*Molality*/mol(kg H_2O)$^{-1}$
glucose	0.056
sucrose	0.054
urea	0.055
NaCl	0.029
K I	0.027
$AgNO_3$	0.030
$K_2Cr_2O_7$	0.017

If these solutions have equal mole fractions of water, then the ionic solutes are dissociated to about the expected number of ions (two for NaCl, KI and $AgNO_3$ and three for $K_2Cr_2O_7$).

3.9.2 Osmotic equilibrium

The gas phase in the experiment just discussed allows the solvent to transfer between dishes, but prevents transfer of the solutes. Many natural and synthetic materials are selective in their ability to allow passage of chemical substances and are used as membranes. For example, the stomach wall allows passage of amino-acids into the blood, but not proteins. Membranes in the kidneys and films of cellulose acetate ("cellophane") allow passage of water, salts and low molecular weight substances like urea and uric acid but hold back sugars, proteins, etc. Low density polythene film wrapping used for meat allows oxygen to reach the surface of the meat while preventing passage of most other contaminants. Hot palladium metal plates can be used to "filter" hydrogen from gas streams.

We wish to draw attention here to those membranes and films which exhibit the ability to allow passage of a solvent, while preventing the transfer of solutes. Such films are called *semipermeable*. The natural process of passage of solvent into a solution through a film like this is *osmosis*. The argument in Chapter 6 shows that the observation can be interpreted as meaning that the solvent has a higher potential when it is pure than when it is present in a solution.

If a greater pressure is applied to the solution than to the solvent using an apparatus like that of Fig. 3.23, the natural transfer can be reduced, or even reversed. The least pressure difference which prevents further entry of the solvent to the solution (and thus keeps the liquid levels constant) is the *osmotic pressure*, π

$$\pi \equiv P_2 - P_1$$

In dilute solutions, there is a simple relationship between the osmotic pressure and the mole fraction x_B of the solute (or solutes).

$$P_2 - P_1 \equiv \pi \simeq x_B RT/\bar{V}_A$$

where $\bar{V}_A$ is the molar volume of the solvent. In most regions where this relationship is used it can be simplified, since the solutions are dilute. The mole fraction can be approximated by the mole ratio,

$$x_B \simeq n_B/n_A$$

This leads to the equations

$$\pi \simeq n_B RT/n_A \bar{V}_A$$

$$\simeq n_B \mathrm{RT}/V$$

$$\simeq c_B RT$$

where c_B is the concentration (in mol m^{-3}). These equations are remarkably similar to those which describe the behaviour of gases at low pressures. When first reported, this excited intense interest and speculation but there is a straightforward explanation (at a theoretical level to which we do not wish to go now). In fact RTx_B is an approximation to $-RT\ln x_A$, which is itself an approximation to the difference of potential between the pure solvent and the solvent in the solution.

Measurements of the osmotic pressure of dilute solutions have high sensitivity. Thus 10^{-3} mol solute dm^{-3} at 300 K yields an osmotic pressure of 2.5 kPa. If measured with a simple U-tube manometer filled with water, that pressure corresponds to a difference of 25 cm in liquid level. Because of this sensitivity, osmotic pressure measurements are used in determining molar masses when only dilute solutions can be used—especially for large molecules, such as occur with proteins.

Osmosis is a common, important, biological phenomenon in the adaptation of many plants and animals to their environments. "Reverse osmosis", the recovery of pure solvent on application of a pressure difference greater than the osmotic pressure, is a potentially useful method for getting fresh water from brackish supplies. We can calculate the minimum operator work (Chapter 6) which we must do in the process and hence the minimum cost in carrying it out. To recover 1 mol H_2O ($\bar{V}$ = 18.0 cm^3 mol^{-1}) from sea water with $\pi = 2.6 \times 10^6$ Pa

$$w_{\text{♀}}(\text{minimum}) = \pi \bar{V}$$

$$= 2.6 \times 10^6 \times 18 \times 10^{-6} \text{ Pa m}^3 \text{ mol}^{-1}$$

$$= 47 \text{ J mol}^{-1}$$

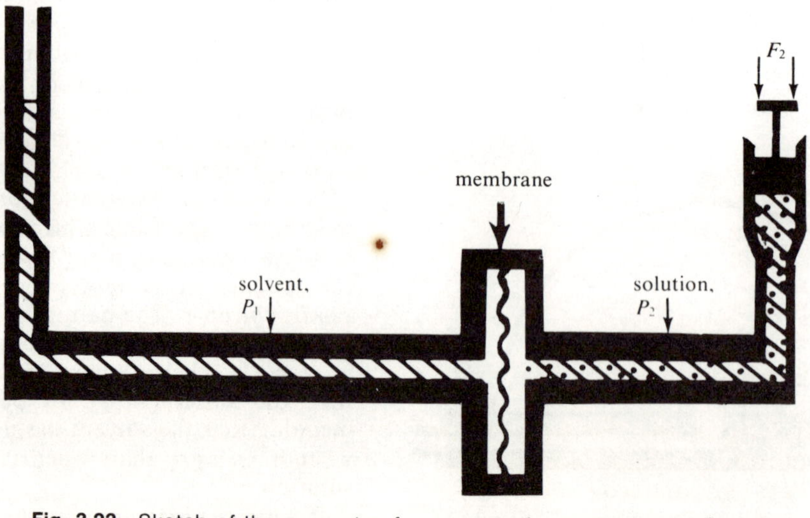

Fig. 3.23 Sketch of the apparatus for an osmotic pressure experiment.

At this complete efficiency, one kilowatt hour (3.6×10^6 J) of energy could recover 1400 kg (approximately 250 gallons) of water. To get useful *rates* of production, such plants are now run at about 20% efficiency, with applied pressures of about 5π.

3.10 Colligative Properties

We have turned up four phenomena in this discussion which appear in dilute solutions and depend principally on the total mole fraction of solutes rather than on the individual chemical properties of these solutes. These are the lowering of partial pressure, the decrease in freezing point, the increase in boiling point, and the osmotic pressure. Together they are known as the *colligative properties.* They all depend on the same thing—the relative number of solute and solvent particles. When a solute is examined which dissociates (for example, a salt present as almost independent ions) or which associates (as many carboxylic acids do in hydrocarbon solvents), there is a change in the colligative properties which reflects the dissociation or association. Note, for examples, the figures in Table 3.1.

Table 3.1
Comparison of Freezing Points of 10^{-2} mol dm^{-3} *Aqueous Solutions*

solute	*freezing point*	*freezing point depression*
sucrose	−0.019°C	0.019K
diethyl ether	−0.019°C	0.019K
NaCl	−0.038°C	0.038K
HCl	−0.038°C	0.038K
$MgCl_2$	−0.057°C	0.057K
Na_2SO_4	−0.057°C	0.057K

Query May we assume from these figures that NaCl, HCl, $MgCl_2$ and Na_2SO_4 are completely dissociated?

Reply Yes, at this concentration. The picture is not quite so simple at higher concentrations.

3.11 The Phase Rule

In the preceding discussion we have repeatedly emphasised that the maps show definite patterns—under some conditions a system may be completely specified by stating the number of phases, sometimes a complete specification requires that some values be assigned to variables like temperature, pressure or composition. From the first discussion in this chapter of the behaviour of carbon dioxide, there has been evidence in the maps that there is some connection between the number of variables required and the number of phases. The *phase rule* is the formal statement of this connection. Basically this rule is a restatement of the conditions which have to be met to achieve equilibrium in a chemical system involving definite numbers of substances and phases, reacting with defined equations. Use of the rule does not "prove" anything. It does give a check for the presence of confused thought in our interpretation of behaviour. In this section we shall give the logical basis (but not the proof) of the phase rule, and some examples of its use.

3.11.1 Logical basis

The logical basis of the phase rule is what we mean by saying that a system is in equilibrium. To be "in equilibrium" normally means:

(a) that there is thermal equilibrium (the temperatures of all the phases are identical);
(b) that there is mechanical equilibrium (the pressures of all the phases are identical in the regions where they make contact);
(c) that some substances reach equilibrium across the boundaries between the phases; and
(d) that when any substance B reaches equilibrium between some phases I and II, and also between the phases I and III, it also reaches equilibrium between II and III—as, indeed, it does with all of the phases which are at equilibrium.

Often there are some phases which are not at equilibrium with others, but which also do not undergo any change anyway. Usually we want reaction vessels to behave like that. This was the kind of restriction we had in making (c) refer to "some" and not "all" substances. Often, too, there are reactions within some phases which link the amounts of some of the substances at equilibrium.

The effect of the above statements is a great reduction in the number of types of independent changes which an experimenter can make in a system at equilibrium without changing the number of phases present. We may call the available *number* of types of independent change, the *number,* $\mathscr{F}$, *of degrees of freedom* of the system. We also count the *number*, $\mathscr{P}$, *of phases*, and the *number*, $\mathscr{B}$, *of substances* which are present in them. In addition, the *number*, $\mathscr{R}$, *of reactions* which link the $\mathscr{B}$ substances is required. It is possible to show, from the concept of chemical potential (Section 6.3), that the number of degrees of freedom is

$$\mathscr{F} = \mathscr{B} + 2 - \mathscr{P} - \mathscr{R}$$

More usually, the *number*, $\mathscr{C}$, *of components* is introduced to mean $\mathscr{B} - \mathscr{R}$, so that the equation becomes

$$\mathscr{F} = \mathscr{C} + 2 - \mathscr{P}$$

Either of these statements is a form of the phase rule. Proofs are given in many standard texts.

Query "Number of degrees of freedom" seems an odd expression for chemistry?
Reply It describes the range of experiments which is possible. Sometimes the number of phases is such that nothing more can be done than *observe*. There is no freedom to experiment, and $\mathscr{F} = 0$. Sometimes a single property can be altered (like the pressure or the temperature) but choice of a particular value of the pressure is enough to define the whole system at equilibrium. Then $\mathscr{F} = 1$. Sometimes one is free to choose values of two properties—for example *both* temperature *and* pressure. Then $\mathscr{F} = 2$, and so on.

Query Does it matter what property is varied?
Reply Not really. Provided the experiment can be done in which the property changes. . . But the *properties* are always ones which are independent of the size of the system, that is, ones like density or molar volume, not like mass or volume. They are measured in a single phase.

Query The pressure is not always constant —for example, in a vertical tube. What does that alter?
Reply There is an effect due to the external field (here gravity). It was left out previously because it is usually small and can then be neglected. If the effect is not small, as in a centrifuge, mechanical equilibrium requires that there be a well defined pressure *gradient* and inclusion of the pressure gradient gives back a modified phase rule. Other types of field can be treated in a similar way.

When discussing the phase rule we count, in $\mathscr{R}$, only those reactions which can get close to equilibrium in the time we take to make an observation.

Example 3.10
Consider a system which contains only pure water, and neglect any reactions (for example, its ionisation).
Then $\mathscr{B} = 1, \mathscr{R} = 0$
and thus $\mathscr{C} = 1$

If the system contains solid H_2O, liquid H_2O and gaseous H_2O, then

$$\mathscr{P} = 3$$

Hence $\mathscr{F} = 1 + 2 - 3 = 0$

Zero degrees of freedom means that the temperature, the density of each phase, and every other property within a phase has a singular, defined value. If we perform an experiment which we think to relate to this system, and find the density of the liquid to be variable, then either we do not have the system we thought (another substance might be present or the system may not be at equilibrium), or the measurements are faulty.

Example 3.11
Consider just the same system, but recognise that the reaction

$$2H_2O = H_3O^+ + OH^-$$

occurs in some or all of the phases ($\mathscr{R} = 1$). To define the composition of any one phase, we now need to state the amounts of *two* of these, so $\mathscr{B} = 2$. Hence when solid H_2O, liquid H_2O and gaseous H_2O are present, $\mathscr{P} = 3$.

$$\mathscr{F} = 2 + 2 - 3 - 1 = 0$$

The agreement with the answer in Example 3.10 may seem surprising. It is due to the fact that we have changed the meaning we give to the term "pure water" between this and the previous discussion. In the first case, although we did not recognise it, the term was used so that it included reactions intrinsic to the substance. The book-keeping operation of the phase rule gave an answer at the same level of understanding as that chosen to describe the system. All our maps for a single substance earlier in this chapter could be drawn without needing to include these "intrinsic" reactions explicitly.

Example 3.12
Suppose that we have water and three phases

again present, but this time some nitrogen is present in the gas phase and is also dissolved in the water and also, presumably, in the solid. Proceeding as in Example 3.11, we have

$$\mathscr{B} = 3$$

$$\mathscr{R} = 1$$

$$\mathscr{P} = 3$$

whence $$\mathscr{F} = 1$$

This implies that, for example, we can alter the relative amount of nitrogen in say the gas phase (the mole fraction), but there are fixed properties of the system for every particular mole fraction of nitrogen in the gas—including a definite solubility in the liquid and in the solid H_2O.

Query Is the idea of *one component* more general than that of one compound?

Reply Yes. The ideas of equilibrium often depend very little on the details of any description of the system. We saw that water could be thought of in two ways in the examples above. Either way it was a one component system.

Query Is this very important?

Reply It often makes conclusions attainable which we could not get otherwise. For example, people have proposed that pure water contains many different molecules, for example, H_2O, $(H_2O)_2$, $(H_2O)_4$. The vapor pressure of the substance "water" does not depend on whether this is true or not. Each additional type of molecule contributes one more substance $\mathscr{B}$ and one more reaction $\mathscr{R}$ and so $\mathscr{C}$ remains unaltered.

PROBLEMS

3.1 The solubility of carbon dioxide in water at 25°C is $1.45\ \mathrm{g\,dm^{-3}}$ when $p(CO_2) = 101.3$ kPa. Calculate the pressure of CO_2 in a bottle of soda-water (which consists only of water and carbon dioxide) at 25°C if the concentration of carbon dioxide in the soda-water is $0.1\ \mathrm{mol\,dm^{-3}}$.

Answer 304 kPa

3.2 Solutions of ammonia in water were shaken with chloroform at 25°C. When the two layers were separated and titrated with hydrochloric acid, the following pairs of solutions were identified as in equilibrium with one another:

	I	II	III
Concentration of NH_3 in aqueous phase/mol $\mathrm{dm^{-3}}$	0.0110	0.085	0.642
Concentration of NH_3 in $CHCl_3$ phase/mol $\mathrm{dm^{-3}}$	0.00044	0.0033	0.0258

(a) What is the distribution coefficient?

Answer $$\frac{c_{NH_3(aq)}}{c_{NH_3(CHCl_3)}} = 25.2$$

(b) The partial pressure of ammonia above aqueous ammonia (0.1 mol $\mathrm{dm^{-3}}$) is 14 mmHg at 25°C. Assuming the applicability of Henry's law, use the above figures to calculate the partial pressure of ammonia over a solution of ammonia, concentration 0.01 mol $\mathrm{dm^{-3}}$, in chloroform at 25°C.

Answer 35.3 mmHg

3.3. Raoult's law may be used to describe the change in vapor pressure of a substance when a solute is present. How should it be modified if there are several solutes present, or if the solute dissociates in the solvent to form several species?

3.4. Vapor pressures at 20°C were determined for the following systems: sulfur, pure carbon disulfide, and a solution of 2.00 g of sulfur in 100 g of carbon disulfide. The values were found to be 3×10^{-4}, 854.0 and 848.9 mmHg respectively. What is the molecular complexity of the sulfur in the solution? The relative molar masses of C and S are respectively 12.01 and 32.06.

Answer The formula is $S_{7.6}$

3.5. Two solutions, one of sulfur, the other of iodine in carbon disulfide, were placed in separate containers in a vacuum desiccator, and allowed to come to equilibrium. Air was then admitted to the desiccator and the solutions were sampled and analysed. The sulfur solution contained 0.0833 g of sulfur associated with 50 g of solvent; the iodine solution contained 0.0824 g of iodine associated with 50 g of solvent. What can you deduce from this data concerning the molecular state of sulfur in solution? (Iodine may be regarded as effectively non-volatile under the conditions of experiment). Compare with the previous answer.

Answer $S_{8.0}$

3.6. Mixtures of chlorobenzene and bromobenzene in any proportions follow Raoult's law for both components. If the vapor pressures of the pure components at 137°C are respectively 115 kPa and 60.4 kPa, calculate both algebraically and graphically:

(a) the composition (in mole fraction) of the liquid mixture which has a normal boiling point of 137°C (that is, which boils at 137°C when the pressure is 101.3 kPa);
(b) the composition of the vapor which is in equilibrium with this liquid at its boiling point; and
(c) the composition of the vapor, and the total pressure over a solution at 137°C containing an equal number of mole of chlorobenzene and bromobenzene.

Answer (a) $x_{C_6H_5Cl} = 0.749$

(b) $x_{C_6H_5Cl(g)} = 0.850$

(c) $x_{C_6H_5Cl(g)} = 0.656$

$P = 87.7$ kPa

3.7. The following data are typical of experimental measurements of freezing points for aqueous solutions of non-electrolytes and weak electrolytes. The *depression* of freezing point ΔT_f is given in each case.

	$-\Delta T_f/K$						
molality/ mol kg^{-1}	0.10	0.20	0.30	0.40	0.50	0.60	0.70
sucrose ($M = 342.3$ g)	0.19	0.37	0.58	0.78	0.98	1.20	1.41
acetic acid ($M = 60.05$ g)	0.18	0.37	0.56	0.74	0.91	1.08	1.26
ethanol ($M = 46.07$ g)		0.36		0.73		1.10	
urea ($M = 60.06$ g)	0.18	0.35	0.54	0.71	0.89	1.07	1.24

Compare the data graphically by plotting on a single sheet, taking care to distinguish the points for different solutes. Is there a general consistency of behaviour?

Find the value of $k_f(H_2O)$ from these experimental data and compare it with the value of $RT_f^2\bar{M}(H_2O)/\Delta\bar{H}$ with $\Delta\bar{H} = 6.009$ kJ mol^{-1}.

3.8. The following data are typical of experimentally determined freezing point depressions for aqueous solutions of strong electrolytes (that is, substances which greatly increase the electrical conductance of water).

	$-\Delta T_f/K$			
molality/ mol (kg $H_2O)^{-1}$	0.05	0.10	0.15	0.20
NaCl	0.16	0.33	0.50	0.67
KBr	0.17	0.34	0.50	0.67
$MgCl_2$	0.25	0.50	0.75	1.00
K_2SO_4	0.24	0.47	0.62	0.83
$MgSO_4$	0.11	0.22	0.30	0.41

Taking the substances in sequence, consider what the data indicate about the behaviour of electrolytes in aqueous solution.

3.9. An aqueous solution of cane sugar ($\bar{M} =$ 342 g mol^{-1}) has an osmotic pressure of 1.5×101.3 k Pa at 18°C. What will be the vapor pressure of this solution at 40°C? If 100 g of this solution is cooled to −3.0°C, what mass of ice will separate out? Vapor pressure of water at 40°C = 55.324 mm Hg.

3.10. Blood freezes at −0.56°C and a solution of 3.0 g of urea ($\bar{M}_r = 60$) in 250 g of water freezes at −0.37°C. What is the osmotic pressure of blood at 37°C?

3.11. When 2.062 g of a compound A ($\bar{M} =$ 213 g mol^{-1}) was dissolved in 43.2 g of benzene, the solution boiled at a temperature 0.605°C higher than that of pure benzene. What is the molar mass of a substance B, 1.839 g of which, when dissolved in 74.7 g of benzene, raises its boiling point by 0.541°C?

3.12. The vapor pressure of a solution of urea is 736.2 mmHg at 100°C. What is the osmotic pressure of this solution at 15°C? At what temperature would the solution begin to freeze? k_f for water $= -1.86$ K mol^{-1} (kg water).

4.

Chemical Reaction and Chemical Equilibrium

4.1 Direction of Change

Preceding chapters have been concerned with the behaviour of materials, but the only changes in chemical composition considered were those associated with the transfer of substances from one phase to another. In this chapter we begin to consider changes in chemical composition due to transformation of one substance into another, that is, *chemical* change in the ordinary sense. We will develop a method which predicts, on the basis of the chemical equation and appropriate values of a new quantity, the "activity" of each substance whether a particular reaction will go forward, backward or be in equilibrium. The reactions for which the method is useful include not only those which we are accustomed to calling "chemical" but also the escape of a substance from one phase to another, and the diffusion of a substance within a liquid, gaseous or solid phase.

We assume that readers are familiar with the use of chemical equations to express the quantitative relationships between the changes in the amounts of reactants and products which accompany chemical reaction.‡ Chemical equations express the fact that atoms are conserved during a chemical reaction. The ordinary chemical equation can be read just as well with the reactants and products interchanged.

‡Convenient aids to revision include Stove, J. D. and Phillips, K. A., *A Modern Approach to Chemistry*, Heinemann, Melbourne, 1963, Chapter 10; Porterfield, W. W., *Concepts of Chemistry*, Norton, New York, 1972, Section 1.8; and Rosenberg, J. L., *Schaum's Outline of College Chemistry*, McGraw-Hill, New York, 1972.

Query Aren't reactions usually written so that they indicate the direction in which change actually happens?

Reply Yes, if we know the answer. However, we will see that the direction in which a given reaction proceeds depends on the conditions.

Query Why are you just talking about "conservation"? Don't equations show what reacts with what?

Reply Life is a bit more difficult than that! Chapter 5 will make some of the difference clearer. At present we are trying to find out what changes can take place. The questions turn out to be different from (and have much simpler answers than) those which deal with the way reactions go.

When we take some particular amounts of each of the reactants and products and bring them together under particular conditions we observe *one of four results*:

1. as time passes there is more of those substances which are shown on the right of the equation ("products") and less of those on the left ("reactants");
2. there are changes, but just the reverse of those in 1;
3. no change is detected; or
4. there are changes but they are not in the proportions which the equation predicts.

The first two observations correspond to the process depicted in the equation going forward, 1, or backward, 2. The third observation is very common. It may mean that, by accident or design, we have hit on the conditions for chemical equilibrium. Usually, it means that we are impatient—there will be change but we have to wait longer or increase the rate of reaction somehow. The fourth observation is also common—we do not really know what is going on, we wrote the wrong equation, or we interpreted the experiment incorrectly. As stated at the beginning of this section, in this chapter we will primarily be concerned with the conditions under which reactions will go forward (result 1), backward (result 2), or be at equilibrium.

4.2 Absolute Activity and Tendency to React

We begin by looking for a quantity to serve as a measure of the "probability" that a substance B may take part in any reaction. Such a quantity is worth the name "absolute activity". We will give it the symbol λ_B. For such a quantity to be useful we need to know

1. how to define it so that it can be measured;
2. how to use it to work out which way a reaction should be able to go; and
3. how to relate it to other quantities we might know or wish to know, such as the amounts or concentrations of substances.

Hence we will be looking for a set of convenient rules
(a) to relate the likely direction of reaction to the activities of individual substances, and
(b) to make numerical estimates of the activities of substances.

4.2.1 Tendency for reaction to occur

The simplest starting point involves the reaction of one substance to form another single substance—a transformation like that in which *trans*-cinnamic acid forms *cis*-cinnamic acid.

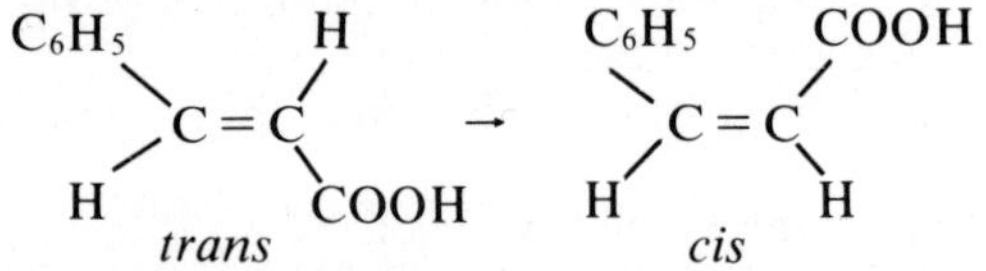

trans *cis*

Suppose we were to start with pure gaseous *trans*-cinnamic acid at some convenient conditions, for example 450 K and 20 kPa. It transforms, at least partly, to the *cis* compound. Now, the tendency for a species (*trans*-cinnamic acid, for example) to take part in chemical change is to be defined and measured by its absolute activity, λ. Let us ask some questions which arise and suggest some suitable answers.

Query How could we state the tendency of *trans*-cinnamic acid to react?

Reply The absolute activity of *trans*-cinnamic acid, $\lambda_{\text{trans-acid}}$, should be the appropriate quantity.

Query How could we state the corresponding tendency of *cis*-cinnamic acid to react?

Reply The corresponding absolute activity of *cis*-cinnamic acid would be $\lambda_{\text{cis-acid}}$.

Query What would be a convenient outcome if

$$\lambda_{\text{trans-acid}} > \lambda_{\text{cis-acid}}$$

Reply From the previous discussion, the above expression implies that the tendency for *trans*-cinnamic acid to react is greater than the tendency for *cis*-cinnamic acid to react. This makes sense if reaction takes place in the forward direction, that is,

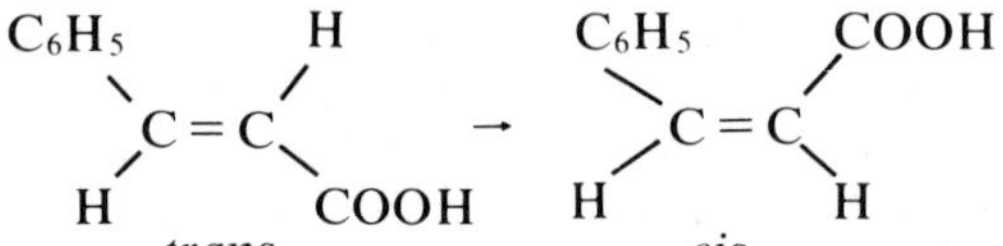

trans *cis*

Query What would be the outcome expected if

$$\lambda_{\text{cis-acid}} > \lambda_{\text{trans-acid}}$$

Reply The reaction would take place in the reverse direction, so that

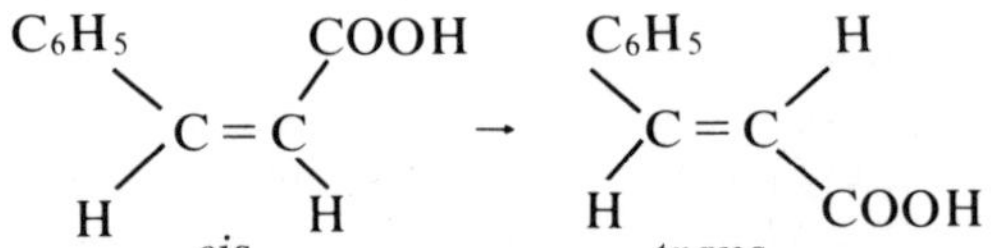

cis *trans*

This is because the tendency for the *cis*-cinnamic acid to undergo reaction is greater than that of the *trans*-cinnamic acid.

Query What would be expected to happen if

$$\lambda_{\text{trans-acid}} = \lambda_{\text{cis-acid}}$$

Reply Nothing! The tendency for *trans*-cinnamic acid to undergo reaction is exactly balanced by the tendency for *cis*-cinnamic acid to undergo reaction. The reaction is in a state of balance or equilibrium.

These criteria for deciding the possible direction of reaction can be summarised as:

$\lambda_{\text{trans-acid}} > \lambda_{\text{cis-acid}}$ reaction will proceed in the forward direction

$\lambda_{trans-acid} < \lambda_{cis-acid}$	reaction will proceed in the reverse direction
$\lambda_{trans-acid} = \lambda_{cis-acid}$	reaction is at equilibrium

Query How do we compare values of absolute activities?

Reply We have two choices. We could compare the *difference*, $\lambda_{cis-acid} - \lambda_{trans-acid}$ with zero; or the *ratio*, $\lambda_{cis-acid} / \lambda_{trans-acid}$ with unity. For activities everyone has chosen to use the *ratio.* In Chapter 6 we will examine an alternative approach to the problem of predicting possible directions of reaction and for predicting the position of equilibrium. There, the quantity used will be the potential for chemical change, and the direction of change will be predicted from the comparison of differences of chemical potential with zero.

Now we take up a more representative type of chemical reaction, one which involves four substances in a gas phase:

$$CO(g) + H_2O(g) = H_2(g) + CO_2(g)$$

The tendency for each of the species present to take part in chemical reaction is still to be defined and measured by values of its absolute activity (λ).

Query What would be the appropriate form to express the tendency of carbon monoxide and water to react together (to produce hydrogen and carbon dioxide)?

Reply The product of the absolute activities of the carbon monoxide and the water, namely, $\lambda_{CO(g)}\lambda_{H_2O(g)}$, is suggested by our earlier choice of a *ratio*.

Query What is the corresponding measure of the tendency of carbon dioxide and hydrogen to react together (to form carbon monoxide and water)?

Reply The corresponding product, namely, $\lambda_{CO_2(g)}\lambda_{H_2(g)}$

Query In a reaction such as $N_2 + 3H_2 = 2NH_3$, what is the expression for the tendency for nitrogen and hydrogen to react to form ammonia?

Reply Since we can write the reaction as

$$N_2(g) + H_2(g) + H_2(g) + H_2(g) = 2NH_3(g)$$

it makes sense if the tendency for forward reaction is

$$\lambda_{N_2(g)}\lambda_{H_2(g)}\lambda_{H_2(g)}\lambda_{H_2(g)} = \lambda_{N_2(g)}\lambda^3_{H_2(g)}$$

Exercise 4.1

Apply the above suggestions to the following cases.

(a) What is the tendency for the ammonia in the question above to decompose to nitrogen and hydrogen?

(b) In the reaction

$$3S_2O_3^{2-} + 4Cr_2O_7^{2-} + 26H_3O^+ = 8Cr^{3+} + 6SO_4^{2-} + 39H_2O$$

(i) What is the tendency for reaction between the thiosulphate, dichromate and hydrogen ions (in the proportions to give chromium(III) and sulphate ions, and water)?

(ii) What is the tendency for reaction between the chromium(III) ions, the sulphate ions and the water (in the proportions to give thiosulphate, dichromate and hydrogen ions)?

We return now to the reaction

$$CO(g) + H_2O(g) = CO_2(g) + H_2(g)$$

and find criteria which enable us to predict (for a given mixture of CO, H_2O, CO_2 and H_2) what direction of reaction might be possible.

Query What would be the expected outcome if

$$\lambda_{CO(g)}\lambda_{H_2O(g)} > \lambda_{CO_2(g)}\lambda_{H_2(g)}$$

Reply This expression is analogous to the expression $\lambda_{trans-acid} > \lambda_{cis-acid}$. It implies that the tendency for the CO and H_2O to react together is greater than the tendency for the CO_2 and H_2 to react together. This makes sense if reaction takes place in the forward direction.

Query What would then be the expected outcome if

$$\lambda_{CO(g)}\lambda_{H_2O(g)} < \lambda_{CO_2(g)}\lambda_{H_2(g)}$$

Reply The reaction would take place in the reverse direction, that is, the reaction

$$CO_2(g) + H_2(g) = CO(g) + H_2O(g)$$

would be able to go forward.

This is because the tendency for the CO_2 and H_2 to react together is greater than the tendency for the CO and H_2O to react together.

Query What happens if

$$\lambda_{CO(g)}\lambda_{H_2O(g)} = \lambda_{CO_2(g)}\lambda_{H_2(g)}$$

Reply Nothing! The tendency for the CO and H_2O to react together is exactly balanced by the tendency for the CO_2 and H_2 to react together. The reaction is in a state of balance, or equilibrium.

These criteria for deciding the possible direction of reaction amount to:

$$\frac{\lambda_{CO_2(g)}\lambda_{H_2(g)}}{\lambda_{CO(g)}\lambda_{H_2O(g)}} \begin{cases} <1 & \text{reaction will proceed in the forward direction} \\ >1 & \text{reaction will proceed in the reverse direction} \\ =1 & \text{reaction is at equilibrium} \end{cases}$$

Thus, for any reaction,

$$\frac{\text{(tendency for the products to react together)}}{\text{(tendency for the reactants to react together)}} \begin{cases} <1 & \text{reaction goes forward} \\ =1 & \text{reaction at equilibrium} \\ >1 & \text{reaction goes backwards} \end{cases}$$

4.2.2 Use of the absolute activity

To illustrate the use of this idea, let us assume that numerical values of absolute activities, designed as above, are available for some experimental situations. We show later (Section 4.7.3) the basis for obtaining these values. For the present we wish to show that such values would be useful.

Example 4.1
A gas mixture consists of:

chlorine ($p_{Cl_2} = 5$ kPa, $\lambda_{Cl_2} = 0.05$),

ethylene ($p_{C_2H_4} = 20$ kPa, $\lambda_{C_2H_4} = 1.4 \times 10^{11}$)

and

1,2-dichloroethane ($p_{C_2H_4Cl_2} = 60$ kPa, $\lambda_{C_2H_4Cl_2} = 9.6 \times 10^{-14}$)

In which direction would you expect the reaction

$$Cl_2 + C_2H_4 = C_2H_4Cl_2$$

to proceed?

$$\frac{\lambda_{C_2H_4Cl_2}}{\lambda_{Cl_2}\lambda_{C_2H_4}} = \frac{9.6 \times 10^{-14}}{5 \times 10^{-2} \times 1.4 \times 10^{11}} = 1.3 \times 10^{-23}$$

This is less than 1. We would expect the reaction to proceed in the *forward* direction.

Exercise 4.2
Use the values given for the absolute activities to determine the direction of reaction in the following examples.

(a) Solid sodium chloride is added to a 1 mol dm^{-3} solution of sodium chloride in water.

reaction of interest: $NaCl(s) = Na^+(aq) + Cl^-(aq)$

under these conditions
$$\lambda_{NaCl(s)} = 1.4 \times 10^{-67}$$
$$\lambda_{Na^+(aq)} = 1.6 \times 10^{-46}$$
$$\lambda_{Cl^-(aq)} = 1.0 \times 10^{-23}$$

(b) Carbon monoxide ($p_{CO} = 100$ kPa) is bubbled through a solution of benzene and benzaldehyde ($x_{C_6H_6} = 0.5$).

reaction: $C_6H_6(l) + CO(g) = C_6H_5CHO(l)$

under these conditions
$$\lambda_{C_6H_6(l)} = 2.1 \times 10^{22}$$
$$\lambda_{CO(g)} = 1.4 \times 10^{-24}$$
$$\lambda_{C_6H_5CHO(l)} = 9 \times 10^{-2}$$

(c) A mixture of carbon monoxide ($p_{CO} = 50$ kPa) and carbon dioxide ($p_{CO_2} = 50$ kPa) is passed over a mixture of iron (III) oxide and iron.

reaction: $Fe_2O_3(s) + 3CO(g) = 2Fe(s) + 3CO_2(g)$

under these conditions
$$\lambda_{Fe_2O_3(s)} = 1.6 \times 10^{-128}$$
$$\lambda_{CO(g)} = 2.1 \times 10^{22}$$
$$\lambda_{CO_2(g)} = 1.4 \times 10^{-69}$$
$$\lambda_{Fe(s)} = 1.$$

4.3 Relative Activity and Tendency to React

We have suggested that we need to be able to relate values of chemical activity to amounts or concentrations of substances. The numbers given in the examples above are not yet related in any simple way to these more familiar quantities. What is needed is a change of scale. The scaled quantity is called the *relative activity*, *a*, and is simply the ratio of the value, λ, of the absolute activity to its value, $\lambda°$, in suitably chosen "standard" conditions.

$$\text{relative activity } a \equiv \frac{\lambda}{\lambda^\circ} \equiv \frac{\text{absolute activity of the substance in a reacting mixture}}{\text{absolute activity of the substance under standard conditions}}$$

The relative activity turns out to be closely related to the relative amount of the substance in the mixture, that is, to the composition of the mixture.

Example 4.2
The absolute activity of CO(*g*) in a reacting mixture ($\lambda_{CO(g)}$) is compared with the absolute activity $\lambda^\circ{}_{CO(g)}$ of CO(*g*) at the same temperature and at some predetermined pressure, P°, usually 101.3 kPa. The value of the standard absolute activity of a substance (i.e., λ°) depends only on the nature of the substance and on the actual reference state; in particular, the solvent, the temperature, the reference concentration and the reference pressure. The value of λ° *does not depend* on the amount of the substance present in the reacting mixture and so is constant as a given reaction advances.

Let us return to the reaction

$$CO(g) + H_2O(g) = H_2(g) + CO_2(g)$$

Query How is the relative activity of carbon monoxide estimated?
Reply It is found that $a_{CO} \simeq p_{CO}/P^\circ$. This is a very simple relationship. A detailed discussion of the estimation of relative activities is given later.
Query How are the relative activities of substances related to the possible direction of reaction?
Reply In the case of the reaction

$$CO(g) + H_2O(g) = CO_2(g) + H_2(g)$$

the possible direction of reaction is determined by

$$\frac{\lambda_{CO_2(g)}\lambda_{H_2(g)}}{\lambda_{CO(g)}\lambda_{H_2O(g)}} \begin{cases} < 1 \text{ reaction goes forward} \\ = 1 \text{ reaction at equilibrium} \\ > 1 \text{ reaction goes backwards} \end{cases}$$

As $\lambda = a\lambda^\circ$, this is equivalent to writing

$$\frac{a_{CO_2(g)}a_{H_2(g)}}{a_{CO(g)}a_{H_2O(g)}} \begin{cases} < \\ = \\ > \end{cases} \frac{\lambda^\circ_{CO}\,\lambda^\circ_{H_2O}}{\lambda^\circ_{CO_2}\,\lambda^\circ_{H_2}}$$

The right hand side of this expression is constant for the reaction. It is called the equilibrium constant of the reaction (K_{eq})

$$K_{eq} \equiv \left(\frac{\lambda^\circ_{CO}\,\lambda^\circ_{H_2O}}{\lambda^\circ_{CO_2}\,\lambda^\circ_{H_2}}\right) \equiv \left(\frac{\lambda^\circ_{CO_2}\,\lambda^\circ_{H_2}}{\lambda^\circ_{CO}\,\lambda^\circ_{H_2O}}\right)^{-1}$$

With this definition of the equilibrium constant for the reaction, relative activities can be used to determine the direction in which reaction is possible, as follows.

$$\frac{a_{CO_2(g)}a_{H_2(g)}}{a_{CO(g)}a_{H_2O(g)}} \begin{cases} < K_{eq} \text{ forward reaction possible} \\ = K_{eq} \text{ reaction at equilibrium} \\ > K_{eq} \text{ reverse reaction possible.} \end{cases}$$

Example 4.3
What is the possible direction of the reaction

$$HSO_4^-(aq) + H_2O = H_3O^+(aq) + SO_4^{2-}(aq), \quad K_{eq} = 1.20 \times 10^{-2}$$

when $a_{H_2O} = 1$; $a_{HSO_4^-} = 0.1$; $a_{H_3O^+} = 1.0$; $a_{SO_4^{2-}} = 0.5$

Answer

$$\frac{a_{H_3O^+}\,a_{SO_4^{2-}}}{a_{H_2O}\,a_{HSO_4^-}} = \frac{1.0 \times 0.5}{1 \times 0.1} = 5 > K_{eq}$$

reaction will take place in the reverse direction, namely,

$$H_3O^+ + SO_4^{2-} = H_2O + HSO_4^-$$

Exercise 4.3
Determine the direction of change in each of the following cases:

(a) $Zn^{2+}(aq) + 4NH_3(aq) = Zn(NH_3)_4^{2+}(aq)$; $K_{eq} = 10^9$ given $a_{Zn^{2+}} = 0.01$; $a_{NH_3} = 1$; $a_{Zn(NH_3)_4^{2+}} = 0.001$.

(b) $Ag_3PO_4(s) = 3Ag^+(aq) + PO_4^{3-}(aq)$; $K_{eq} = 10^{-16}$ given $a_{Ag_3PO_4} = 1$; $a_{Ag^+} = 10^{-5}$; $a_{PO_4^{3-}} = 10^{-2}$

(c) $HCN + H_2O = H_3O^+ + CN^-$; $K_{eq} = 6 \times 10^{-10}$ given $a_{H_3O^+} = 10^{-12}$; $a_{CN^-} = 10^{-1}$; $a_{HCN} = 1.7 \times 10^{-4}$; $a_{H_2O} = 1$.

It reduces the effort of writing these expressions (but adds nothing to our knowledge of them) if we write them in a more compact form. To do this we rearrange the ordinary equation for any reaction:

$$aA + bB + cC + \ldots = xX + yY + zZ$$

to the form:

$$0 = xX + yY + zZ - aA - bB - cC$$

and express it more compactly as:

$$0 = \sum_B \nu_B B$$

In the compact form, we add together the product of the formula of the substance B and its stoichiometric coefficient (ν_B) for all the substances involved in the reaction (ν is negative for reactants and positive for products). To determine the direction in which reaction is possible we evaluate the quantity

$$\frac{a_X^x a_Y^y a_Z^z}{a_A^a a_B^b a_C^c} = a_X^x a_Y^y a_Z^z a_A^{-a} a_B^{-b} a_C^{-c}$$

The latter expression can be written in the symbolic form

$$\prod_B a_B^{\nu_B}$$

where the activity of each substance (a_B) is raised to the power of its stoichiometric coefficient (ν_B) and the results are multiplied (symbol Π) together.

The general criteria for direction of reaction then become

$$\prod_B a_B^{\nu_B} \begin{cases} < K_{eq} \text{ forward reaction} \\ = K_{eq} \text{ reaction at equilibrium} \\ > K_{eq} \text{ reverse reaction} \end{cases}$$

Example 4.4
Determine the criterion for the reaction

$$NH_3 = \frac{1}{2}N_2 + \frac{3}{2}H_2$$

to proceed in the forward direction.

Answer
The criterion is $\prod_B a_B^{\nu_B} < K_{eq}$

For the reaction $\nu_{NH_3} = -1$; $\nu_{N_2} = \frac{1}{2}$; $\nu_{H_2} = \frac{3}{2}$

hence, for forward reaction

$$\prod_B a_B^{\nu_B} = a_{N_2}^{1/2} a_{H_2}^{3/2} a_{NH_3}^{-1} < K_{eq}$$

Exercise 4.4
Determine the criteria (in terms of relative activities and equilibrium constants) for the following cases:

(a) the reaction $CO(g) + 2H_2(g) = CH_3OH(g)$ to proceed in the forward direction;
(b) the reaction $2NO_2(g) = N_2O_4(g)$ to proceed in the reverse direction;
(c) the reaction $CH_3COOH(aq) + H_2O = CH_3COO^-(aq) + H_3O^+(aq)$ to be at equilibrium;
(d) the reaction $Hg_2Cl_2(s) = Hg_2^{2+}(aq) + 2Cl^-(aq)$ to be at equilibrium;
(e) the reaction $H_2O_2(aq) + 3I^-(aq) + 2H_3O^+(aq) = I_3(aq) + 4H_2O$ to be at equilibrium.

4.4 The Estimation of Relative Activity ("activity")

The numerical values of a relative activity (often, loosely, just called activity) will depend on the actual standard state used. The standard states for particular substances are chosen to be as convenient as possible; that is, in order that the relationship between the activity and the concentration be as simple as possible. To achieve this, different classes of substance need different treatments, which we shall now present.

4.4.1 Gases at low pressures

For a gaseous component of a reacting mixture it is found that, at low pressures, the *absolute activity* of the gas becomes directly proportional to its partial pressure, that is,

$$\lambda_g \simeq kp_g$$

where k is a proportionality constant.

The activity of the gas is usually determined relative to its value when pure at a pressure $P°$ of 101.3 kPa and at the same temperature.

Query What is the partial pressure of the gas in its reference state?
Reply The partial pressure, p_g, is defined as the product of the mole fraction x_g in the gas phase and the total pressure P_{total}.

$$p_g \equiv x_g P_{total}$$

In the reference state, the gas is pure ($x_g = 1$) and the total pressure is $P°$. So, in the reference state, $p_g = P°$.

Query What is the absolute activity of the gas in the reference state?
Reply At low pressures $\lambda_g \simeq kp_g$ therefore in the reference state

$$\lambda_g^\circ \approx kP°$$

Query What is the relative activity of the gas in the reacting mixture?
Reply

$$a_g \equiv \lambda_g / \lambda_g^\circ$$

that is,

$$a_g \simeq p_g / P°$$

Example 4.5
Estimate the activity of oxygen in a gas mixture in which $p_{O_2} = 80$ kPa.

Answer

$$a_{O_2} \simeq p_{O_2}/P^\circ = 80\ \text{kPa}/101.3\ \text{kPa}$$

that is, $$a_{O_2} \approx 0.79$$

Exercise 4.5
(a) What is the relative activity of oxygen in air, where its mole fraction is 0.21, if the pressure of the air is 20 kPa?
(b) What is the relative activity of mercury in air at 100 kPa, when the partial pressure of the mercury is 10^{-2} Pa?

4.4.2 Solutes in dilute solutions

In very dilute solutions in a given solvent it is found that the absolute activity of each solute species B is proportional to its concentration:

$$\lambda_B \simeq k c_B$$

For solutes, there is no convenient experimental state available for use as a reference. The way round this problem is to define the proportionality constant in the above expression as λ°/c°, that is

$$\lambda_B \simeq \lambda_B^\circ\, c_B/c^\circ$$

with the standard concentration $c^\circ = 1$ mol dm^{-3}. For dilute solutions, the corresponding expression for the relative activity of a solute is

$$a_B \simeq c_B/c^\circ.$$

Query Does this mean that the absolute activity of a solute is equal to λ° when its concentration is 1 mol dm^{-3}.

Reply No. Such a solution is not "very dilute". At such concentrations of solutes, absolute activities and concentrations are no longer proportional. The standard activity, λ_B°, for a solute, is often described as "the value that λ_B would have at the standard concentration if the simple proportionality relation did apply up to the standard concentration".

Query Why have absolute activities of solutes in dilute solutions been compared with concentrations, c_B, rather than molalities, m_B, or mole fractions, x_B?

Reply For dilute solutions the three common measures of concentration are proportional to one another. We have a choice of writing

$$\lambda_B = \lambda_B^\circ\, c_B/c^\circ, \quad \text{with } c^\circ = 1\ \text{mol dm}^{-3}$$

$$\text{or } \lambda_B = \lambda_B^\circ\, m_B/m^\circ \quad \text{with } m^\circ = 1\ \text{mol (kg solvent)}^{-1}$$

$$\text{or } \lambda_B = \lambda_B^\circ\, x_B/x^\circ \quad \text{with } x^\circ = 1.$$

The choice is made according to which concentration scale we find most satisfactory for our work. We chose volume concentration as the most familiar scale.

Query Surely the λ_B°s for the three scales are not the same?

Reply Correct. The numerical value of λ_B° for a solute depends on the concentration scale to which it applies.

Query Does this mean that the numerical values of relative activities also depend on which concentration scale is used?

Reply Yes. When molalities are used to specify concentrations of solute species in *dilute solutions* we can write $a_B = m_B/m^\circ$. When mole fractions are used we can write $a_B = x_B/x^\circ$.

Query What else will depend on which concentration scale is used?

Reply K_{eq}, because the values of λ° of the solutes are different for the different concentration scales.

Query How large are the differences in practice?

Reply For dilute *aqueous* solutions the differences between m_B/m° and c_B/c° are negligible. In all other cases the differences are important.

Example 4.6
Estimate the relative activity of urea for each scale of concentration for an aqueous solution which can be described as
(a) 0.025 mol urea dm^{-3};
(b) 0.025 mol urea (kg water)$^{-1}$;
(c) $x_{urea} = 0.00045$.

Answer
(a) concentration scale:

$$a_{urea} \approx (c/c^\circ)_{urea} = 0.025\ \text{mol dm}^{-3}/1\ \text{mol dm}^{-3} = 0.025$$

(b) molality scale:

$$a_{urea} \approx (m/m^\circ)_{urea} = \frac{0.025\ \text{mol (kg water)}^{-1}}{1\ \text{mol (kg water)}^{-1}} = 0.025$$

(c) mole fraction scale:

$$a_{urea} = (x/x^\circ)_{urea} = 0.00045/1.000 = 0.00045$$

Exercise 4.6
Estimate the relative activity of glucose in solutions which have
(a) 10^{-4} mol glucose dm^{-3},
(b) 2×10^{-3} mol glucose (kg water)$^{-1}$,
(c) $x_{glucose} = 10^{-7}$

Answer
(a) 10^{-4};
(b) 2×10^{-3};
(c) 10^{-7}.

Equilibrium constants listed in, say, *The Handbook of Chemistry and Physics* use the molality scale for concentrations of solutes. Most of the entries refer to aqueous solutions at 298K and may be used, with less than 0.1 per cent error, as the values of equilibrium constants for the volume concentration scale.

Many practical calculations of amounts of solutes in chemical systems at equilibrium are carried out using volume concentrations. In the remainder of this chapter we use only the volume concentration scale for solutes which are involved in chemical equilibria.

4.4.3 Ionic species in dilute solutions

It is not possible to vary the concentrations of all ionic species independently, even in very dilute solutions. However, it is found that the patterns of equilibria for reactions involving ionic species in *very dilute* solutions correspond to simple proportionality between absolute activities and volume concentrations for each ionic species. We may write

$$\lambda_{\text{ionic species}} \equiv (\lambda^\circ a)_{\text{ionic species}} = (\lambda^\circ c/c^\circ)_{\text{ionic species}}$$

at low total concentrations of solutes. In this respect the behaviour of ionic species is just like that of non-ionic species such as glucose or urea. An important difference is that the restriction of "very dilute" is much more severe. In "less dilute" solutions the relative activity and the relative concentration can no longer be equated. We have only the approximate relationship:

$$a \approx (c/c^\circ)$$

To interrelate the relative activity and the relative concentration we define the "activity coefficient", y_i, such that, for an ionic species, i, in a solution of given composition

$$a_i \equiv (yc/c^\circ)_i$$

Example 4.7
What is the relative activity of chloride ion in 0.30 mol dm^{-3} solution of NaCl in water? $y_{Cl^-} = 0.70$.

Answer

$$a_{Cl^-} = (yc/c^\circ)_{Cl^-} = 0.70 \times 0.30 \text{ mol dm}^{-3} / 1 \text{ mol dm}^{-3} = 0.21$$

Exercise 4.7
What are the relative activities of the positive ions in the following aqueous solutions
(a) 0.8 mol HBr dm^{-3}; $y_{H^+} = 0.8$;
(b) 0.1 mol $CdCl_2$ dm^{-3}; $y_{Cd^{2+}} = 0.3$
(c) 0.2 mol $Al_2(SO_4)_3$$dm^{-3}$; $y_{Al^{3+}} = 10^{-4}$?

In the earlier part of this century, experimenters realised that the activity coefficient of a given ionic species depends mainly on the concentrations and electric charges of the ions present in the medium. A useful relationship for estimating values has been obtained by Davies from theoretical models developed by Debye and Hückel (1923). For an ionic species, i, with charge, z_i,

$$\log_{10} y_i = -A z_i^2 \frac{\sqrt{I}}{1 + Da\sqrt{I}} + BI$$

where I is "the ionic strength of the solution", calculated from the formula

$$I = \tfrac{1}{2} \sum_{\text{all ions}} (z^2 c/c^\circ)_i$$

In this equation, a is the ionic diameter which, for most ions, is of the order of 3×10^{-10}m. A and D are theoretical constants which can be calculated from the known properties of the pure solvent. For aqueous solutions at 25°C, $A = 0.51$ and $D = 0.3 \times 10^{10}$ m^{-1}. B is a coefficient which can be assigned the approximate value 0.1. Thus, at 25°C for ions in aqueous solution the "Davies equation for activity coefficients" takes the form

$$\log_{10} y_i \approx -0.51 z_i^2 \frac{\sqrt{I}}{1 + 1.0\sqrt{I}} + 0.1 I$$

According to the Davies equation, activity coefficients of different ionic species in the same solution are related via their charges. If y_1 is the activity coefficient of the species of unit charge, then

$$\log y_i \approx z_i^2 \log y_1$$

or

$$y_i \approx y_1^{z_i^2}$$

The Debye, Hückel, Davies expressions for activity coefficients of individual ionic species emerge when the principles of mechanics and electrostatics are used to find the most probable distributions of ions near a given ion in dilute solution. They account for the observed behaviour of dilute solutions of salts. Any salt solution must contain more than one type of ionic species.

Example 4.8

(a) For sodium chloride at 0.10 mol dm^{-3}

$$z^2_{Na^+} = 1, \qquad (c/c^\circ)_{Na^+} = 0.10$$

$$z^2_{Cl^-} = 1, \qquad (c/c^\circ)_{Cl^-} = 0.10$$

Thus $I = \frac{1}{2}(0.10 + 0.10)$

$= 0.10$

(b) For sodium sulfate at 0.10 mol dm^{-3}

$$z^2_{Na^+} = 1, \qquad (c/c^\circ)_{Na^+} = 0.20$$

$$z^2_{SO_4^{2-}} = 4, \qquad (c/c^\circ)_{SO_4^{2+}} = 0.10$$

Thus $I = \frac{1}{2}(1 \times 0.20 + 4 \times 0.10)$

$= 0.30$

(c) Use the Davies equation to estimate activity coefficients of Cu^{2+} and Cl^- in a 2×10^{-2} mol dm^{-3} solution of copper (II) chloride in water at 25°C.

Answer

The ionic strength of the solution is

$$I = \tfrac{1}{2}(2^2 \times 0.02 + 1^2 \times 0.04)$$

$$= 0.06$$

For the Cl^- ion

$$\log_{10} y \approx -0.51 \frac{0.06}{1 + 1.0\sqrt{0.06}} + 0.1 \times 0.06$$

$$y_{Cl^-} \approx y_1 = 0.80_5$$

For the Cu^{2+} ion

$$\log_{10} y_2 \approx y_1^4$$

$$y_{Cu^{2+}} \approx 0.42$$

The above example illustrates the general finding that activity coefficients of ionic species cannot be treated as unity except in solutions of very low ionic strength.

4.4.4 Non-ionic solutes in less dilute solutions

The activities and concentrations of uncharged solutes in less dilute solutions can be interrelated by using activity coefficients, y_B. These are defined in just the same way as for ions (Section 4.4.3):

$$a_B \equiv (yc/c^\circ)_B$$

However, the activity coefficients of non-ionic solutes remain close to unity up to much higher values of ionic strength.

Query Is there a formula for estimating the activity coefficient of an uncharged solute?

Reply Yes. We find

$$\log_{10} y \approx BI$$

where I is the ionic strength and B is a coefficient which is found by experiment.

Values of B are most often in the range 0 to 0.2. They are somewhat dependent on which ionic species are contributing to the ionic strength. The value 0.1 which appears in the Davies equation (Section 4.4.3) is useful for making approximate estimates.

Exercise 4.8

Estimate the relative activity of ether, 1 mol dm^{-3}, in water

(a) as the only solute;

(b) with NaCl, 2 mol dm^{-3}, also present. Assume that $B = 0.1$.

Answer

$a_{ether} = (yc/c^\circ)_{ether}$

(a) $y = 1.00$, $a = 1.00$

(b) $\log_{10} y \approx 0.1 \times 2.0$

$y \approx 1.6$, $a \approx 1.6$.

4.4.5 Solvents in dilute solutions

In a very dilute solution, the activity of the solvent (A) is found to be proportional to the mole fraction of solvent in the solution:

$$\lambda_A \simeq k\, x_A$$

Hence, for the reference state

$$\lambda_A^\circ = k \times 1$$

Query Then what is the relative activity of the solvent in a very dilute solution?

Reply $a_A = \lambda_A / \lambda_A^\circ$

$\simeq x_A$

Example 4.9

Estimate the relative activity of water in a 0.1 mol (kg water)$^{-1}$ solution of sucrose.

Answer

$$a_{H_2O} \simeq x_{H_2O} = \frac{55.5}{55.6} = 0.998 \simeq 1$$

In most problems involving solutions the activity of the solvent differs from unity by a few per cent at most. Hence, in practical calculations, it is usually quite sufficient to write the activity of the solvent, a_A, as 1.

Exercise 4.9

(a) Estimate the relative activity of water in a solution in which its mole fraction is 0.95.

(b) A solution of potassium chloride of concentration 1 mol (kg H_2O)$^{-1}$ has a mole fraction of water of 0.965. Estimate the relative activity of the water. How much error is likely to be involved in writing this activity as unity?

Answer

(a) 0.95; (b) 0.965; $<4\%$.

4.4.6 Pure solid substances

Frequently one or more of the substances which may participate in chemical change are present in essentially pure solid form. The absolute activity of a solid substance is referred to the value for the pure solid phase at the same temperature. It is nearly always satisfactory to approximate the absolute activity of a pure solid by this reference value, that is,

$$\lambda_s = \lambda_s^\circ$$

Thus the relative activity of such a solid is unity

$$a_s \equiv \lambda_s / \lambda_s^\circ = 1$$

Query That means the relative activity of pure solid water, or pure solid copper, is unity?

Reply Yes.

Query What would be a reasonable guess for the relative activity of copper in brass?

Reply To treat it just in the same way as a liquid solution—it is a solid solution. We saw above that the mole fraction of a solution is a useful approximation to the relative activity for liquid solutions. The mole fraction is also a reasonable approximation for copper in some brasses.

The relative activity of a pure solid is not unity if it is not in mechanical equilibrium—as when part of a metal rod is under stress, or has been cooled very quickly. Another aspect which can arise is the degree of subdivision. We find that the direction of chemical change is not normally dependent on how much surface of a solid reagent is exposed, though the rate may be greatly affected. Only for extremely small particles does relative activity depend on their size.

4.4.7 Gases at higher pressures

At higher pressures, the simple proportionality relationship between the absolute activity of a gas and its partial pressures may become too inaccurate to use.

Query What do we do when this happens?

Reply We define the activity coefficient of the gas in terms of the ratio of its activity (the desired quantity) to its partial pressure (the readily measured quantity), through the equation

$$a_{g(\text{higher pressures})} = f_g p_g / P^\circ$$

Query Are there any gases for which f_{gas} is not unity at $p_g = P^\circ$?

Reply Yes, a few, notably HF. They get treated the way we treated solutes. That is, we use the result that, at sufficiently low pressures,

$$\lambda_g = k p_g$$

and define

$$k \equiv \lambda_k^\circ / P^\circ$$

giving the general low pressure result

$$a \simeq p / P^\circ$$

4.4.8 Solvents in more concentrated solutions

As we might expect, in more concentrated solutions the simple relationship

$$a_{\text{solvent}} \simeq x_{\text{solvent}}$$

may no longer be sufficiently accurate. As before, we use an activity coefficient to relate the mole fraction of the solvent to its relative activity,

$$a_{\text{solvent}} = f_{\text{solvent}}\, x_{\text{solvent}}$$

Query How can one judge when the value of the activity coefficient will differ significantly from unity?

Reply For the types of solution normally encountered (say $c_B < 2$ mol dm^{-3}) setting the activity coefficient of an *uncharged solute* equal to unity will seldom introduce errors of more than 10%. *Ionic solutes* on the other hand, particularly divalent and trivalent ions, have activity coefficients which differ that much from unity at quite low concentrations of ions (e.g. 0.01 mol dm^{-3}). For gases, there is no problem until there is serious departure from the equation

$$P\bar{V} = RT$$

For most cases that takes us up to a few times P°—say, to five times atmospheric pressure—before the difference is likely to be more than 10%.

Query You have only given *rules* for measuring activities. Why don't you prove them?

Reply If by "prove" you mean "show by algebra that these choices are right", it cannot be done. What we can do is show that the choices we made link up with the behaviour we find. We will give three cases.

(a) *Gas at low pressure.* Consider an experiment, sketched in Fig. 4.1, in which a gas *mixture* in one vessel is separated from another vessel which contains one of its constituent gases in a pure state. If we can find a special sort of membrane which lets only this special gas pass, we can investigate the conditions at equilibrium. We find that, provided the gas pressures are not too high, the special gas reaches equilibrium when its pressure in the vessel where it is pure is equal to its *partial pressure* in the vessel where it is part of the mixture. So

$$P_{B(\text{pure gas})} = x_{B(\text{mixture})}\, P_{\text{total(mixture)}} \equiv p_{B(\text{mixture})}$$

These are conditions of equilibrium, so

$$\lambda_{B(\text{pure gas})} = \lambda_{B(\text{mixture})}$$

This is exactly the kind of behaviour we had in mind in saying that λ for a gas at low pressures is directly proportional to its partial pressure. Partial pressure is a useful physical property to be defined because of the existence of such relationships

(b) *Gases which dissolve slightly.* The second fairly simple case is the one we met as Henry's law (in Section 3.8.6). It involves equilibria of the kind

$$B(g) = B(solution)$$

and has the experimental relationship

$$p_B \simeq k\, x_{B(\text{solution})}$$

when p_B and $x_{B(\text{solution})}$ are not too high. We can treat this equilibrium in terms of activities and write the *absolute activities as*

$$\lambda_{B(g)} = \lambda_{B(\text{solution})}$$

$$a_{B(g)}\, \overset{\circ}{\lambda}_{B(g)} = a_{B(\text{solution})}\, \overset{\circ}{\lambda}_{B(\text{solution})}$$

If we make substitution for the relative activities of the gas, B, and for the solute, B, this becomes

$$\frac{p_B\, \overset{\circ}{\lambda}_{B(g)}}{P^\circ} = f_{B(\text{solution})}\, x_{B(\text{solution})}\, \overset{\circ}{\lambda}_{B(\text{solution})}$$

$$\simeq x_{B(\text{solution})}\, \overset{\circ}{\lambda}_{B(\text{solution})}$$

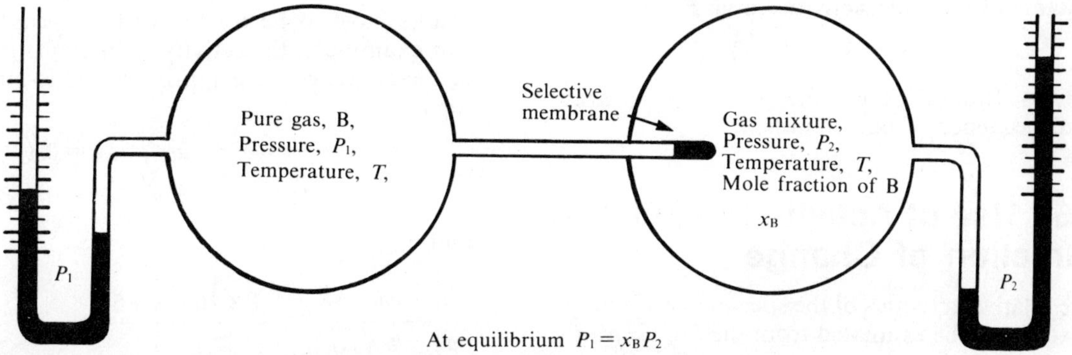

Fig. 4.1 Equilibrium between a gas mixture and a pure gas.

Putting all the constant factors together, this may be written as

$$p_B = \frac{\lambda^{\circ}_{B(solution)} P^{\circ}}{\lambda^{\circ}_{B(g)}} x_{B(solution)}$$

This is just the same as Henry's law.

(c) *Solvents in dilute solutions.* For these situations, equilibrium with the gas phase has been described (Section 3.8.5) in terms of Raoult's law. The equilibrium may be written as

$$A\,(g) = A\,(solution) \qquad (1)$$

and the law expressed by

$$p_A \simeq p^{\circ}_A\, x_{A(solution)} \qquad (2)$$

where p°_A is the "vapor pressure" for the equilibirum

$$A\,(g) = A\,(l, pure) \qquad (3)$$

Note: we use the symbol A to denote a solvent and the symbol B to denote a solute. The mole fraction of A will normally be greater than 0.5 while the mole fraction of B will normally be rather less than 0.5.

To use activities, we may again begin with λ and express the equilibrium in (1) as

$$\lambda_{A(g)} = \lambda_{A(solution)}$$

$$a_{A(g)}\lambda^{\circ}_{A(g)} = a_{A(solution)}\lambda^{\circ}_{A(solution)}$$

Thus, approximately, for dilute solutions

$$\frac{p_A\, \lambda^{\circ}_{A(g)}}{P^{\circ}} \simeq x_{A(solution)}\lambda^{\circ}_{A(solution)} \qquad (4)$$

For the pure substance A, equation (3) may be analysed in just the same way and we find

$$\frac{p^{\circ}_A\, \lambda^{\circ}_{A(gas)}}{P^{\circ}} \simeq \lambda^{\circ}_{A(solution)} \qquad (5)$$

On combining (4) and (5), we eliminate the standard absolute activities and P°.

$$p_A \simeq x_{A(solution)}\, p^{\circ}_A$$

Thus Raoult's law emerges as a natural consequence of our choices.

4.5 Use of Activity to Find the Direction of Change

The relative activities of the species in a reacting mixture can be estimated from their concentrations and activity coefficients. The likely direction of change can then be determined.

Example 4.10
For the reaction

$$CH_3COOH + H_2O = CH_3COO^- + H_3O^+ \; (aq, 298\ K),$$

$$K_{eq} = 1.8 \times 10^{-5}$$

what is the probable direction of change in an aqueous solution containing

$$CH_3COOH,\ 0.1\ mol\ dm^{-3},\ y_{CH_3COOH} \approx 1;$$

$$CH_3COO^-,\ 0.5\ mol\ dm^{-3},\ y_{CH_3COO^-} \approx 0.7;$$

$$H_3O^+,\ 10^{-7}\ mol\ dm^{-3},\ y_{H_3O^+} \approx 0.7?$$

Answer
The probable direction of reaction is determined by comparing with K_{eq} the value of the expression

$$\frac{a_{H_3O^+}\, a_{CH_3COO^-}}{a_{CH_3COOH}\, a_{H_2O}}$$

Now $a_{H_3O^+} = 0.7 \times 10^{-7}$;

$a_{CH_3COOH} = 0.1$;

$a_{H_2O} \approx 1$;

and $a_{CH_3COO^-} = 0.35$

that is

$$\frac{a_{H_3O^+}\, a_{CH_3COO^-}}{a_{CH_3COOH}\, a_{H_2O}} \approx \frac{0.7 \times 10^{-7} \times 0.35}{0.1 \times 1}$$

$$\approx 2.4 \times 10^{-7}$$

$$< K_{eq}$$

so reaction may proceed in the forward direction.

Query What if we don't have values for the activity coefficients?

Reply We approximate them by unity. The results have to be treated with due caution. For example, if the activity coefficients in the last example were all approximated by unity, then

$a_{H_3O} \approx 1 \times 10^{-7}$; $a_{CH_3COOH} \approx 0.1$

$a_{H_2O} \approx 1$; $a_{CH_3COO} \approx 0.5$

and

$$\frac{a_{H_3O^+}\, a_{CH_3COO^-}}{a_{H_2O}\, a_{CH_3COOH}} = \frac{1 \times 10^{-7} \times 0.5}{0.1 \times 1}$$

$$= 5 \times 10^{-7},\ \text{(an error of 100\%)}$$

Even so, $5 \times 10^{-7} < K_{eq}$ and hence the prediction of the direction of reaction is still correct.

Exercise 4.10
Determine the probable direction of reaction under the following conditions.

(a) A gaseous mixture of PCl_5, PCl_3 and Cl_2, such that $p_{PCl_5} = 90$ kPa; $p_{PCl_3} = 3$ kPa; $p_{Cl_2} = 3$ kPa react according to the equation

$$PCl_5(g) = PCl_3(g) + Cl_2(g)$$

for which $K_{eq} = 1.2$

(b) A gaseous mixture of NH_3 and H_2S above solid NH_4HS reacts according to the equation

$$NH_4HS(s) = NH_3(g) + H_2S(g)$$

$p_{NH_3} = 32$ kPa; $p_{H_2S} = 32$ kPa; $K_{eq} = 0.11$

(c) An aqueous solution of H_2S, $ZnCl_2$ and HCl which react according to the equation

$$Zn^{2+}(aq) + H_2S(aq) + 2H_2O = ZnS(s) + 2H_3O^+(aq),$$

$$K_{eq} = 6.2 \times 10^{-2}$$

under conditions in which there are:

H_2S, 7.5×10^{-2} mol dm^{-3}, $y_{H_2S} \approx 1$

Zn^{2+}, 0.01 mol dm^{-3}, $y_{Zn^{2+}} \approx 0.75$

H_3O^+, 0.01 mol dm^{-3}, $y_{H_3O^+} \approx 0.9$

(d) Will 20 mg (1.4×10^{-4} mol) of AgCl dissolve in a tank containing
(i) 5 dm^3 water,
(ii) 5 dm^3 water containing 5 mol of $NaNO_3$?

(In the sodium nitrate solution $y_{Ag^+} \approx y_{Cl^-} \approx 0.4$; $K_{eq} = 1.7 \times 10^{-10}$.)

(e) A 500 cm^3 flask contains 1.5 mmol NO(g), 1 mmol $Cl_2(g)$ 2.5 mmol ClNO(g). The temperature is 503 K. The gases react according to the equation

$$2NO(g) + Cl_2(g) = 2ClNO(g)$$

$K_{eq} = 0.89$

In which direction will change occur?

4.6 Reactions at Equilibrium: a Useful, Special Case

In a reacting system *at equilibrium*

$$\Pi_B a_B^{\nu_B} = K_{eq}$$

Therefore a knowledge of the activities of some of the species in a reacting mixture at equilibrium enables us to determine others. In particular, because of the close relationship between the activity and the concentration, the condition that a reaction be at equilibrium may be used to

(a) determine the concentrations and hence amounts of the species present at equilibrium;
(b) determine the extent to which a given reacting mixture will proceed before equilibrium is reached, and hence determine the yield of products to be expected;
(c) control the concentration of one species by judicious choice of the activities of the other species involved in the reaction.

Query What is the strategy for using the equilibrium condition in solving such problems?

Reply

(a) write down the equation for the reaction, for example,

$$H_2(g) + Cl_2(g) = 2HCl(g)$$

(b) write down the equilibrium condition for the reaction, at equilibrium,

$$a_{HCl(g)}^2 \, a_{H_2(g)}^{-1} \, a_{Cl_2(g)}^{-1} = K_{eq}$$

(c) replace the activity by the appropriate concentration and activity coefficient,

$$(f_{HCl} p_{HCl} / P^\circ)^2 (f_{H_2} p_{H_2} / P^\circ)^{-1} (f_{Cl_2} p_{Cl_2} / P^\circ)^{-1} = K_{eq}$$

(d) substitute values for the appropriate terms,

$$f_{HCl} = f_{H_2} = f_{Cl_2} = 1$$

Note: in practice, it may be impossible to evaluate the activity coefficients until all the concentrations are known. In such cases the problem is solved approximately, the activity coefficients determined, and the calculation repeated.

(e) examine the problem in order to determine any relationships between the quantities involved. In many problems, approximate values may be used for some quantities:
(f) the problem is then solved;
(g) all approximations made in (e) are then checked for consistency using the answer,

and any necessary improvement in approximations fed back into the calculations.

Example 4.11

(a) Estimate the concentration of H_3O^+ in an aqueous solution containing sodium acetate and undissociated acetic acid

CH_3COOH 0.1 mol dm^{-3}, $y_{CH_3COOH} \approx 1$,

CH_3COO^- 0.01 mol dm^{-3}, $y_{CH_3COO^-} \approx 0.83$,

H_3O^+ $y_{H_3O^+} \approx 0.83$

$f_{H_2O} \approx 1$

For the reaction

$$CH_3COOH + H_2O = CH_3COO^- + H_3O^+ \ (aq, 298\,K),$$

$$K_{eq} = 1.8 \times 10^{-5}$$

Answer

At equilibrium

$$a_{CH_3COO^-}\, a_{H_3O^+}\, a_{H_2O}^{-1}\, a_{CH_3COOH}^{-1} = K_{eq}$$

This is rewritten in terms of concentrations as

$$(yc/c^\circ)_{CH_3COO^-}\,(yc/c^\circ)_{H_3O^+}\,(fx/x^\circ)_{H_2O}^{-1}\,(yc/c^\circ)_{CH_3COOH}^{-1} = 1.8 \times 10^{-5}$$

At low concentrations $x_{H_2O} \approx 1$. The known values are substituted into the equation for equilibrium, to give:

$$(0.01 \times 0.83)\left(\frac{c_{H_3O^+}}{c^\circ} \times 0.83\right)(1)^{-1}(0.1 \times 1)^{-1} = 1.8 \times 10^{-5}$$

$$\therefore \quad c_{H_3O^+} = 2.6 \times 10^{-4} c^\circ$$

$$= 2.6 \times 10^{-4}\ \text{mol dm}^{-3}$$

(b) Estimate the partial pressure of $H_2S(g)$ above a solid sample of ammonium hydrogen sulphide contained in an open beaker at 25°C. For the reaction

$$NH_4HS(s) = NH_3(g) + H_2S(g),$$

$$K_{eq}(298\ K) = 0.096$$

Answer

At equilibrium

$$a_{NH_3}\, a_{H_2S}\, a_{NH_4HS}^{-1} = K_{eq}$$

and

$$a_{NH_3} = f_{NH_3}\, p_{NH_3}/P^\circ$$

$$a_{H_2S} = f_{H_2S}\, p_{H_2S}/P^\circ$$

$$a_{NH_4HS} = 1$$

For a gas at low pressure, $f_{NH_3} \approx f_{H_2S} \approx 1$ and from stoichiometry $n_{NH_3} = n_{H_2S}$, thus $p_{NH_3} = p_{H_2S}$.

When we substitute these relations into the condition for equilibrium we obtain

$$(p_{H_2S}/P^\circ)^2 = 0.096$$

or

$$p_{H_2S} = 0.31\ P^\circ$$

$$= 0.31 \times 101.3\ \text{kPa}$$

$$\therefore \quad p_{H_2S} = 31\ \text{kPa}$$

(c) Estimate the solubility of barium sulphate in an aqueous solution of sodium sulphate in which

$$c_{SO_4^{2-}} = 0.1\ \text{mol dm}^{-3}$$

$$y_{Ba^{2+}} = y_{SO_4^{2-}} = 0.08$$

For the reaction

$$BaSO_4(s) = Ba^{2+}(aq) + SO_4^{2-}(aq),\ K_{eq} = 1 \times 10^{-10}$$

Answer

At equilibrium

$$a_{Ba^{2+}}\, a_{SO_4^{2-}}\, a_{BaSO_4}^{-1} = K_{eq}$$

or, in terms of concentrations,

$$(c_{Ba^{2+}}\, y_{Ba^{2+}}/c^\circ)\,(c_{SO_4^{2-}}\, y_{SO_4^{2-}}/c^\circ)\,(1)^{-1} = 1 \times 10^{-10}$$

Let the solubility of barium sulphate in the solution be S mol dm^{-3}

then

$$c_{Ba^{2+}}/c^\circ = S$$

$$c_{SO_4^{2-}}/c^\circ = 0.1 + S \approx 0.1$$

that is,

$$(S \times 0.08)\,(0.1 \times 0.08) = 1 \times 10^{-10}$$

$$\therefore \quad S = 1.6 \times 10^{-7}$$

The approximation that $0.1 + S \approx 0.1$ can now be checked. Indeed $0.1 + 1.6 \times 10^{-7} \approx 0.1$. Thus the solubility of barium sulphate in a 0.1 mol dm^{-3} solution of sodium sulphate is 1.6×10^{-7} mol dm^{-3}.

Comment These calculations appear to be very complicated.

Reply The illusion that equilibrium calculations are complex can only be dispelled by the familiarity which is gained by practice. There appear to be three areas in which practice is required. First, identifying and setting up problems. Second, making the appropriate approximations. Skill in this area grows as familiarity with the basic chemistry of the substances involved increases. The third area for practice is the arithmetic skills required for solving simple linear, quadratic, maybe cubic, equations, and manipulating logarithms. Some examples for practice are given below. Further practice examples may be found in many texts of general chemistry. Two suitable texts are Porterfield, W.W., *Concepts of Chemistry*, Norton, New York, 1972, Chapter 10; and Rosenberg, J.L., *Schaum's Outline of College Chemistry*, McGraw-Hill, New York, 1972, Chapters 16, 17 and 18. In almost all texts tne activity coefficients of the species present are approximated by unity. A result of this approximation is that, in some cases, the answers to the calculations may be too inaccurate to be useful. We recommend the habit of going through steps (a)-(d) listed on p. 67. By doing this you will always be aware of the nature of the approximations you are making.

Exercise 4.11

(a) Estimate the partial pressure of carbon dioxide in equilibrium with calcium carbonate and calcium oxide in a lime kiln at 1100 K if, for the reaction,

$$CaCO_3(s) = CaO(s) + CO_2(g),$$
$$K_{eq}(1100\ \text{K}) = 0.22$$

If the total pressure of gas inside the kiln is 100 kPa, estimate the mole fraction of carbon dioxide in the gas.

(b) Estimate the solubility of silver bromate in water at 20°C if, for the reaction

$$AgBrO_3(s) = Ag^+(aq) + BrO_3^-(aq),$$
$$K_{eq}(292\text{K}) = 6\times10^{-5}$$

(under these conditions $y_{Ag^+} = y_{BrO_3^-} \simeq 1$)

(c) Estimate the hydrogen ion concentration of an aqueous solution of 0.1 mol dm^{-3} sodium carbonate, for which

$y_{H_3O^+} \approx 0.6$, $y_{OH^-} \approx 0.6$, $y_{CO_3^{2-}} \approx 0.08$,

$y_{HCO_3} \approx 0.6$, $f_{H_2O} \approx 1$

For the reaction

$$HCO_3^- + H_2O = CO_3^{2-} + H_3O^+(aq)$$
$$K_{eq} = 4.7\times10^{-11}.$$

and

$K\{2H_2O = H_3O^+ + OH^-(aq)\}_{298K} = 10^{-14}$

(d) Estimate the hydrogen ion concentration of the above solution, approximating the values of the activity coefficients by unity.

(e) (A more challenging problem.) You are given an initial mixture of N_2 and H_2 in a sealed glass flask such that $p_{N_2} = 50$ kPa and $p_{H_2} = 100$ kPa which undergoes the reaction:

$$N_2(g) + 3H_2(g) = 2NH_3(g),\ K_{eq} = 0.1$$

At equilibrium, what are the values of:

(i) p_{N_2}; (ii) p_{H_2}; (iii) p_{NH_3} and (iv) p_{total}?

Comment pH and pK are also common ways of writing some of these quantities.

Reply Yes. But in practice the meaning of the "p" is not as similar in those two cases as Sorensen (who devised this way of reporting results) would have wanted. Take the second case first. Equilibrium constants have a very wide range of numerical values, at least from 10^{-50} to 10^{+50}, and we often only need a rough value for solution of our problem. The obvious way of proceeding is to use exponential notation and to note only the changes in the exponent. Or, equivalent to that, to take the logarithm of our equilibrium constant (for example $\log_{10} K_{eq}$). Either way we compress the scale to one of familiar numerical size. But many of the people who need to use such numbers get scared of mathematical terms like "$\log_{10}$". Sorensen was mainly concerned with numbers less than unity (negative values of the logarithm). He proposed (and the proposal was accepted) that, for any quantity, Q, the negative logarithm to the base 10 of Q be called pQ, that is,

$$pQ \equiv -\log_{10}Q$$

so that, for example,

$$pK_{eq} = -\log_{10} K_{eq}$$

Example 4.12

For the reaction at 298 K,

$$H_2(g) + Cl_2(g) = 2HCl(g),\ K_{eq} = 1.2\times10^{33}$$

so $pK_{eq} = -\log_{10}(1.2\times10^{33}) = -33.08$

The acidity or alkalinity of solutions is an important element in their reactivity and one which needs to be controlled in a very great number of processes, from bread making to waste disposal. In solutions in water, the concentration of hydrogen ion, $c_{H_3O^+}$, may lie anywhere from about 10 to 10^{-15} mol dm^{-3}, so the pQ notation is appropriate. It is near enough to say that

$$pH = -\log a_{H_3O^+}$$

Comment You imply some doubt about this relationship?

Reply Yes. pH has a precise legal definition which results in it having no precise fundamental meaning. In practice, the definition says that pH is what may be measured on a particular sort of instrument used in a particular way. When someone describes the pH at which you can make a kind of cheese or grow healthy cabbages, that is what he means and, given the recipe, any of us can make the test without needing to know the chemistry. At moderate concentrations the legal definition and the relationship above agree with one another.

Query Where can I find the definition?

Reply In the journal of the International Union of Pure and Applied Chemistry, *Pure and Applied Chemistry*, 1970, *21*, p.33.

4.7 Determination of the Equilibrium Constant of a Reaction

Five main methods are used to determine equilibrium constants.

4.7.1 From known values of equilibrium constants

From combination of the *known values of* the *equilibrium constants for other reactions* we get many which are not readily measured directly. Consider the equations in the following examples.

(a) $Cu(s) + 2H^+(aq) = H_2(g) + Cu^{2+}(aq)$ (1)

$Cu^+(aq) + H^+(aq) = \frac{1}{2}H_2(g) + Cu^{2+}(aq)$ (2)

$Cu(s) + H^+(aq) = \frac{1}{2}H_2(g) + Cu^+(aq)$ (3)

Note that equation (1) = equation (2) + equation (3)

$$K_{eq}(1) = K_{eq}(2) \times K_{eq}(3)$$

Proof

$$K_{eq(1)} = \frac{\lambda^{\circ}_{Cu(s)} \lambda^{\circ 2}_{H^+(aq)}}{\lambda^{\circ}_{H_2(g)} \lambda^{\circ}_{Cu^{2+}(aq)}}$$

$$K_{eq(2)} K_{eq(3)} = \frac{\lambda^{\circ}_{Cu^+(aq)} \lambda^{\circ}_{H^+(aq)}}{\sqrt{\lambda^{\circ}_{H_2(g)}}\, \lambda^{\circ}_{Cu^{2+}(aq)}} \times \frac{\lambda^{\circ}_{Cu(s)} \lambda^{\circ}_{H^+(aq)}}{\sqrt{\lambda^{\circ}_{H_2(g)}}\, \lambda^{\circ}_{Cu^{+}(aq)}}$$

$$= \frac{\lambda^{\circ}_{Cu(s)} \lambda^{\circ 2}_{H^+(aq)}}{\lambda^{\circ}_{H_2(g)} \lambda^{\circ}_{Cu^{2+}(aq)}}$$

$$= K_{eq}(1)$$

(b) $HCOOH + H_2O = H_3O^+ + HCOO^-$; (1)

$K_{eq}(1) = 1.8 \times 10^{-4}$

$2H_2O = H_3O^+ + OH^-$; $K_{eq}(2) = 10^{-14}$ (2)

What is the equilibrium constant for:

$HCOOH + OH^- = H_2O + HCOO^-$? (3)

Answer
equation (3) = equation (1) − equation (2)

$$\therefore \quad K_{eq}(3) = K_{eq}(1) / K_{eq}(2)$$
$$= 1.8 \times 10^{-4} \times 10^{14}$$
$$= 1.8 \times 10^{10}$$

4.7.2 From measurements of relative activities

If we *measure the relative activities* of the various species present in a reaction at equilibrium, we can calculate K_{eq}. For example, for

$$CO + H_2O = CO_2 + H_2$$

at equilibrium $$\frac{a_{CO_2} a_{H_2}}{a_{CO} a_{H_2O}} = K_{eq}$$

So, if we can observe the four relative activities in a reaction at equilibrium, we can calculate K_{eq} from them.

Example 4.13
A mixture of H_2, I_2 and HI was found to be in equilibrium when

$$a_{H_2} = 5.62 \times 10^{-2};$$
$$a_{I_2} = 0.0594 \times 10^{-2};$$
$$a_{HI} = 1.27 \times 10^{-2}$$

What is K_{eq} for the reaction $H_2 + I_2 = 2HI$?

Answer
At equilibrium $$\frac{a_{HI}^2}{a_{I_2} a_{H_2}} = K_{eq}$$

$$\therefore \quad K_{eq} = \frac{(1.27 \times 10^{-2})^2}{5.94 \times 10^{-4} \times 5.62 \times 10^{-2}}$$
$$= 4.8$$

4.7.3 From values of standard absolute activity

Values of *standard absolute activity*, $\lambda°$, for the substances involved can be used in calculation of K_{eq}.

Example 4.14
At 298 K, $\lambda^{\circ}_{NO_2(g)} = 8.6 \times 10^8$
and $\lambda^{\circ}_{N_2O_4(g)} = 1.5 \times 10^{17}$

Therefore, for the reaction

$$2NO_2(g) = N_2O_4(g)$$
$$K_{eq}(298K) = (\lambda^{\circ}_{N_2O_4(g)} / \lambda^{\circ 2}_{NO_2(g)})^{-1}$$
$$= \frac{(8.6 \times 10^8)^2}{1.5 \times 10^{17}}$$
$$= 4.9$$

Lists of standard absolute activities are published, but not under that name. Indeed, although we find the concept of absolute activities to be a useful one, only a few others have used it as more than a stepping stone to the relative activity.

Like any other scale, the scale of absolute activities needs to have a beginning and we can choose one. A sensible choice seems to us to be that chosen for many other chemical properties, for example, enthalpies of formation. So we assign the value *unity* to the *absolute activity* of the *most stable state of each elementary substance* at the temperature of the experiments and at the standard pressure, $P°$.

Query Does this mean that the standard values of the absolute activities of lead and gold are the same?

Reply Yes.

Comment But lead is much more chemically active than gold.

Reply That is so. However, in practice we will never wish to compare the absolute activities of lead and gold directly. What we may wish to compare are the values of $\frac{\lambda° Au}{\lambda° AuCl}$ and values of $\frac{\lambda° Pb}{\lambda° PbCl_2}$, for example, to determine the equilibrium constant for the reaction

$$2Au + PbCl_2 = 2AuCl + Pb$$

Such ratios are not affected by our decision to make $\lambda°_{Au} = \lambda°_{Pb} = 1$. It is outside the limits of usefulness of the ideas of absolute activity to seek to determine the value of

$$\frac{\lambda° Au}{\lambda° Pb}$$

which would represent the equilibrium constant of the non-chemical transmutation reaction

$$Pb(s) = Au(s) \quad !$$

With the above choice for $\lambda°$ it is easy to show that, for any substance, $\lambda°$ is equal to the reciprocal of the equilibrium constant, K_f, for its formation from the elements.

$$\lambda° = K_f^{-1}$$

Thus $\lambda°$ for NO_2 at 298 K is K_{eq}^{-1} for the reaction:

$$0.5N_2(g) + O_2(g) = NO_2(g) \text{ at } 298 \text{ K}$$

Many tables, such as *JANAF Thermochemical Tables*, Nat. Stand. Ref. Data Ser., Nat. Bur. Stand. (U.S.), 1965, 1971, or *Handbook of Chemistry and Physics*, Chemical Rubber Co., list values of the logarithm of these equilibrium constants, normally identified as $\log_{10} K_f$, or $-pK_f$.

The pK for any reaction is the sum of pK_f values taken according to the equation for the reaction. Thus for

$$A + B = C$$

$$pK = pK_f(C) - pK_f(A) - pK_f(B)$$

or $$K = K_f(C)\, K_f(A)^{-1}\, K_f(B)^{-1}$$

4.7.4 Two other methods

The *remaining two methods* for determining

equilibrium constants involve tabulated values of $\Delta \bar{G}_f^\circ$ and E°. The definitions of these quantities and their use for the present purpose are discussed in Section 6.7.1.

Exercise 4.12
Use the given experimental information to determine the equilibrium constants for the following reactions.

(a) $H_2(g) + I_2(g) = 2HI(g)$ at 458°C

if, at equilibrium,

$p_{H_2} = 29.7$ kPa, $p_{I_2} = 3.14$ kPa,
$p_{HI} = 67.2$ kPa.

(b) $COCl_2(g) = CO(g) + Cl_2(g)$

if, at equilibrium,

$x_{COCl_2} = 0.33$, $x_{CO} = 0.25$,
$x_{Cl_2} = 0.42$, $P = 62.9$ kPa.

(c) $2NO_2(g) = N_2O_4(g)$.

A bulb to which 0.05 mol of N_2O_4 is added is found at equilibrium to contain 0.01 mol NO_2 and be at a pressure of 14 kPa.

(d) $2H_2O(g) = 2H_2(g) + O_2(g)$

if, at 1000°C and 101 kPa, the fraction of water dissociated is 24×10^{-6}.

(e) $H_2O(g) = H_2(g) + \tfrac{1}{2}O_2(g)$

if, for $CO(g) + H_2O(g) = CO_2(g) + H_2(g)$,

$K_{eq} = 0.36$; and, for

$CO(g) + \tfrac{1}{2}O_2(g) = CO_2(g)$, $K_{eq} = 9.8 \times 10^5$.

(f) $4HCl(g) + O_2(g) = 2H_2O(g) + 2Cl_2(g)$

if, for $H_2(g) + Cl_2(g) = 2HCl(g)$,

$K_{eq} = 1.2 \times 10^{33}$; and, for

$2H_2(g) + O_2(g) = 2H_2O(g)$, $K_{eq} = 3.8 \times 10^{13}$.

Question
How is K_{eq} related to K_p and K_c?

Answer
For gases at low pressures for example,

$$NH_3 = \tfrac{1}{2}N_2 + \tfrac{3}{2}H_2$$

$$K_{eq} \equiv a_{N_2}^{1/2} a_{H_2}^{3/2} a_{NH_3}^{-1}$$

$$= (f_{N_2} p_{N_2}/P^\circ)^{1/2} (f_{H_2} p_{H_2}/P^\circ)^{3/2} (f_{NH_3} p_{NH_3}/P^\circ)^{-1}$$

$$= p_{N_2}^{1/2} p_{H_2}^{3/2} p_{NH_3}^{-1} (1/P^\circ)^{1/2+3/2-1}$$

K_p for this reaction is defined as

$$K_p \equiv p_{N_2}^{1/2} p_{H_2}^{3/2} p_{NH_3}^{-1}$$

It follows that

$$K_{eq} \equiv \Pi_B (fp/P^\circ)_B^{\nu_B}$$
$$\equiv (\Pi_B f_B^{\nu_B}) K_p (1/P^\circ)^{\Sigma \nu_B}$$

That is, K_p must be converted to the scale of relative pressures and multiplied by the stoichiometric product of activity coefficients to obtain the equilibrium constant for a reaction which involves gaseous substances. The use of K_p is thus less general than the use of K_{eq}, but most entries in tables, under the heading of K_p, are actually values of K_{eq}. Values of K_p will be *constant* only over ranges of conditions in which the activity coefficients are *constant*. The relationship to K_{eq} is direct only when their value is unity.

Note also that K_p is a quantity which has units. The values therefore change with change in the unit used for pressure, whereas K_{eq} is a number (without units) and so is the same no matter what pressure unit is chosen.

A similar relationship exists between K_{eq} and K_c for reactions in solution. The activity of each *solute* species is given by

$$a_B \equiv y_B c_B / c^\circ$$

If we approximate the solvent activity, wherever it appears, by unity we have

$$K_{eq} \equiv K_c (1/c^\circ)^{\Sigma \nu_B} (\Pi_B y_B^{\nu_B})$$

Query What are K_a, K_{sp}, K_b and K_w?
Reply These are important classes of K_{eq}:
K_a is K_{eq} for dissociation of acids;
K_{sp} is K_{eq} for dissolution of ionic substances (and thus equal to the product of the activities of the ions in saturated solutions of such substances);
K_b is K_{eq} for dissociation of bases; and
K_w is K_{eq} for dissociation of water.

When possible, we prefer to write the chemical equation whenever we introduce an equilibrium constant. We think this reduces the ambiguities significantly. It happens to make these particular symbols for classes much less necessary—though they are useful in tables.

Query Is that your excuse for using K_f?

Reply Yes. The range of usefulness of tables of K_f seems to justify that class name. It is also one which gets messy to write out in words or chemical equations time after time.

4.8 Temperature Change and Equilibrium Constant

There is an astonishing regularity about the way equilibrium constants change with changing temperature. Whenever the value of the equilibrium constant increases with increasing temperature, the reaction belongs to one class when examined at almost constant temperature. When K_{eq} decreases with increasing temperature, it belongs to another. The comparison experiment which generates the two classes to which we refer is simple, but it needs to be stated rather carefully. Suppose that we observe what happens when equilibrium is *nearly* established in one of these reactions, and choose to make the observations in some vessel which is insulated from thermal interactions with its surroundings. This might be a Dewar vessel or a foam plastic beaker. In reactions of the first class, the temperature will be found to be *falling* as equilibrium is approached. In reactions of the second class, it will be *rising*.

This is quite a general observation. The most simple examples to illustrate it are probably the changes in the solubility of substances with changing temperature. Suppose we take sodium sulfate and dissolve it in an aqueous solution which is already nearly saturated with sodium sulfate. This is an experiment which we can do with different results according to the solid salt with which we start, and the experimental temperature. If we start with solid $Na_2SO_4 \cdot 10H_2O$ and a temperature of 10° C, we will find

(a) that we can dissolve more salt at higher temperatures, and
(b) that the temperature tends to fall in the dissolution process.

If we start with solid Na_2SO_4 and temperatures above 32.38° C, we find

(a) we can dissolve rather less salt at higher temperatures, and
(b) the temperature tends to *rise* in the dissolution process.

Figure 4.2 shows the phase diagram for this system—it is only slightly different from those we met in Section 3.6.

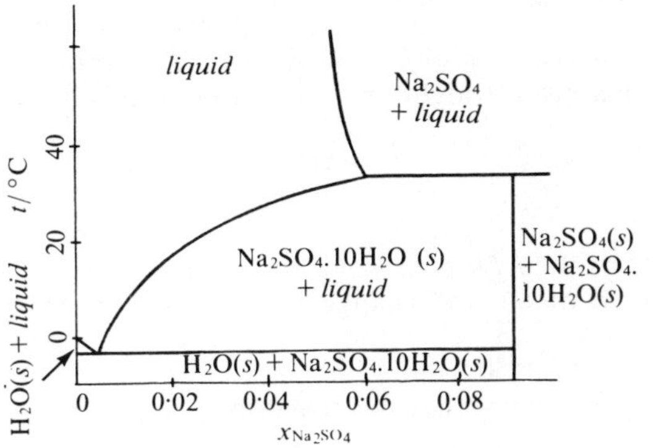

Fig. 4.2 Phase equilibria for sodium sulfate and water.

Closer examination of this regularity shows that it involves a relationship between the *logarithm* of the equilibrium constant, ln K_{eq}, and the standard enthalpy change, $\Delta \bar{H}°$.
In fact, a graph of ln K_{eq} against T has the slope

Thus

$$\frac{d \ln K_{eq}}{dT} = \frac{\Delta \bar{H}°}{RT^2}$$

The relationship proves to be an exact one and is as useful for determining values of $\Delta \bar{H}°$ as of K_{eq}. It was established by J.H. van't Hoff in 1884.

Query Does $\Delta \bar{H}°$ change much with temperature?

Reply No. If we need to know values of K_{eq} which are accurate to, say, 5%, then we almost never have a temperature range big enough to make the change in $\Delta \bar{H}°$ significant. If we do need to make a correction, there is a straightforward way to do it.

Comment Then, if $\Delta \bar{H}°$ is near enough to constant, the equation can be reorganised to cover substantial ranges in temperature.

Reply Yes. It involves a simple integration.

$$\int_{T_1}^{T_2} \mathrm{d}\ln K_{eq} = \int_{T_1}^{T_2} \frac{\Delta \bar{H}^\circ}{RT^2}\mathrm{d}T$$

that is,

$$\ln K_{eq}(T_2) = \ln K_{eq}(T_1) - \frac{\Delta \bar{H}^\circ}{R}\left(\frac{1}{T_2} - \frac{1}{T_1}\right)$$

or,

$$\log_{10} K_{eq}(T_2) = \log_{10} K_{eq}(T_1) - \frac{\Delta \bar{H}^\circ}{2.303\,R}\left(\frac{1}{T_2} - \frac{1}{T_1}\right)$$

That is, a plot of log K_{eq} vs. $1/T$ is approximately a straight line. The slope of this line is

$$-\frac{\Delta \bar{H}^\circ}{2.303\,R}$$

Exercise 4.13

(a) Estimate the change in K_{eq} between 25°C and 37°C for a reaction with $\Delta \bar{H}^\circ = 50$ kJ mol^{-1}.

(b) Assuming that the reaction

$$3C_2H_2(g) = C_6H_6(g),\ \Delta \bar{H}^\circ = -574 \text{ kJ mol}^{-1}$$

can be maintained at equilibrium, what alterations of temperature and pressure would be expected to improve the yield of benzene?

PROBLEMS

Relative activities

4.1. Write down the relationship between relative activity and
   (a) the partial pressure, p, of a gas in a mixture of gases;
   (b) the concentration, c, of a solute in a solvent;
   (c) the mole fraction, x, of a substance B in a liquid A;
   (d) the molality, m, of a solute in a solvent.

4.2. A solution of NO_2 in $CCl_4(l)$ is prepared with concentration $c_{NO_2} = 0.19$ mol dm^{-3} and activity coefficient $y_{NO_2} = 0.98$. What is the relative activity of NO_2 in the solution?

4.3. The mole fraction of water in a sample of air ($P = 100$ kPa) is 0.02 at 25°C. Estimate the activity of the water.

Approach to equilibrium

4.4. Determine the criteria (in terms of relative activities and equilibrium constant) for the reaction

$$2NO_2(g) = N_2O_4(g)$$

to proceed in the reverse direction.

4.5. The equilibrium constant for the reaction

$$H_2(g) + I_2(s) = 2HI(g)$$

at 25°C is $K_{eq} = 0.35$. If the relative activities of the substances involved are:

$$a_{H_2(g)} = 0.10;\ a_{I_2(s)} = 1.0;\ a_{HI(g)} = 0.20$$

is it possible for the reaction to produce more HI? Briefly justify your answer.

4.6. Determine the direction of change in the reaction

$$Zn^{2+}(aq) + 4NH_3(aq) = Zn(NH_3)_4^{2+}(aq)$$

given that

$$a_{Zn^{2+}} = 0.01;\ a_{NH_3} = 1.0;\ a_{Zn(NH_3)_4^{2+}} = 0.001$$

and $K_{eq} = 9.0$.

4.7. Determine the direction of change in the reaction

$$Zn^{2+}(aq) + H_2S(aq) + 2H_2O = ZnS(s) + 2H_3O^+(aq),\ K_{eq} = 10^{4.15}$$

$$c_{H_2S} = 7.5 \times 10^{-2} \text{ mol dm}^{-3}$$

$$c_{Zn^{2+}} = 10^{-2} \text{ mol dm}^{-3};\ y_{Zn^{2+}} = 0.75$$

$$c_{H_3O^+} = 10^{-2} \text{ mol dm}^{-3};\ y_{H_3O^+} = 0.9$$

4.8. Describe in terms of relative activities the position of equilibrium for the reaction

$$H_2O_2(aq) + I_3^-(aq) + 2H_3O^+(aq) = 3I^-(aq) + 4H_2O$$

4.9. Write expressions for the equilibrium constants for the following reactions:
   (a) in terms of the equilibrium values of the relative activities of the components;
   (b) in terms of the appropriate relative concentrations and activity coefficients at equilibrium.

   (i) $CH_4(g) + 2O_2(g) = CO_2(g) + 2H_2O(g)$
   (ii) $Fe^{3+}(aq) + NCS^-(aq) = FeNCS^{2+}(aq)$
   (iii) $CaCO_3(s) = CaO(s) + CO_2(g)$.

4.10. Determine the value of the equilibrium constant of the reaction

$$H_2(g) + I_2(g) = 2HI(g)$$

if at equilibrium

$$p_{H_2} = 29.7 \text{ kPa}$$

$$p_{I_2} = 3.14 \text{ kPa}$$

$$p_{HI} = 67.2 \text{ kPa}$$

Applications of the equilibrium law

4.11. If the equilibrium constant for the reaction

cis–2–butene = *trans*–2–butene

in liquid cyclohexane at 25°C is 1.5, what will be the concentration of the *trans*-isomer at equilibrium after 0.1 mol of the *cis*-isomer is added to 1 dm^3 of CCl_4? (In such a solution the activity coefficients may be taken as 1.0.)

4.12. Estimate the pH of an aqueous solution which contains 0.5 mol dm^{-3} NH_3. Assume that

$$y_{NH_4^+} = y_{H_3O^+} = 0.90$$

$$pK_a(NH_4^+) = 9.2$$

4.13. For benzoic acid, C_6H_5COOH, in aqueous solution at 20°C, K_a is 6.30×10^{-5}. The pH of a saturated aqueous solution is observed to be 3.0. Estimate the solubility of this acid, in g dm^{-3}.

4.14. For acetic acid, $pK_a = 4.76$. Estimate the $c_{H_3O^+}$ of a solution containing 0.05 mol acetic acid and 0.05 mol sodium acetate per 1 dm^3. The activity coefficients of the various species present may be taken as:

$y_{HAc} = 1.0$; $y_{Ac^-} = 0.8$; $y_{H_3O^+} = 0.8$;

$y_{Na^+} = 0.8$; $f_{H_2O} = 1.0$.

4.15. Determine the approximate pH at 25°C of a solution which contains 0.02 mol Na_2HPO_4 and 0.04 mol NaH_2PO_4 in 500 cm^3 of aqueous solution. The ion activity coefficient of HPO_4^{2-} is 0.60, and that for $H_2PO_4^-$ is 0.78. For the reaction:

$$H_2PO_4^- + H_2O = HPO_4^{2-} + H_3O^+;$$

$$K_{eq}(25°C) = 6.3 \times 10^{-8}$$

Answer 6.8

4.16. Determine the solubility of slaked lime (calcium hydroxide) in water. For the reaction:

$$Ca(OH)_2(s) = Ca^{2+}(aq) + 2OH^-(aq);$$

$$K_{eq}(25°C) = 4 \times 10^{-6}$$

The activity coefficients may be taken as

$y(Ca^{2+}) = 0.69$; $y(OH^-) = 0.90$.

Answer 1.2×10^{-2} mol dm^{-3}

4.17. Carbon dioxide at a partial pressure of 101 kPa is bubbled through 0.001 mol (0.1g) of calcium carbonate suspended in 100 cm^3 of water. The reaction which occurs can be described by the equation:

$$CaCO_3(s) + CO_2(g) + H_2O(l)$$
$$= Ca^{2+}(aq) + 2HCO_3^-(aq)$$

$$K_{eq} = 2.5 \times 10^{-6}$$

Take $y(Ca^{2+}) = 0.7$; $y(HCO_3^-) = 0.9$. Will all the calcium carbonate dissolve?

Answer Yes

4.18. For the following processes, indicate the direction of chemical change, with reasons for your decision.

(a) Formic acid reacts with water according to the equation

$$HCOOH(aq) + H_2O(l)$$
$$= HCOO^-(aq) + H_3O^+(aq)$$

Such a solution is diluted by adding an equal volume of water.

(b) A gaseous mixture contains acetylene and benzene in equilibrium at 2000K. The equation is

$$3C_2H_2(g) = C_6H_6(g);$$

$$\Delta\bar{H}° = -600 \text{ kJ mol}^{-1}$$

The mixture is cooled to 500 K.

4.19. The equilibrium constant for equilibrium between Ag_2SO_4 in aqueous solution and $Ag_2SO_4(s)$ is

$$K\{Ag_2SO_4(s) = 2Ag^+ + SO_4^{2-}(aq)\}_{298K}$$
$$= 2 \times 10^{-5}$$

Estimate the solubility of Ag_2SO_4 in water at 298 K.

4.20. Consider the solution equilibrium:

$$Ag^{+}(aq) + Cl^{-}(aq) = AgCl(s)$$

What would occur if some solid AgCl were added to this system? Give reasons.

4.21. A minimum amount of a relatively concentrated solution of silver nitrate ($1\ mol\ dm^{-3}$) was added to a solution of sodium chloride ($0.01\ mol\ dm^{-3}$) in water until the solution began to appear "cloudy". The solubility product of silver chloride is 2×10^{-10}. Estimate the concentration of silver ion left in solution,

(a) if you have no knowledge of the values of the activity coefficients;

(b) given that in such a solution

$$y_{Ag^{+}} = y_{Cl^{-}} = 0.75 \pm 0.05$$

4.22. $CO(g) + H_2O(g) = CO_2(g) + H_2(g)$;

$$K_1 = 0.36$$

$CO(g) + \frac{1}{2}O_2(g) = CO_2(g)$;

$$K_2 = 9.8 \times 10^5$$

Use this data to find the equilibrium constant for the decomposition of steam

$$H_2O(g) = H_2(g) + \frac{1}{2}O_2(g)$$

Effect of temperature on K_{eq}

4.23. For the reaction

$$CO(g) + 2H_2(g) = CH_3OH(g)$$

the value of K_{eq} is 2×10^4 at 298 K, and $\Delta \bar{H}°$ is $-90.6\ kJ\ mol^{-1}$.
Estimate K_{eq} at 400 K.

4.24. For the reaction

$$2H_2O = H_3O^{+} + OH^{-}(aq)$$

K_{eq} at 298 K is 1.0×10^{-14}. At 273 K the value of K_{eq} is only 1.13×10^{-15}.
Estimate $\Delta \bar{H}°$ for this reaction.

4.25. Consider a gaseous system containing carbon monoxide, hydrogen and methanol.

(a) What is the immediate effect on the relative activities if the volume of the system is halved without change of temperature?

(b) If helium is added to the above system without change of temperature or volume what is the effect on the relative activities of each original substance?

(c) For the reaction

$$CO + 2H_2 = CH_3OH(g)$$

$\Delta \bar{H}°$ is $-90\ kJ\ mol^{-1}$. What will be the effect on the equilibrium constant of increasing the temperature?

(d) If, in the original system the reaction

$$CO + 2H_2 = CH_3OH(g)$$

was at equilibrium, what would be the effects on the amounts of these substances of changes (a) — (c)?

5.

Rates of Reactions

5.1 The Timescales of Chemistry

In Chapter 4 we developed a method to determine in which direction chemical change is possible. However this method does not tell us anything about the time it takes for a reaction to get to equilibrium (note the examples in Table 5.1), or how it occurs. This chapter takes up the timescales of chemical change and their variation with conditions—the study known as *chemical kinetics.*

Chemists expend a lot of effort on the study of the timescales of chemical change. It may involve search for methods to slow down undesired reactions such as the corrosion of metals and alloys, the oxidation of motor tyres or the evaporation of water from a storage reservoir. Alternatively, it may involve speeding up desired reactions, as in bread making or the decay of garbage. Often such a search aims to find conditions which achieve the maximum yield of desired products at the same time as the minimum yield of undesired ones. In complex cases like petroleum refining, the processes need to be alterable (within limits) so that the balance between products can be altered with alteration in market demand. Each product (gas, particular grades of fuel, etc.) in such a case needs to be sufficiently defined in chemical terms for successful operation of the industry.

Control like this requires knowledge and understanding of the factors which alter products and production rates. So we will first present the methods which are used to define, describe and measure rates of chemical change and the ways in which they are affected by the prevailing conditions.

The results of a study like this also provide some of the clues in one of the great detective games in chemistry—the determination of the path by which change occurs. Exploration of this area occupies the later parts of the chapter.

Table 5.1
Timescales for some Highly Possible Reactions

Reaction	*Typical reaction times*, 298K	$K_{eq,298K}$
$H^+ + OH^- \rightarrow H_2O(aq)$	$<<10^{-6}$s	10^{14}
$H_2O \rightarrow H^+ + OH^-(aq)$	seconds	10^{-14}
$2H_2(g) + O_2(g) = 2H_2O(l)$	many years	3×10^{41}
$2H_2O_2(aq) = 2H_2O + O_2(g)$	years	10^{41}
$Zn(s) + 2H^+(aq) = Zn^{2+}(aq) + H_2(g)$	minutes	6×10^{25}
$4Al(s) + 3O_2(g) = 2Al_2O_3(s)$	centuries?	10^{277}

5.2 Defining the Rates of Chemical Change

5.2.1 Monitoring change

The first part of getting at the rate of a chemical change involves finding what the change is, and how much of it has occurred. We will assume that we know the appropriate chemical equations. In practice a variety of methods are available for determining the amount of change resulting from chemical reaction. Some will be more convenient or accurate than others. The following examples give an impression of the choices which are likely to be available.

(a) If we are concerned with the corrosion of the zinc coating on galvanised steel equipment in contact with an acidic fruit juice, the chemical equation is

$$Zn(s) + 2H_3O^+(aq) = Zn^{2+}(aq) + H_2(g) + 2H_2O$$

We may choose to monitor either the volume of hydrogen produced, or the mass of zinc which remains, with results like those sketched in Fig. 5.1a.

(b) If we are to examine the "burning" of limestone to calcium oxide,

$$CaCO_3(s) = CaO(s) + CO_2(g)$$

the most likely approach is to "trap" the carbon dioxide in an alkaline solution. As an alternative, we could use a chemical method of analysis for $CaCO_3$ or for CaO in the presence of the other. Another option is simply to measure the mass of the remaining solid. These measurements will be related (Fig. 5.1b) to one another.

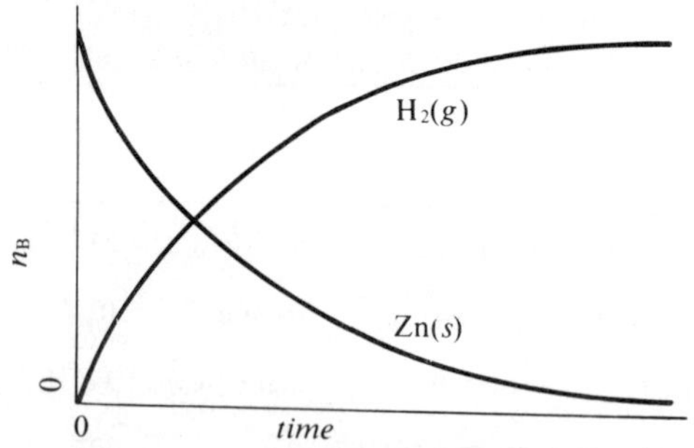

(a) Corrosion of a zinc coating.

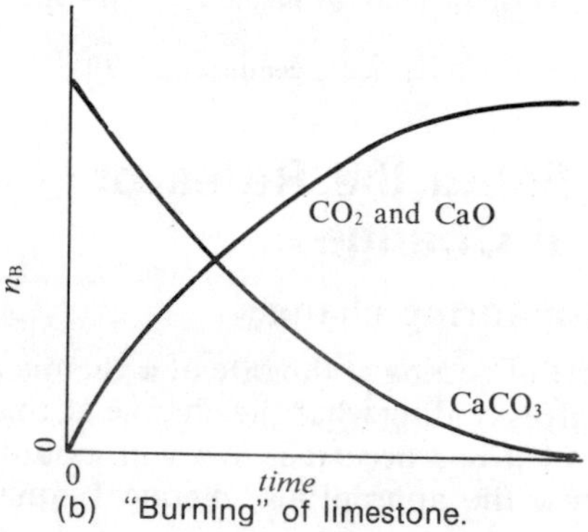

(b) "Burning" of limestone.

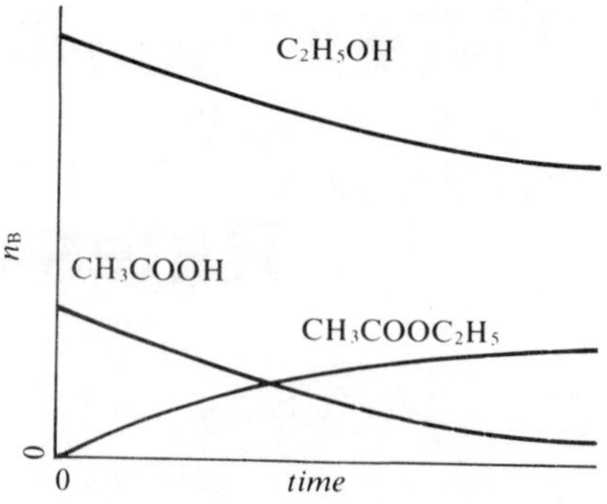

(c) Ester forming in maturing wine.

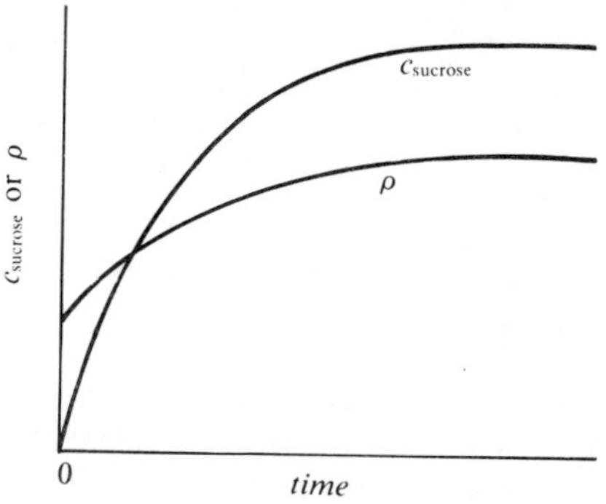

(d) Formation of sugar syrup.

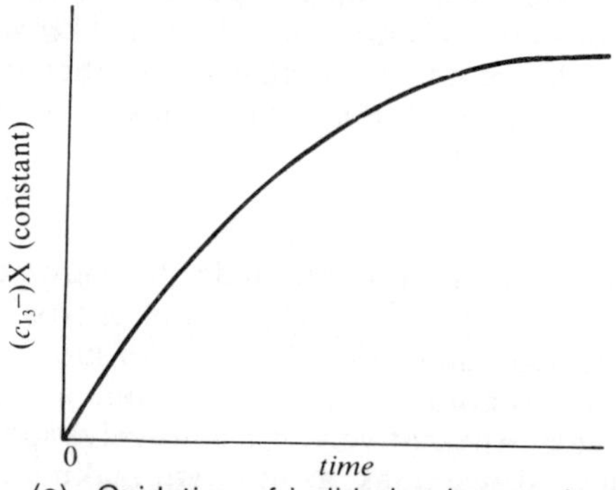

(e) Oxidation of iodide ion by aquoiron (III).

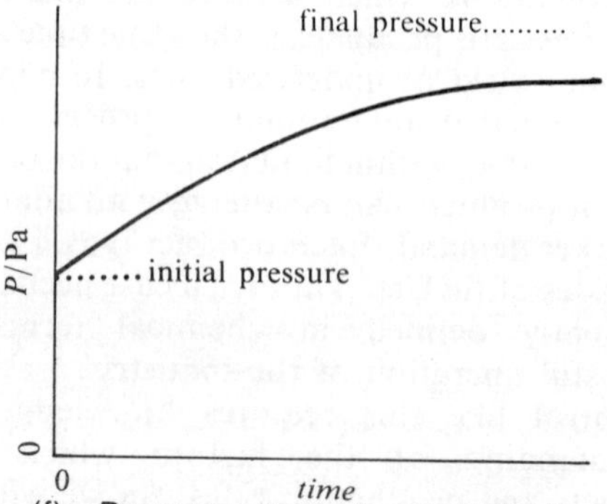

(f) Decomposition of acetaldehyde.

Fig. 5.1 Changes accompany reaction.

(c) One important group of chemical changes which occur in the maturing of wine (Fig. 5.1c) involves the formation of esters, for example

$$CH_3COOH + C_2H_5OH = CH_3COOC_2H_5 + H_2O$$

The extent of these changes may be followed by removing small samples and titrating the acid; this shows the general relationships between the changes which occur.

(d) The rate at which sugar (sucrose) dissolves to form "syrup"

$$C_{12}H_{22}O_{11}(s) = C_{12}H_{22}O_{11}(aq)$$

is most easily monitored by measuring either the density of the solution or its rotation of polarised light (Fig. 5.1d), which is proportional to the concentration. One will not choose to measure the mass of undissolved sugar because quantitative separation from the solution is too difficult.

(e) Oxidation of iodide ion by iron (III)

$$3I^-(aq) + 2Fe^{3+}(aq) = I_3^-(aq) + 2Fe^{2+}(aq)$$

is most easily followed by measuring changes in the light absorption due to formation of I_3^- (Fig. 5.1e). This is proportional to the concentration of I_3^-.

(f) In the gas phase, the decomposition of acetaldehyde

$$CH_3CHO(g) = CH_4(g) + CO(g)$$

can be followed by measuring the increasing pressure in an experiment in which the volume and the temperature are kept constant (Fig. 5.1f).

Some of the methods of measurement suggested are ordinary methods of chemical analysis and some involve monitoring of properties which depend on the amounts or the proportions of the substances present. The latter have the advantage that they can often be measured continuously without altering the reacting system, and so more (or more rapid) measurements can be taken. Chemical analyses tend to have greater selectivity, at the expense of lower convenience.

Exercise 5.1

For each of the following reactions suggest a method which is likely to be convenient for measurement of the rate of reaction.

(a) $N_2O_5(g) = N_2O_4(g) + \frac{1}{2}O_2(g)$

(b) $CH_3COOCH_3 + OH^- = CH_3COO^- + CH_3OH$ in aqueous solution.

(c) $CH_3COOCH_3 + H_2O = CH_3COOH + CH_3OH$ in acidic aqueous solution.

(d) $\underset{\text{(colourless)}}{Fe(CN)_5OH_2^{3-}} + C_6H_5NO = \underset{\text{(violet)}}{Fe(CN)_5(ONC_6H_5)^{3-}} + H_2O$ in aqueous solution.

(e) Crystallisation of $CaF_2(s)$ from supersaturated aqueous solution.

What is the reason for each choice?

5.2.2 Definition of reaction rate

The average rate of production of a substance in a reaction is defined as the increase in amount of that substance divided by the time interval involved.

$$\textit{Average rate of accumulation of substance B} \equiv \frac{\Delta n_B}{\Delta t}$$

The amounts of the various substances produced in any particular reaction are related through the extent of reaction, ξ, as previously shown (Section 1.4.2). Consequently the average *rate of reaction* can be defined by the change in ξ, that is, $\Delta\xi$, divided by the time interval:

$$\textit{Average rate of reaction} \equiv \frac{\Delta\xi}{\Delta t}$$

Thus for the example of Fig. 5.1a

$$\frac{\Delta\xi}{\Delta t} = \frac{\Delta n(Zn^{2+})}{\Delta t} = \frac{\Delta n(H_2)}{\Delta t} = -\frac{\Delta n(Zn)}{\Delta t} = -\frac{1}{2}\frac{\Delta n(H_3O^+)}{\Delta t}$$

Exercise 5.2

If 0.10 mol HI is produced in 10 minutes by the reaction

$$H_2 + I_2 = 2HI$$

what are the average values of

(a) rate of production of HI;

(b) rate of production of H_2;

(c) rate of production of I_2;

(d) rate of reaction?

Answer
(a) 1.7×10^{-4} mol HI s^{-1}

(b) -8.3×10^{-5} mol H_2 s^{-1}

(c) -8.3×10^{-5} mol I_2 s^{-1}

(d) $+8.3 \times 10^{-5}$ mol ($H_2 + I_2 = 2HI$) s^{-1}

The graphs in Fig. 5.1 show that the rates of those reactions are not constant since their slopes are not constant. They depend on how far the reaction has progressed. Hence the rate of reaction at any stage in any one of these reactions should be defined in terms of the tangent to the appropriate curve.

$$\textit{Rate of reaction} \equiv \frac{d\xi}{dt}$$

or in Fig. 5.1a,

$$\textit{rate of reaction} = -\frac{dn(\text{Zn})}{dt}_{\text{at time t}}$$

Figure 5.2 shows two values of the rate of reaction (at times t_1 and t_2) and an average rate, for this particular case, as based on measurements of the hydrogen formed.

Often we are concerned with systems in which more than one reaction is in progress. For instance, the reaction of carbon with oxygen normally generates carbon monoxide as well as carbon dioxide and thus two chemical equations must be written:

$$2C + O_2 = 2CO \qquad (1)$$

$$2CO + O_2 = 2CO_2 \qquad (2)$$

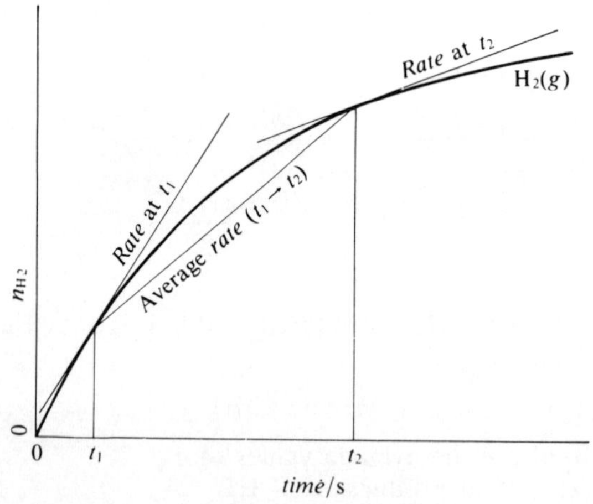

Fig. 5.2 Rate of reaction and average rate.

In consequence, there are two rates of reaction to be considered:

$$\textit{rate of reaction (1)} = \frac{d\xi_1}{dt}$$

$$\textit{rate of reaction (2)} = \frac{d\xi_2}{dt}$$

The rate of production of the substances then represents a balance between the two rates, for example, the rate of production of carbon monoxide is

$$\frac{dn_{CO}}{dt} = 2\frac{d\xi_1}{dt} - 2\frac{d\xi_2}{dt}$$

Exercise 5.3
If the rate of reaction (1) is 1 mole per second and the rate of reaction (2) is 1.5 mole per second, what is the rate of production of oxygen?

Query Surely we want to know the rate of production or consumption of particular substances more often than we want the rate of reaction?

Reply Probably. We want to know lots of things and they are all tied together. This is the most convenient way we know for relating them.

PROBLEMS

5.1. The rate of the reaction

$$2C_8H_{18} + 25O_2 = 16CO_2 + 18H_2O$$

can be defined in terms of the rate of change of amount of any one of the substances involved or in terms of the rate of change of amount of reaction.

Write the five definitions and the relationships between them.

5.2. A malfunctioning internal combustion engine burns 1.00 mol iso-octane. The exhaust gas has the composition:

0.02 mol C_8H_{18}
6.40 mol CO_2
1.44 mol CO
8.82 mol H_2O
0.01 mol NO_2

(a) Write chemical equations to describe the reactions involved (three are needed).
(b) Evaluate the amount of each reaction.
(c) What tests of consistency are available from the given data?
(d) If this engine uses 1.00 mol iso-octane every 400 seconds, what are the corresponding reaction rates?

5.2.3 Rates involving a homogeneous medium

Whenever a reaction occurs in which some, at least, of the reactants or products are present in a homogeneous medium (a well-mixed liquid or gas), it is likely to be most convenient to base measurements of the rate of reaction on measurements of the concentrations of these substances, as illustrated by the following examples.

(a) In an aqueous solution in which chloride ion is being formed, we might determine the concentration of chloride in the solution by titrating small samples of known volume with a standard solution of silver nitrate.
(b) In an aqueous solution in which glucose is being consumed, we might be measuring the optical rotation of the solution and using calibration data to convert these values to ones of c_{glucose} in solution.
(c) In the extraction of coffee we might determine the concentration (g solids dm^{-3}) of the coffee by measuring the density of the solution.

For these types of situations the rate of reaction is related to changes in the concentration of a particular substance (B) as follows:

$$\textit{rate of reaction} \equiv \frac{d\xi}{dt}$$

$$\equiv \frac{1}{\nu_B}\frac{dn_B}{dt}$$

$$\equiv \frac{1}{\nu_B}\frac{d(c_B V)}{dt}$$

If, and only if, the volume of the medium is not changing significantly with time, we can rewrite this expression as

$$\textit{rate of reaction} = \frac{V\text{(homogeneous medium)}}{\nu_B}\frac{dc_B}{dt}$$

It is common practice to divide both sides of this equation by the volume of the homogeneous phase and so obtain the new quantity "rate over volume".

$$(\textit{rate}/V) = \frac{1}{\nu_B}\frac{dc_B}{dt}$$

In the simplest case, the reaction is occurring wholly in the homogeneous medium, for example the reaction

$$CH_3Cl(aq) + OH^-(aq) = CH_3OH(aq) + Cl^-(aq)$$

However the same treatment is convenient if the reaction is not in the homogeneous medium, but perhaps at the boundary of this phase with some other. The extraction of coffee is such an example. The primary measurements will be of (*rate*/*V*), although we might wish later to relate these rates to, say, the area of the surface of the coffee grounds.

Many chemists have limited their thinking to reactions which occur entirely within a single well-mixed solution and refer to what we (following I.U.P.A.C.) call (*rate*/*V*) simply as the "rate" or "velocity" of reaction. This gives no problems until one meets a system where the volume of the homogeneous medium changes significantly as the reaction proceeds, or a system containing more than one phase. In these cases the general approach is more useful.

Query Why isn't the quantity (*rate*/*V*) called "the rate of reaction per unit volume"?

Reply Such a name would imply that the rate of reaction in a small sample of volume (δV) would be

$$\textit{Rate} = (\textit{rate}/V)\,\delta V$$

There is a question as to how large δV may be. When the gas or liquid medium is homogeneous, this question is trivial. When the liquid or gaseous medium is non-uniform but the density at any point remains constant, a separate treatment is needed. This condition is often met in practice in continuous chemical reactors, burning gas jets, reactions of pollutants being discharged at a steady rate into rivers or the atmosphere, etc.

For systems such as these the total rate of reaction,

$$\frac{d\xi}{dt}$$

will simply be the local values of $(rate/V)\delta V$ summed over the total volume.

$$Rate = \int_{\text{volume}} (rate/V)\,\mathrm{d}V$$

It is the *local* values of $(rate/V)$ which are simply related to the *local* conditions (of concentration, temperature, pressure), but it is probably the total rate which is economically significant.

Exercise 5.4
Suppose that an adequately stirred neutralising tank is receiving, through the drains from a research laboratory, a steady trickle ($0.1\ \mathrm{dm^3\ min^{-1}}$) of dilute hydrochloric acid ($0.5\ \mathrm{mol\ dm^{-3}}$) and that it contains 200 kg $CaCO_3$. The overflow liquid is at pH5, so substantially complete removal of HCl occurs in the tank.
(a) What is the rate of HCl addition in $\mathrm{mol\ s^{-1}}$?
(b) What reaction equation is involved?
(c) What is the rate of this reaction?
(d) If the flow continues as described, is it satisfactory to add another 200kg $CaCO_3$ to the tank each Monday?
$\bar{M}(CaCO_3) = 100\ \mathrm{g\ mol^{-1}}$

Answer
(a) $8 \times 10^{-4}\ \mathrm{mol\ s^{-1}}$
(b) $2HCl + 2CaCO_3 = Ca(HCO_3)_2 + CaCl_2$
(c) $4 \times 10^{-4}\ \mathrm{mol\ s^{-1}}$
(d) No!

Exercise 5.5
The reaction

$$3A + 2B = C$$

occurs in a well-mixed solution.
Write down three expressions for "$(rate/V)$" and the relationships between them.

Answer

$$Rate/V = \frac{dc_C}{dt} = -\frac{1}{3}\frac{dc_A}{dt} = -\frac{1}{2}\frac{dc_B}{dt} = \frac{1}{V}\frac{d\xi}{dt}$$

Exercise 5.6
The reaction

$$BrO_3^- + 6H_3O^+ + 5\,Br^- = 3\,Br_2 + 9H_2O$$

is occurring in an aqueous solution at 298K. If a 20 $\mathrm{cm^3}$ sample, taken at a known time, contained 0.0014 mol Br_2 and another, taken 5 minutes later, contained 0.0020 mol Br_2, what is the average value of $(rate/V)$?

Answer

$$Rate/V = 3.3 \times 10^{-4}\ \mathrm{mol\ dm^{-3}\ s^{-1}}$$

5.2.4 Initial rates

Sometimes it happens that the method of measurement of a rate of reaction is sufficiently sensitive that the rate can be evaluated over a very small fraction of reaction (e.g. less than 5%). In a properly formulated reaction solution, the rate of reaction (and $rate/V$) will be constant over such a fraction of reaction. These quantities are called "initial rates".

Query What does the term "initial rate" mean?
Reply The rate of the reaction at the beginning of the reaction.
Query Why bother?
Reply At the beginning of the reaction, the composition and other properties are defined by the way we make up the reaction mixture. This cuts down the uncertainties about other processes.

5.3 A Particular Reaction

Many of the features of the kinetics of reactions in solution can be developed in a case study of the reaction between hydrogen peroxide and iodide ion, forming triiodide, I_3^-, and water, as it occurs in acidified aqueous solution,

$$H_2O_2 + 3I^- + 2H_3O^+ = I_3^- + 4H_2O \qquad (1)$$

Progress of this reaction with passage of time was almost the first quantitative approach to rates. It was studied by Harcourt and Esson about 1860. The reaction is conveniently studied in homogeneous solution, and the rate of reaction may be defined by the relationships:

$$Rate/V \equiv \frac{dc_{I_3^-}}{dt} \equiv -\frac{dc_{H_2O_2}}{dt} \equiv -\frac{1}{3}\frac{dc_{I^-}}{dt} \equiv -\frac{1}{2}\frac{dc_{H_3O^+}}{dt}$$

Consequently measurements of the rate of reaction could be based on analysis for any one of these.

In practice, analysis for I_3^- has been employed, through the reaction with thiosulfate, $S_2O_3^{2-}$:

$$I_3^- + 2S_2O_3^{2-} = 3I^- + S_4O_6^{2-}$$

Starch is a convenient, sensitive, indicator in this analysis because of the formation of blue starch—iodine complexes when more than very low concentrations of I_3^- are present. Neither thiosulfate nor tetrathionate, $S_4O_6^{2-}$, alters the rate of reaction at the concentrations which are necessary. These considerations are typical of those required in picking a method for study of a rate of reaction.

Two main simple procedures have been adopted.

(a) Small samples *taken from* an experiment at known times may be titrated with a standard thiosulfate solution to establish $c_{I_3^-}$ at that time.

(b) Very small known amounts of thiosulfate may be *added to* a reaction experiment, together with starch indicator, and the *times* measured at which sufficient I_3^- has been generated to just exceed what is consumed by the $S_2O_3^{2-}$.

Method (a) is of more restricted usefulness than (b) because of a competing reaction between iodine and hydrogen peroxide to form oxygen, which interferes unless the concentrations of iodide and acid are moderately high (probably greater than 10^{-2} mol dm^{-3}). In method (b), the concentration of I_2 remains very low until the indicator colour change appears. (The equilibrium, $I_2 + I^- = I_3^-$, is established very quickly). When both methods are appropriate they give equally satisfactory results. Figure 5.3 shows typical behaviour, observed in experiments with varying initial hydrogen peroxide concentrations.

Table 5.2
Values of Initial *rate*/ *V* for the Reaction
$H_2O_2 + 3I^- + 2H_3O^+ = I_3^- + 4H_2O$
in aqueous solution at 298K.

$c_{H_2O_2}$ / mol dm^{-3}	c_{I^-} / mol dm^{-3}	$c_{H_3O^+}$ / mol dm^{-3}	10^6 *Rate*/ *V* / mol dm^{-3} s^{-1}
0.0010	0.10	0.10	2.8
0.0020	0.10	0.10	5.8
0.0040	0.10	0.10	11.5
0.0010	0.10	0.10	2.8
0.0010	0.20	0.10	5.8
0.0010	0.40	0.10	11.5
0.0010	0.10	0.10	2.8
0.0010	0.10	0.20	4.7
0.0010	0.10	0.30	6.3
0.0010	0.10	0.40	8.2
0	0.10	0.10	0.0

5.4 Rate Laws

5.4.1 Rate laws and concentrations

The expression "rate law" in chemical kinetics means an equation which expresses the dependence of the rate of a chemical reaction on the concentrations (or sometimes, the amounts or the surface areas) of substances.

To start looking at the dependence of rate on concentration we present some values which correspond to the use of Harcourt and Esson's method in study of the hydrogen peroxide–iodide reaction. The numbers correspond to those in much later work (Liebhafsky and Mohammad, 1933), with some rounding off for

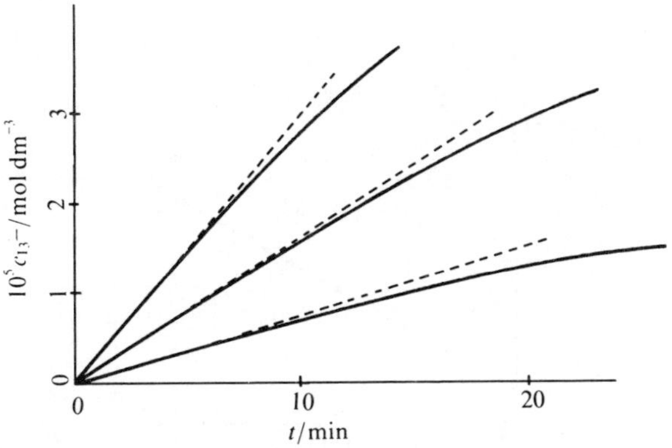

Fig. 5.3 Progress of reaction of hydrogen peroxide and iodide ion. The dotted lines are the extensions of the straight lines which best fit the experimental data (solid lines) in the initial stages of reaction (small values of the time, t).

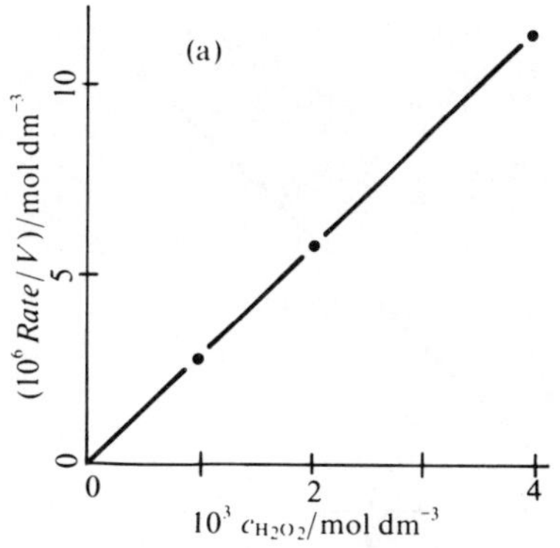

Fig. 5.4 Variation of initial rate of reaction of hydrogen peroxide
(a) with initial hydrogen peroxide concentration.

ease in manipulation. In Fig. 5.4 we display the values of initial *rate*/ *V* in terms of the three concentrations which vary in Table 5.2, viz., $c_{H_2O_2}$, c_{I^-} and $c_{H_3O^+}$. It is clear that the main relationships involved are:

initial *rate*/ *V* increases in proportion to $c_{H_2O_2}$ (Fig. 5.4a),

$$\text{initial } rate/V \approx \alpha c_{H_2O_2}, \text{ at fixed } c_{I^-} \text{ and } c_{H_3O^+}$$

initial *rate*/ *V* increases in proportion to c_{I^-} (Fig. 5.4b),

$$\alpha \approx \beta\, c_{I^-}, \text{ at fixed } c_{H_3O^+}$$

initial *rate*/ *V* increases linearly with $c_{H_3O^+}$ (Fig. 5.4c)

$$\beta \approx \gamma + \delta\, c_{H_3O^+}$$

and in the last case zero $c_{H_3O^+}$ appears not to correspond to zero *rate*/ *V*.

These observations imply that the experimental relationship between (*rate*/ *V*) and the concentrations of reagents appears to have the form,

$$\text{initial } rate/V = ((\gamma + \delta\, c_{H_3O^+})\, c_{I^-})\, c_{H_2O_2}.$$

In chemistry, it is traditional to use a "*k*" to represent a coefficient in a rate equation. So we write

$$\text{initial } rate/V = k c_{H_2O_2} c_{I^-} + k' c_{H_2O_2} c_{I^-} c_{H_3O^+} \quad (1)$$

The best reported values of the *rate coefficients* are $k = 0.0115\ \text{dm}^3\ \text{mol}^{-1}\ \text{s}^{-1}$ and $k' = 0.173\ \text{dm}^6\ \text{mol}^{-2}\ \text{s}^{-1}$ at 298K.

Exercise 5.7
Verify the values of the rate coefficients by calculating the gradients that would be expected for the graphs of Fig. 5.4 a-c and the expected intercept for Fig. 5.4c.

Answer
In Fig. 5.4a,

$$c_{I^-} = 0.10;\ c_{H_3O^+} = 0.10$$

Hence

$$rate/V = [(0.0115 + 0.173\, c_{H_3O^+}) c_{I^-}]\, c_{H_2O_2}$$

and predicted gradient is 2.88×10^{-3}.

In Fig. 5.4b,

$$c_{H_2O_2} = 0.0010;\ c_{H_3O^+} = 0.10$$

Hence

$$rate/V = (0.0115 + 0.173\, c_{H_3O^+})\, c_{H_2O_2} c_{I^-})$$

and the predicted gradient is 2.88×10^{-5}.

In Fig. 5.4c,

$$c_{H_2O_2} = 0.0010;\ c_{I^-} = 0.10$$

Hence

$$rate/V = 1.15 \times 10^{-6} + 1.73 \times 10^{-5}\ c_{H_3O^+}$$

and the predicted gradient is 1.73×10^{-5}, while the predicted intercept is 1.15×10^{-6}.

The experimental (*rate*/ *V*) relationship given above can be described in words. It is a "rate

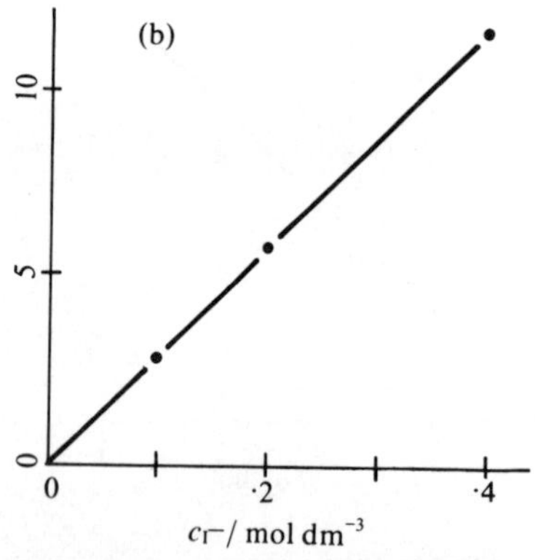

Fig. 5.4 (b) with initial iodide ion concentration.

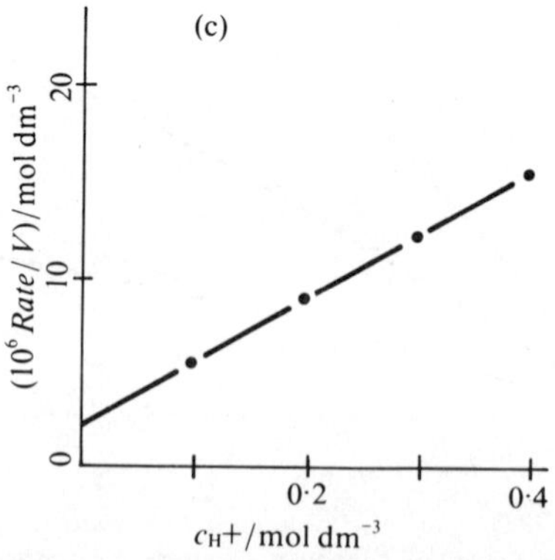

Fig. 5.4 (c) with hydrogen ion concentration.

law in terms of concentrations". It consists of two "terms", both of which involve concentration of hydrogen peroxide to the first power and the concentration of iodide to the first power. We say that both terms are "first order" with respect to hydrogen peroxide and "first order" with respect to iodide ion. The first term is then "zeroth order" with respect to hydrogen ion and the second is "first order" with respect to hydrogen ion. The first term is "second order overall" because it contains two concentrations, each to the "first order", while the second term is "third order overall" because the sum of powers of concentrations in this term is three.

5.5 Temperature Dependence of Rate Coefficients

5.5.1 Observed behaviour

Rates of reaction, and therefore rate coefficients, almost always increase when the temperature is raised. A fair example is the value of k for the first term in the hydrogen peroxide-iodide reaction (Fig. 5.5). The increase is dramatic. For most reactions it amounts to a factor of 2 or 3 for each 10K rise of temperature at about 300K. Such a change is enough to explain the impression we often get that chemical reactions (like those of cookery) just do not happen until some minimum temperature is reached.

The simplest and most useful mathematical representation of such dependence is

$$\frac{\mathrm{d}\ln k}{\mathrm{d}T} \equiv \frac{E_A}{RT^2} \quad (1)$$

A little manipulation gives the alternative relationships

$$\frac{\mathrm{d}\ln k}{\mathrm{d}(1/T)} \equiv - E_A/R \quad (2)$$

and

$$k \equiv \mathscr{A} e^{-E_A/RT} \quad (3)$$

Each of these is known as the "Arrhenius equation" in honour of the original proponent. The parameters which appear are

E_A — the "Arrhenius activation energy" or "activation energy", defined by (1),

and $\mathscr{A}$ — the "Arrhenius preexponential factor", defined by (3).

Values of E_A are usually obtained from the gradients of graphs of $\log_{10}k$ versus the reciprocal of the Kelvin temperature (equation (2); for example, Fig. 5.6). In practice, the Arrhenius activation energy is usually almost constant over a wide range of temperature. There are theoretical reasons to expect that $\mathscr{A}$ should change slowly with temperature. However, there is very little experimental data of sufficient range and accuracy for changes in $\mathscr{A}$ to be apparent.

To estimate the rate coefficient at one temperature from that at another we usually rearrange equation (3) to:

$$\log_{10}\left(\frac{k(\text{at } T_2)}{k(\text{at } T_1)}\right) \simeq \frac{-E_A}{2.303R}\left(\frac{1}{T_2} - \frac{1}{T_1}\right) \simeq \frac{-E_A}{2.303R}\left(\frac{T_1 - T_2}{T_1 T_2}\right)$$

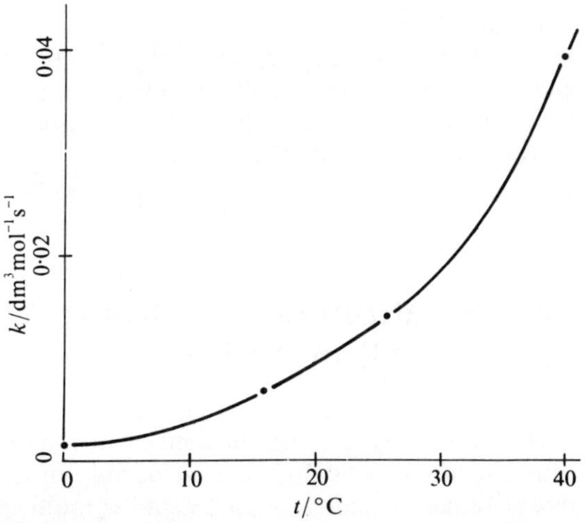

Fig. 5.5 Variation with temperature of rate coefficient for H_2O_2 - I^- reaction.

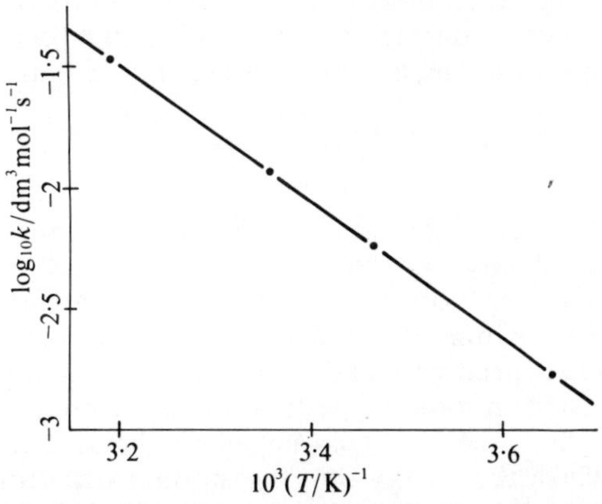

Fig. 5.6 "Arrhenius plot" of rate coefficients for H_2O_2 - I^- reaction.

From Fig. 5.6, E_A for k is obtained as 56.0 kJ mol^{-1}. A similar treatment for k' gives $E_A' = 43.7$ kJ mol^{-1}.

Exercise 5.8

(a) For the reaction of H_2O_2 and I^- at 298 K at very low salt concentrations,

$$k = 0.0110 \text{ dm}^3 \text{ mol}^{-1} \text{ s}^{-1}$$

$$k' = 0.317 \text{ dm}^6 \text{ mol}^{-2} \text{ s}^{-1}$$

Using the values of E_A, calculate the values of $\mathscr{A}$ and $\mathscr{A}'$. Estimate the rate coefficients at 90°C.

Answer

$$\mathscr{A} = 7.1 \times 10^7 \text{ dm}^3 \text{ mol}^{-1} \text{ s}^{-1}$$

$$\mathscr{A}' = 1.43 \times 10^7 \text{ dm}^6 \text{ mol}^{-2} \text{ s}^{-1}$$

At 90°C, $k = 0.63 \text{ dm}^3 \text{ mol}^{-1} \text{ s}^{-1}$

$$k' = 7.4 \text{ dm}^6 \text{ mol}^{-2} \text{ s}^{-1}$$

(b) The reaction

$$O_3 + NO_2 = NO_3 + O_2$$

is thought to be significant in the chemistry of the stratosphere. The rate coefficients at 286−302 K are reported as

$$k = 5.9 \times 10^9 \, e^{-29000/RT} \text{dm}^3 \text{ mol}^{-1} \text{ s}^{-1}$$

What rate coefficient is predicted for the stratospheric region in which the temperature is 210 K?

Answer 360 dm^3 mol^{-1} s^{-1}

5.5.2 Interpretation of Arrhenius equation

The formulation of the Arrhenius equation was partly based on prior work by Van't Hoff on the temperature dependence of equilibrium constants, from which it may be concluded that

$$\frac{\text{dln } K_{eq}}{\text{d}T} \equiv \frac{\Delta \bar{H}^\circ}{RT^2}$$

In this equation $\Delta \bar{H}^\circ$ may be measured independently as the standard change of enthalpy per mole of reaction and is not greatly different from the molar energy difference between products and reactants. Arrhenius suggested, and the suggestion is still reasonable, that the value of E_A is related to the amount by which the energy of a transitional structure exceeds the energy of the reactants. Independent measurement of the energies of such short-lived complexes is not yet possible. The values of E_A (usually positive and between 50 and 100 kJ mol^{-1}) correspond to the energy of the transition structure being very considerably higher than the average energy of the reactant species.

On a molecular view, we suggest that the reactant molecules have their normal structure distorted by stretching or compressing the distances between atoms to a transitional arrangement from which spontaneous collapse into product molecules is possible.

Query How could molecules get that much extra energy?

Reply We believe that there is always a small fraction of molecules with high energy in the normal state of the material, due to the unequal exchanges of energy which occur in the continual movement and jostling of molecules. According to the model developed by Ludwig Boltzmann in 1896, the number, N_1, of molecules of one energy, ϵ_1, compared with the number, N_2, of another energy, ϵ_2, should depend on the exponential of the ratio of energy difference to temperature, that is, N_2/N_1 should be proportional to $e^{-(\epsilon_2-\epsilon_1)L/RT}$, where L is the Avogadro number (6.023×10^{23} mol^{-1}).

In the simplest model, the fraction of molecules which has energy *greater* by more than some value ϵ^* turns out to be approximately $e^{-\epsilon^* L/RT}$. If the model fits and $E_A \approx \epsilon^* L$, it follows that $e^{-E_A/RT}$ should indicate roughly the fraction of molecules for which reaction is possible.

Exercise 5.9

We have shown that E_A for the $H_2O_2 + I^-$ reaction is 56.0 kJ mol^{-1}. What indication does this give of the fraction of the molecules capable of reaction in the "simplest model"?

Answer 1.5×10^{-10} at 298 K

5.6 Relationships between Concentration and Time

So far we have studied and interpreted *rates* of reactions as they depend on concentrations in solution: our practical interest is usually in the time it takes to remove a particular amount of one substance or produce an amount of another—not small changes in extent of reaction but large ones. In a suitably designed experiment we can deal with this question.

Consider again our particular reaction

$$H_2O_2 + 3I^- + 2H_3O^+ = I_3^- + 4H_2O$$

for which the rate law has been found to be

$$Rate/V \simeq k c_{H_2O_2} c_{I^-} + k' c_{H_2O_2} c_{I^-} c_{H_3O^+} \quad (1)$$

It is possible to design experiments so that c_{I^-} and $c_{H_3O^+}$ are much larger than $c_{H_2O_2}$. Then, even if complete consumption of H_2O_2 is examined, c_{I^-} and $c_{H_3O^+}$ will be almost constant during decay of $c_{H_2O_2}$ in a given batch. So equation (1) becomes the even simpler equation (2).

$$Rate/V = k_{batch} c_{H_2O_2} \quad (2)$$

with $\quad k_{batch} = k c_{I^-} + k' c_{I^-} c_{H_3O^+}$

Using one of the equivalent expressions for $(rate/V)$,

$$-d\, c_{H_2O_2}/dt = k_{batch} c_{H_2O_2} \quad (3)$$

Query Are there other limitations to this equation?

Reply Yes. In the experiment which we are now to propose the amounts of products will build up. This could change the rate law through entry into the rate law of terms involving product concentrations.

5.6.1 First order decay

"If X is a quantity which decays at a rate proportional to its current value, what is the relation between X and time?" Such a question arises from equation (3) above, with $X = c_{H_2O_2}$. We have

$$\frac{-dX}{dt} = kX \quad (4)$$

and we let $X = X_0$ at time $t = 0$.

It is necessary, first, to separate the variables to obtain

$$-X^{-1}\, dX/dt = k$$

and $\quad -\int_0^t X^{-1}(dX/dt)dt = \int_0^t k dt$

that is, $\quad -\int_{X_0}^{X} X^{-1}\, dX = \int_0^t k dt$

The two definite integrals are then solved

$$-\ln X \Big|_{X_0}^{X} = kt \Big|_0^t$$

that is $\quad \ln(X_0/X) = kt \quad (5)$

or $\quad X = X_0 e^{-kt} \quad (6)$

In the present case we have $X = c_{H_2O_2}$, and $k = k_{batch}$, so that the equation should be

$$c_{H_2O_2} = (c_{H_2O_2})\, e^{-k_{batch} t} \quad (7)$$

The equations (5) and (6) are known as the first order decay relationships. Figure 5.7 shows the variation of $c_{H_2O_2}$ with time. The marker lines indicate the times taken for $c_{H_2O_2}$ to reach $\frac{1}{2}c_{H_2O_2,o}$, $\frac{1}{4}c_{H_2O_2,o}$, $\frac{1}{8}c_{H_2O_2,o}$. The times for successive halvings of the amount are equal and are often reported as the "half life", $t_{1/2} = \ln 2/k$. Inspection to see if the half life is constant is a quick test for first order decay of an experimentally-monitored quantity.

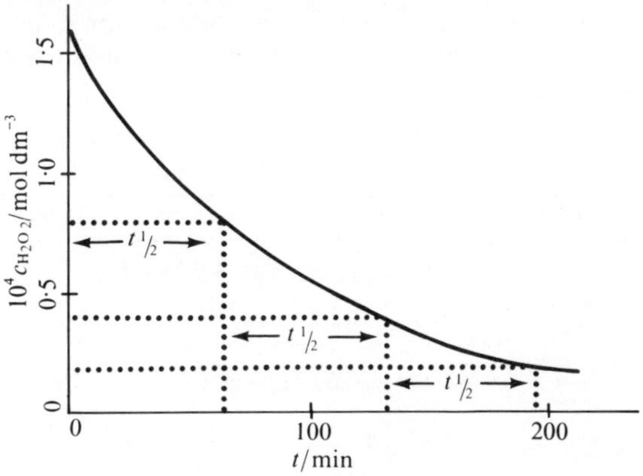

Fig. 5.7 Decay of hydrogen peroxide concentration with time in H_2O_2 - I^- batch reaction with $c_{H_2O_2}$ c_{I^-}, $c_{H_3O^+}$.

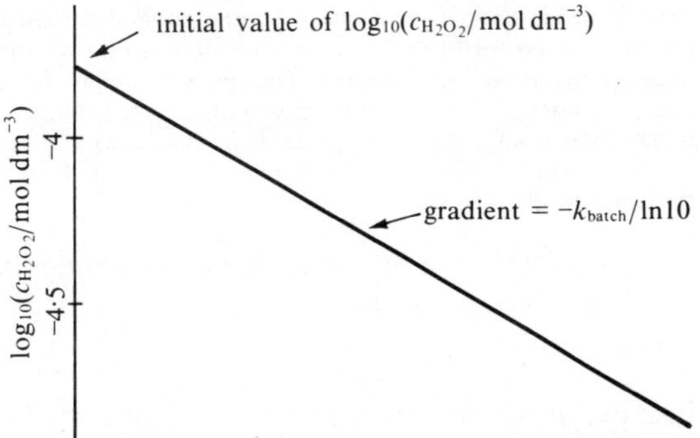

Fig. 5.8 "First order decay" plot for H_2O_2 -I^- batch reaction with $c_{H_2O_2} \ll c_{I^-}$, $c_{H_3O^+}$.

In order to test the goodness of fit of the first order decay relation and to obtain the best estimate of the first order decay coefficient, we go back to equation (5) and prepare the graph shown in Fig. 5.8. The straight line confirms that first order decay of $c_{H_2O_2}$ continues throughout the measured range of reaction. The first order rate coefficient for the run is obtained from the gradient of the straight line of best fit.

Comment The slope in Fig. 5.8 looks a much more accurate way to get the rate coefficient than the slopes of the initial (*Rate*/ *V*) plots in Fig. 5.3.

Reply It is. Almost everyone uses equations like (5) to get values of k if they can. The differential rate equations relating (*Rate*/ *V*) and concentrations of species may be easier to write. They are simple to test experimentally and they do make fewer assumptions, but accurate values of k are hard to get from them.

Table 5.3
Simple Batch Decay Relationships*

$-dX/dt$ equals	order	X versus time	comment
k	zero	$1-(X/X_0) = kt/X_0$	$t_{1/2} = X_0/2k$
kX	first	$\ln(X_0/X) = kt$	$t_{1/2} = \ln 2/k$
kX^2	second	$(X_0/X) - 1 = X_0kt$	$t_{1/2} = 1/kX_0$
kX^3	third	$(X_0/X)^2 - 1 = 2X_0^2kt$	$t_{1/2} = 3/2kX_0^2$
kXY with $X_0 - X = Y_0 - Y$	second, being first order in X	$\frac{1}{X_0-Y_0} \ln \frac{X}{X_0}\frac{Y_0}{Y} = kt$	general
	and first order in Y	$\ln \frac{X_0}{X} \simeq Y_0kt$	if $Y_0 >> X_0$

*The term "batch" refers to one lot of materials in a closed system with no further input or output of matter after the initial preparation of the system. The overall rate law for a chemical reaction is likely to be more complicated than the batch decay law for one of its reactants. A special case arises for decay which is first order in X and first order in Y with X_0 equal to Y_0.

Most people try to design their experiments so that what they measure is a "first order decay" process. We have developed the first order decay relationship in general form because it applies to many other phenomena, such as the depletion of the value of a loan if the capital is used to pay the interest, the loss of electric charge from a capacitor, the cooling of hot bodies, etc. The decay of radioisotopes is a particularly well-known example. In this case, X represents the number of remaining radioactive nuclei of a particular type. Constancy of the time taken to halve the distance from equilibrium is a convenient criterion of first order decay. Such decay is often characterised by the "half life", $t_{1/2}$, equal to $\ln 2/k$.

5.6.2 Other common relationships

Depending on the design of the experiment, more than one concentration may vary significantly during a reaction. This gives rise to a number of different relationships between concentration and time. Some of the more common ones are collected in Table 5.3. More extensive lists are given in specialised monographs.

5.7 Mechanisms and Rates of Reaction

5.7.1 Rate laws, structures and mechanisms

The temperature dependence of rate coefficients has been interpreted in terms of a molecular picture of the energies of the reactants and of the transition structure. We now proceed to use molecular ideas to interpret the experimental rate laws and the fact that they involve only some of the reactants. The key idea is that an experimentally observed reaction may be the overall result of a sequence of "elementary reaction steps". Each such "elementary step" has its own transitional complex with structure between one set of stable molecular structures and another. The products of one elementary step include the reactants of the next. Within such a sequence, the rates of one or two steps may limit the supply of reactant for later steps and thereby determine the rate of the whole (experimentally observed) reaction. The formulae of the transitional complexes of these kinetically significant steps can be inferred from the limiting forms of the terms in the experimentally determined rate law.

A scheme of elementary reactions proposed to explain an observed reaction is known as a "reaction mechanism". Most frequently we cannot detect and measure the amounts of the intermediate species in a suggested mechanism. We call these species "reactive intermediates". The sequence of steps is the result of using our imagination, intuition and experience on the clues provided by

(a) comparing the structures of initial reactant molecules with structures of the final product molecules; and
(b) the formula of the transitional species in the rate determining step, as indicated by the experimental rate law.

Thus in the $H_2O_2 + I^-$ reaction it is possible to think of reaction (1) as the overall result of several steps

$$H_2O_2 + 3I^- + 2H_3O^+ = I_3^- + 4H_2O \qquad (1)$$

The mechanism suggested as corresponding to the first term, $k c_{H_2O_2} c_{I^-}$, in the expression for $(rate/V)$ is

(a) $H_2O_2 + I^- \rightarrow HOI + HO^-$
"rate determining", via $(H_2O_2I^-)\ddagger$

(b) $H_3O^+ + OH^- \rightarrow 2H_2O$
"fast", kinetically insignificant

(c) $HOI + I^- \rightarrow I_2 + HO^-$
"fast", kinetically insignificant

(d) $H_3O^+ + OH^- \rightarrow 2H_2O$
"fast", kinetically insignificant

(e) $I_2 + I^- \rightarrow I_3^-$
"fast", kinetically insignificant

Steps (a) – (e) add up to the experimentally observed reaction

$$H_2O_2 + 3I^- + 2H_3O^+ = I_3^- + 4H_2O \qquad (1)$$

Query Why are arrows used in writing the steps of the mechanism when equality signs appear in the overall equation?

Reply There are several reasons:
(a) to distinguish them from the experimentally observed reaction;
(b) because we think that these steps are "elementary molecular transformations" with no other intermediate molecules; and
(c) because we wish to think of the steps as going in the indicated direction. In writing a mechanism we treat the reverse of each step as a separate reaction step. (Near to equilibrium the reverse steps must become important.)

Query How do we proceed to show that this mechanism would lead to a rate law corresponding to the term that it is supposed to explain?

Reply The assumption is made that an elementary reaction step, having no stable intermediate, will follow a rate law corresponding to the reactants that come together. That is, the rate of a step of the type

$$A + B \rightarrow P + Q$$

should be given by

$$Rate/V = k c_A c_B$$

The term "rate determining step" implies that the rate of the overall reaction is sensitive to alterations in the concentrations of reactants for that step. Thus

$$Rate(1)/V \simeq rate(a)/V$$
$$= k(a)\, c_{H_2O_2}\, c_{I^-}$$

Query What is meant by the term "fast step"?

Reply A fast step, occurring after the rate determining step, transforms an intermediate into another intermediate or a final product, almost as rapidly as it is formed. Very sensitive methods would be needed to detect the presence of such reactive intermediates. Often the best available methods do not succeed.

Query If the fast steps cannot be directly verified by monitoring their reactants and products, what is the basis for writing them in the mechanism?

Reply Educated guess-work, mostly based on our ideas of molecular structure. Each step given above involves only a minimal structural change—no more than the breaking of one chemical bond and the making of another. For example,

(a)

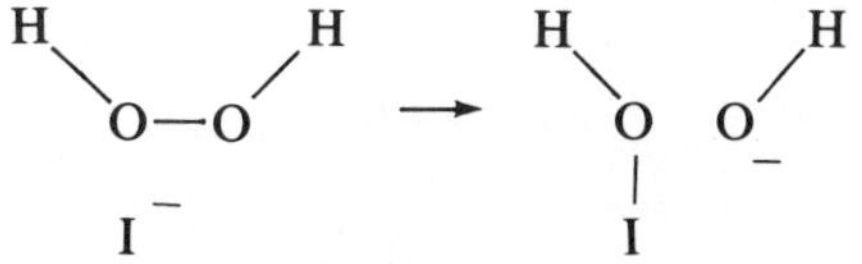

(b)

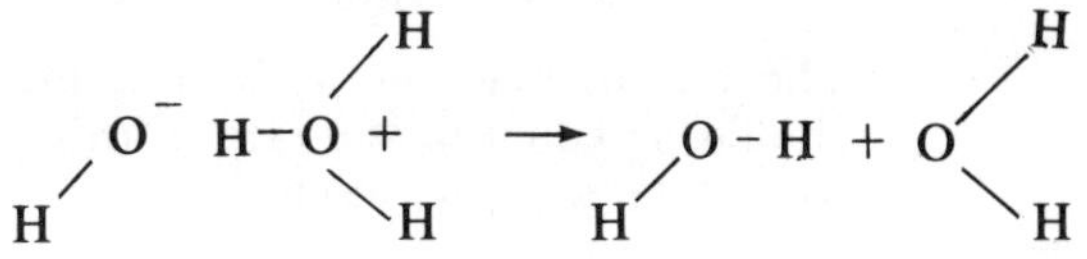

(c)

(d) As illustrated in (b) above.

(e) $I-I \quad I^- \rightarrow I-I^--I$

In the present case, reaction steps (b) and (e) can be studied by supplying OH^- or I_2, in the absence of hydrogen peroxide and are known to be rapid. This independent verification is not always available, but in the present case leaves only one step which is likely to limit the reaction rate to that observed.

5.7.2 Parallel pathways for reaction

The rate law for the hydrogen peroxide–iodide reaction contains a second term, $k'c_{H_2O_2}c_{I^-}c_{H_3O^+}$, which must also be interpreted.

Query How can this be accounted for in terms of a mechanism similar to the first term?

Reply To explain the occurrence of this expression we suggest the steps

$$\left.\begin{array}{ll}(a') & H_2O_2 + H_3O^+ \rightarrow H_3O_2^+ + H_2O \\ (-a') & H_3O_2^+ + H_2O \rightarrow H_2O_2 + H_3O^+\end{array}\right\}$$

$$(b') \quad H_3O_2^+ + I^- \rightarrow HOI + H_2O \quad \text{rate determining}$$

followed by the fast steps

$$(c) \quad HOI + I^- \rightarrow I_2 + HO^-$$

$$(d) \quad H_3O^+ + OH^- \rightarrow 2H_2O$$

$$(e) \quad I_2 + I^- \rightarrow I_3^-$$

as before.

If steps (a′) and (−a′) are in balance, as suggested, we have an equilibrium

$$(a') \quad H_2O_2 + H^+ \rightleftharpoons H_3O_2^+$$

with $\quad a_{H_3O_2^+} = K(a')\, a_{H_2O_2} a_H$

or $\quad c_{H_3O_2^+} \simeq K(a')c_{H_2O_2}c_{H^+}/c^\circ$

The rate of reaction (1) by this mechanism should be equal to the rate of (b′)

$$Rate\,(1)/V \simeq Rate\,(b')/V$$

$$= k(b')c_{H_3O_2^+}c_{I^-}$$

$$\simeq k(b')K(a')c_{H_2O_2}c_{H^+}c_{I^-}/c^\circ$$

$$\simeq k'c_{H_2O_2}c_{H^+}c_{I^-},$$

as required by experiment. We describe this mechanism as a second pathway for the overall reaction. It runs parallel to the mechanism developed to explain the first term in the rate law. The second step is supposed to be rate determining. The first step must therefore be in equilibrium. We call it a "pre-equilibrium". The experimental rate coefficient k' is then a combination of the rate coefficient $k(b')$ for the rate determining step and the equilibrium constant $K(a')$ for the pre-equilibrium

$$k' \simeq K(a')\, k(b')$$

Query Would not the single step

$$H_2O_2 + H_3O^+ + I^- \rightarrow HOI + 2H_2O$$

be just as acceptable as steps (a′) and (b′) above?

Reply There are often several mechanisms that would follow a given rate law. The "termolecular" reaction step would certainly be expected to follow the observed rate relationship. However, it would require that sets of three molecules or ions meet at the "same moment" in time. Estimates of the frequencies of simultaneous termolecular collisions suggest that such reaction steps must be very slow. Hence the two step scheme, involving only "unimolecular" and "bimolecular" steps, is preferred.

5.7.3 Other common patterns

Our case study does not easily lead us into the other common patterns which appear in reaction mechanisms. We therefore give very brief mention to the most common.

Consecutive reactions show up when an intermediate in sequences such as those we have been examining has sufficient stability for its concentration to build up to measurable levels. Figure 5.9 shows the kind of dependence on

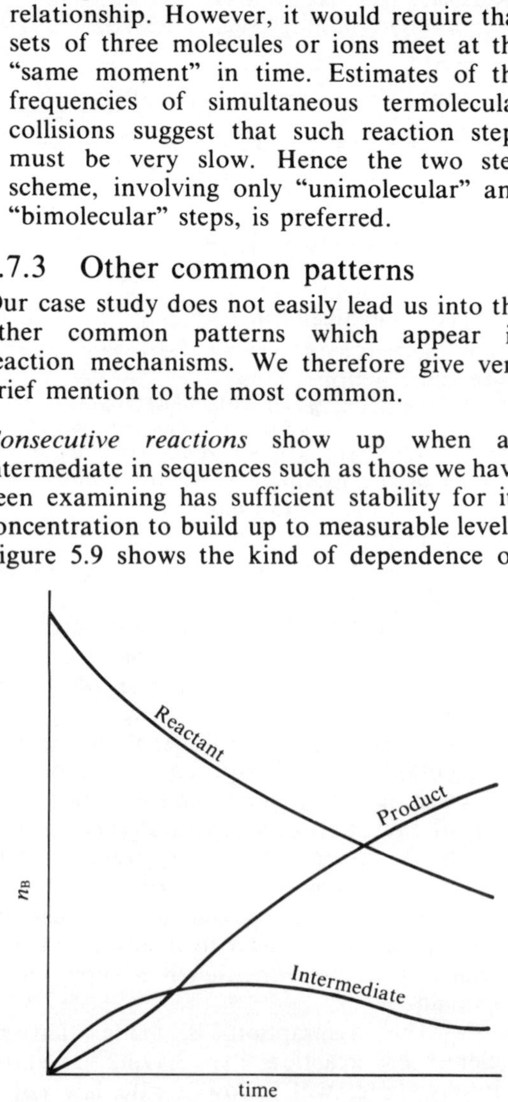

Fig. 5.9 Variation in amount of reactant, intermediate and product during reaction.

time of the concentrations of reactants, intermediates and eventual products which are to be expected in such a case. Examples include hydrolysis of hydroxylamine -N, O-disulfonate

$$O_3SNHOSO_3^{2-} + H_2O \rightarrow H_2NOSO_3^- + HSO_4^-$$

$$H_2NOSO_3^- + H_2O \rightarrow H_2NOH + HSO_4^-$$

or of other molecules from which successive groups can be split such as diethylphthalate and the formation of large (polymer) molecules as the terminal process in a sequence which involves many intermediate molecules of smaller molecular mass.

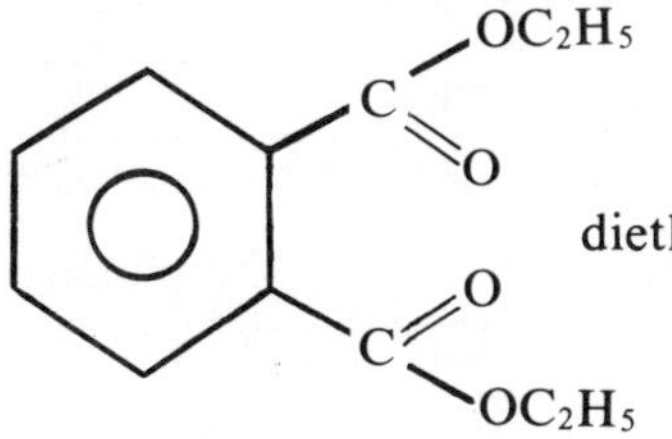

diethylphthalate

Catalysis is a term used to describe the faster rates which are observed in the presence of various substances which, at first sight, ought not to alter the rate. A more precise definition is that a *catalyst* is a substance which appears in a term in the rate law of a reaction to higher order than its stoichiometric coefficient in the equation for the reaction. Consider the decomposition of hydrogen peroxide in aqueous solution (1).

$$2H_2O_2(aq) = 2H_2O + O_2(g) \qquad (1)$$

This reaction is observed to occur more rapidly when aqueous solutions of iron (III) salts, solid manganese dioxide, finely divided solid platinum, pieces of liver, or many other materials, are added.

To take a particular example, the rate law for reaction (1) in the presence of added acid and bromide ion is often dominated by the term

$$Rate/V = kc_{H_2O_2}c_{H_3O^+}c_{Br^-}$$

We can suggest a mechanism similar to that for the hydrogen peroxide–iodide ion reaction

$$H_2O_2 + H_3O^+ \rightleftharpoons H_3O_2^+ + H_2O$$
"pre-equilibrium"

$$H_3O_2^+ + Br^- \rightarrow HOBr + H_2O$$
rate determining

followed by a series of fast steps.

Query But the products don't include Br_2 or HOBr, if the equation for the reaction is (1)?

Reply So the fast steps need to include ones which reform Br^- from HOBr. In these steps $O_2(g)$ is to be formed. One way of achieving this is to write the reaction

$$H_2O_2 + HOBr \rightarrow O_2 + H_3O^+ + Br^-$$

Query Does HOBr react rapidly with H_2O_2 and is $O_2(g)$ formed?

Reply Yes. And that tests the proposal.

With the developments which have occurred in ideas about rate laws and mechanisms it has become increasingly difficult, and perhaps unnecessary, to give a single statement to define "catalysis". This case meets our proposal in terms of "order". Increasingly we tend to say of reactions that they have mechanisms, that these mechanisms may, or may not, include species which are in the equation to describe the total reaction. Then we look for patterns in the mechanisms for different reactions. The notion of "catalyst" doesn't prove to be very useful as defining *one* pattern of mechanisms.

Query What about solid "catalysts" like $MnO_2(s)$ or $Pt(s)$?

Reply We believe that such materials bind reactant molecules ("adsorb" them) at their surfaces. In the process, we believe that other bonds are weakened or broken. In consequence reactions may be facilitated between such "adsorbed" species, and the reaction products be removed without permanent effect on the solid material. Solid catalysts are attractive because they can often be separated easily from the products when reaction is complete.

5.8 Models for Elementary Reactions

To get a more detailed account of the path of chemical change, we need to enquire about the changes which are involved within an elementary reaction. This we shall do in terms of two models, the collision or encounter model, and a model which assumes that transition species are in equilibrium with reactants.

5.8.1 The Collision Model

The collision model of elementary reactions attempts to justify the assumption that the rate law for an elementary reaction can be written down solely on the basis of its chemical equation by describing elementary reactions in terms of effective meeting of the reactants in a collision.

Consider a "displacement process" in homogeneous solution such as

$$A + BC \rightarrow AB + C$$

We are assuming that

$$Rate/V = kc_A c_{BC}$$

One way of visualising what happens would be to think of colliding pairs of molecules, A and BC. If they meet with sufficient velocity their structure may be altered enough to reach the transition structure, (ABC). The frequency of collision would then be a factor in the rate of transformation. Thus we expect (*Rate*/*V*) to be proportional to the number of A per unit volume and also to the number of BC per unit volume. This means that we should expect the expression for (*Rate*/*V*) to contain the product of concentrations, $c_A c_{BC}$.

Other factors, which will appear in the rate coefficient as we have written it, should be the cross-sectional areas of the molecules (with a shape factor to allow for the requirement that A should strike BC at the B-end rather than the C-end), the average speeds of travel and the Boltzmann factor, $e^{-\epsilon^* L/RT}$. The last is given approximately by $e^{-E_A/RT}$, and represents the fraction of collisions which involve kinetic energy greater than a critical value, ϵ^*. Detailed calculations for gaseous systems and typical molecular masses and diameters give

$$k \approx p \times 10^{11}\, e^{-E_A/RT}\, dm^3\, mol^{-1}\, s^{-1}$$

where p is the shape (or steric) factor, and should be no greater than unity. The Arrhenius pre-exponential factors should thus have values up to about $10^{11} dm^3 mol^{-1} s^{-1}$.

In practice, this is largely true though there are a few reactions for which $\mathcal{A}$ is as much as 10^{12} to $10^{13} dm^3 mol^{-1} s^{-1}$.

5.8.2 Transition species equilibrium model

The collision model does not give rate laws in terms of activities and activity coefficients, at least in its simpler forms. This is because it treats molecules as independent of one another between collisions. Activity coefficients change from unity only when the average behaviour of one molecule is affected by that of another. Some insight is gained by adopting a different picture of the events which lead up to molecular transformation in an elementary reaction. Arrhenius, and after him many others, worked on the idea that some kind of equilibrium is involved between reactant molecules and activated molecular complexes whose atomic arrangements have been distorted to the range of transitional structures from which collapse into product molecules is probable. Our development of the transition complex model is based on the work of Henry Eyring, 1935. We treat a reaction step such as

$$A + BC \rightarrow AB + C$$

as if there is an equilibrium

$$A + BC = (ABC)^{\ddagger}$$

with equilibrium constant, $K_{\ddagger}$. The passage of $(ABC)^{\ddagger}$ into product molecules is imagined as occurring

$$(ABC)^{\ddagger} \rightarrow AB + C$$

with the rate law:

$$Rate/V = k_{\ddagger} c_{(ABC)^{\ddagger}} \qquad (1)$$

Since

$$K_{\ddagger} = \frac{a_{(ABC)^{\ddagger}}}{a_A\, a_{BC}}$$

$$= \frac{c_{ABC^{\ddagger}}\, y_{(ABC)^{\ddagger}}\, c^{\circ}}{c_A c_{BC}\, y_A y_{BC}} \qquad (2)$$

this leads to

$$Rate/V = k_{\ddagger} K_{\ddagger}\, c_A c_{BC} y_A y_{BC} / y_{(ABC)^{\ddagger}} c^{\circ}$$

$$= k^{\circ} \frac{a_A a_{BC} c^{\circ 2}}{y_{(ABC)^{\ddagger}}} \qquad (3)$$

The rate coefficient, k°, is therefore equal to

$$k_{\ddagger} K_{\ddagger} / c^{\circ}$$

Query Where did that c° come from?

Reply We put it in to make the units right. There is a difference in the conventions for equilibria and for reaction rates and it erupted just then. We see three ways out of this problem:

(a) write rate laws in terms of *relative* concentrations (c/c°) so that the dimensions of rate coefficients are constant;

(b) abandon the convention for equilibrium constants;

(c) tuck the occasional c° into the rate equation.

Of these, (b) is a step backward; (a) seems rational but premature—nobody does it; and (c) puts the problem where it is likely to be recognised clearly.

We choose (c) for the present.

A similar treatment applied to the dissociation reaction

$$AB \rightarrow A + B$$

would give

$$Rate/V = k^\circ\, a_{ABC^\circ}/y_{(AB)\ddagger}$$

This equilibrium transition model gives the same forms for ($Rate/V$) in terms of activities as appear empirically.

Query Equation (1) assumes that the rate of passage of transition species to products is proportional simply to their numbers or concentrations. Is this done "to get the right answer"?

Reply Yes. Application of quantum theory suggests that the value of $k_\ddagger$ may well be

$$RT/Lh$$

with

L = Avogadro's number

h = Planck's constant

This makes $k_\ddagger$ at 300 K approximately $10^{13}\,s^{-1}$ and would give any particular transition structure a lifetime of about 10^{-13} second.

5.8.3 Rate laws and activities

The rate coefficients which are found for reactions in solution frequently depend on other properties of the solution—particularly when the reactions involve ions. This fact became clear early this century, about the time that the concept of activity was being discovered. The question then raised was "if equilibrium constants are expressed in terms of activity, and not of concentration, should rate laws also be expressed in terms of activity?" The answer to this question was "Yes—provided the resulting expression for the rate law is divided by a guessed activity coefficient". This coefficient was given a value corresponding to a hypothetical species having the charge, and about the size, of the sum of everything in the rate law. Thus, for the term $ka_{H_2O_2}a_{I^-}$, y for something like ($H_2O_2I^-$) was required. For the term $k'a_{H_2O_2}a_{I^-}a_{H_3O^+}$, y for something like ($H_2O_2IH_3O$), without nett charge was required.

Table 5.4
Rate Coefficients for the Reaction of H_2O_2 *and* I^- *at* 298K in the Presence of Salts.

$c_{\text{added salt}}$ mol dm^{-3}	*First term* k	k/k°	*Second term* k'	k'/k'°
0	0.0110	1.00	0.317	1.00
0.021 KCl	0.0110	1.00	0.250	0.79
0.052 KCl	0.0112	1.02	0.203	0.64
0.092 KCl	0.0113	1.03	0.188	0.59
0.202 KCl	0.0118	1.07	0.153	0.48
0.25 $NaClO_4$	0.0130	1.18	0.175	0.55
0.45 $NaClO_4$	0.0142	1.29	0.170	0.54
0.85 $NaClO_4$	0.0167	1.52	0.172	0.54
1.70 $NaClO_4$	0.0222	2.02	0.180	0.57
3.7 $NaClO_4$	0.0377	3.43	0.175	0.55.

In about half the cases, the use of activity values without $y_\ddagger$ leads to calculated rate coefficients which are *less* constant than those based on concentrations.

Query Is there a problem about units? Activities are numbers and concentrations have units.

Reply Yes. To "write the rate law in terms of activities" means that we substitute c_B by $c^\circ a_B \equiv c_B y_B$.

On substituting values of activity, our standard reaction has the rate law

$$Rate/V = k^\circ \frac{a_{H_2O_2}\, a_{I^-}\, c^{\circ 2}}{y_{1\ddagger}} + k'^\circ \frac{a_{H_2O_2}\, a_{I^-}\, a_{H_3O^+} c^{\circ 3}}{y_{2\ddagger}}$$

The previous values of the rate coefficients are related to those in this equation by the equations

$$k = k^\circ \frac{y_{H_2O_2}\, y_{I^-}}{y_{1\ddagger}}$$

$$k = k'^\circ \frac{y_{H_2O_2}\, y_{I^-}\, y_{H_3O^+}}{y_{2\ddagger}}$$

The rate constants k°, k'° are the limiting or standard values of the rate coefficients which are found in very dilute solution. At higher concentrations, it would be expected that k and k' will change as the activity coefficients deviate from unity.

Table 5.4 lists values of k and k' found in solutions in which varying concentrations of sodium perchlorate or potassium chloride were present. It will be seen that *both* rate coefficients change, though k' changes only at the lower end of the concentration range. This is probably the result of accidental cancellation of opposed variations.

Comment This all looks like a case of fiddling.
Reply It certainly is. The object of the exercise was to find the simplest way of representing the empirical dependence of the rate of reaction on the properties of the solution. Eyring's model is used because it yields the same result.

PROBLEMS

5.3. (a) State the most general definition for the rate of a chemical reaction and its basic unit.
(b) Suggest appropriate units for measuring rates in the following practical situations, and explain your choice.
(i) the production of urea in a fertiliser factory;
(ii) the continuous production of ammonia in a catalytic reaction;
(iii) the hydrolysis of an amide in acidified aqueous solution;
(iv) the production of ethylene by cracking petroleum hydrocarbons on the surface of a solid catalyst;
(v) the decomposition of gaseous ethylene chloride in a research study of the rate of this reaction;
(vi) general comparison of the rates of homogeneous reactions.

5.4. The following results were obtained for concentration of a reactant as a function of time.

time/min	0	10	20	40	80	120	160
c/mmol dm^{-3}	33.3	22.2	16.7	11.1	6.67	4.76	3.70

(a) Evaluate the time for half reaction and the time for three-quarters reaction (graphically).
(b) Evalute *Rate*/*V* at each time from the graph. Comment on the suitability of these data for the graphical determination of rate.
(c) Plot log (graphical *rate*) versus log (*concentration*) and estimate the order of the reaction run reported.

5.5. At 298 K, the vapor pressure of solid N_2O_5 is 414.0 mmHg. Gaseous N_2O_5 decomposes according to the equation

$$N_2O_5 = N_2O_4 + \tfrac{1}{2}O_2$$

The following data were obtained in the presence of excess solid N_2O_5 at 298 K (1 mmHg = 133 Pa):

t/s	0	1200	1800	2400	3000	3600
P/mmHg	414.0	435.0	449.0	465.3	474.6	487.5

(a) What is the relationship between the rates of change of concentrations of N_2O_5, N_2O_4 and O_2?
(b) Show that the reaction is zero order with respect to N_2O_4 and O_2.

5.6. In the absence of solid N_2O_5, the reaction in problem 5.5 yielded the following data at 318 K:

$10^{-3}t$/s	0	1.0	2.0	3.0	4.0	5.0	6.0	7.0	8.0
$p_{N_2O_5}$/kPa	46.4	28.1	17.1	10.4	6.3	4.0	2.4	1.6	1.0

Determine the order of the reaction with respect to N_2O_5, write down the rate law for this run, and evaluate the rate coefficient.

5.7. At 25°C and pH = 5 the "inversion" of sucrose (i.e. sucrose = *d*-fructose + *d*-glucose) proceeds with a constant half life of 500 min. At this same temperature, but at pH = 4, the half life is constant at 50 min. What must be the values of the exponents a and b, in the rate law

$$dc_{sucrose}/dt = -kc^{a}_{sucrose}c^{b}_{H_3O^+} \text{ ?}$$

5.8. The reaction

$$Co(NH_3)_5F^{2+} + H_2O = Co(NH_3)_5(H_2O)^{3+} + F^-$$

follows the rate law

$$(Rate/V) \equiv \frac{dc_{Co(NH_3)_5F^{2+}}}{dt} = kc^{a}_{Co(NH_3)_5F^{2+}}c^{b}_{H_3O^+}$$

The times for half and three quarters of the complex to react are given below for the indicated temperatures and initial concentrations.

Concentrations of

$Co(NH_3)_5F^{2+}$ / $mol\ dm^{-3}$	H_3O^+ / $mol\ dm^{-3}$	T / °C	$t_{1/2}$ / min	$t_{3/4}$ / min
0.10	0.010	25	60	120
0.10	0.020	25	30	60
0.10	0.010	35	30	60

(a) Establish values for the exponents a and b.
(b) Calculate the value of the rate coefficient, k.
(c) Calculate the Arrhenius activation energy.

5.9. The kinetics of the reaction

$$2N_2O = 2N_2 + O_2$$
(g, 986 K, constant volume)

were found to be second order with respect to N_2O. Given that the initial rate of increase of total pressure for a system, starting with pure N_2O at standard atmospheric pressure, was 0.063 mmHg s^{-1}, what is the value of the rate coefficient expressed in $dm^3\ mol^{-1}\ s^{-1}$?

5.10. For the reaction

$$H_3O^+ + OH^- \rightarrow 2H_2O\ (aq, 25°C)$$

Eigen and de Maeyer have established the rate law

$$\frac{-dc_{H_3O^+}}{dt} \equiv \frac{-dc_{OH^-}}{dt}$$

$$= (1.3 \pm 0.2) \times 10^{11}\ c_{H_3O^+}c_{OH^-}\ dm^3\ mol^{-1}$$

Estimate the time required to complete half reaction between H_3O^+ and OH^- when both are initially present at 10^{-4} mol dm^{-3}.

5.11. For the reaction

$$FCH_2COOEt + OH^- = FCH_2CO_2^- + EtOH$$
(aqueous solution, 25°C),

$$(Rate/V) = kc_{OH^-}c_{FCH_2COOEt}$$
with $k = 14.0\ dm^3\ mol^{-1}\ s^{-1}$

The Arrhenius activation energy at this temperature is reported to be 38.4 kJ mol^{-1}.

(a) Write down the equation for rate of change of concentration of FCH_2COOEt at 298 K in a solution in which c_{OH^-} is held continuously at 1.00×10^{-4} mol dm^{-3}. What would be the time taken for half of the FCH_2COOEt to be removed under these conditions?
(b) Estimate the rate coefficient for this reaction at 35°C.

5.12. The reaction

$$S_2O_3^{2-} + BrCH_2CO_2^- = S_2O_3CH_2CO_2^{2-} + Br^-$$
(aqueous solution, 25°C)

is found to have $(Rate/V) = k\ c_{S_2O_3^{2-}}\ c_{BrCH_2COO^-}$ where the rate coefficient k is constant for any given run. By reference to the "transition state equilibrium model" as developed by Eyring, deduce the formula of the transition species in this reaction.

What is the ratio of the rate coefficient, k, to the rate constant, $k°$, corresponding to estimates of activity coefficients for the ionic species, as $y(B^-) = 0.90$, $y(B^{2-}) = 0.81$ and $y(B^{3-}) = 0.73$?

5.13. Suggest a mechanism for the formation of butadiene from cyclobutene.

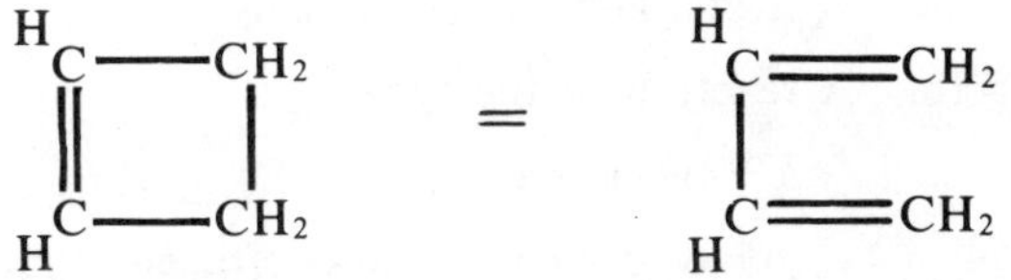

cyclobutene butadiene

The rate law is reported to be

$$Rate/V = kc_{cyclobutene}$$

5.14. The formation of 3-hydroxybutyraldehyde from acetaldehyde

$$2CH_3CHO = CH_3CH(OH)CH_2CHO$$
(aq, alkaline)

in dilute alkaline solution is reported to have the rate law

$$Rate/V = kc_{CH_3CHO}c_{OH^-}$$

The mechanism proposed for the reaction is

$$CH_3CHO + OH^- \underset{k_1}{\rightarrow} {}^-CH_2CHO + H_2O$$

$$^-CH_2CHO + H_2O \underset{k_{-1}}{\rightarrow} CH_3CHO + OH^-$$

$$CH_3CHO + {}^-CH_2CHO \underset{k_2}{\rightarrow} CH_3CH(O^-)CH_2CHO$$

$$CH_3CH(O^-)CH_2CHO + H_2O \underset{k_3}{\rightarrow} CH_3CH(OH)CH_2CHO + OH^-$$

(a) Write down equations for the rate of each step.

(b) What conditions must be fulfilled if this mechanism is to explain the observed rate law?

5.15. Suggest mechanisms for the following reactions, taking account of the rate laws

(a) $CH_2{-}CH_2$ (with O bridging, ethylene oxide) $+ H_2O = HOCH_2CH_2OH$ (*aq*, acidic)

$$Rate/V = k_1\, c_{CH_2-CH_2\,(O)} + k_2\, c_{H_3O^+}\, c_{CH_2-CH_2\,(O)}$$

(b) $CH_3COCH_3 + I_2 + OH^-$

$= CH_3COCH_2I + H_2O + I^-$

(*aq*, alkaline)

$$Rate/V = k c_{CH_3COCH_3}\, c_{OH^-}$$

5.16. A reaction of the type

$$A + B = C + D; \quad (Rate/V) = k c_A\, c_B$$

is found to have a rate coefficient equal to $0.10\ dm^3\ mol^{-1}\ s^{-1}$ at 27°C and an activation energy of $40\ kJ\ mol^{-1}$. Evaluate the rate of reaction immediately after solutions of A and B, each being $0.20\ mol\ dm^{-3}$, are mixed in equal volumes at 60°C.

5.17. Measurements carried out on the following reactions under controlled conditions in aqueous solution have revealed experimental rate laws as indicated:

1. $H_2O_2 + 2I^- + H_3O^+ = I_2 + 4H_2O(aq)$
   $(Rate/V) = k_1\, c_{H_2O_2}\, c_{I^-}$
2. $2CH_3CHO = CH_3CH(OH)CH_2CHO(aq)$,
   $(Rate/V) = k_2\, c_{CH_3CHO}\, c_{OH^-}$
3. $CH_3COOC_2H_5 + H_2O = CH_3COOH + C_2H_5OH(aq)$,
   $(Rate/V) = k_3\, c_{CH_3COOC_2H_5}$
4. $NH_4^+ + NCO^- = CO(NH_2)_2(aq)$,
   $(Rate/V) = k_4\, c_{NH_4^+}\, c_{NCO^-}$
5. $2ClNO = Cl_2 + 2NO(g)$,
   $(Rate/V) = k_5\, c^2_{ClNO}$

(a) Give a reasoned assessment of the possibility, in each case, that the reaction could be elementary, i.e. a simple one-stage process.

(b) Which, if any, of these reactions would be said to be catalysed?

(c) In a single run involving the determination of the variation of reactant concentration with time, which reactions would show first order behaviour?

(d) The condition that was most carefully controlled during the experimental studies was the temperature. Why?

ADDITIONAL PROBLEMS

The problems below attempt to follow particular reactions through from initial observations to some conclusions. They are genuine examples, some with unexpected conclusions which may later prove false. They are not all of equal difficulty.

5.18. In aqueous solution, the ethyl ester of iodoacetic acid, $ICH_2COOC_2H_5$, reacts with sodium hydroxide to form iodoacetate ion, the ethyl ester of hydroxyacetic acid, iodide ion and ethanol.

(a) Write equations to describe the chemical changes which occur in this situation.

(b) How is the rate of formation of hydroxide ion related to the rates of the reactions in (a) and to the rate of reaction of ethyl iodoacetate?

(c) What different information could be obtained from a study of the rates of formation of iodide ion?

(d) The rates of formation of hydroxide ion are found to be related by the equation

$$-dc_{OH^-}/dt = k c_{OH^-}\, c_{ICH_2COOC_2H_5}$$

At 298 K, k is 20 $dm^3\ mol^{-1}\ s^{-1}$. If the hydroxide ion concentration is held constant at 10^{-5} mol dm^{-3}, how long can a solution of $ICH_2COOC_2H_5$ be stored if its concentration must not drop below half the initial value (0.01 mol dm^{-3})?

(e) It is found that the rate of formation of iodide ion in this situation at 298 K is always 10% of the rate of hydroxide ion consumption. What are the rate laws and the rate coefficients for formation of ICH_2—COO^- and of $HOCH_2COOC_2H_5$?

(f) The Arrhenius activation energies are 50 kJ mol^{-1} and 100 kJ mol^{-1} for formation of ICH_2COO^- and of $HOCH_2COOC_2H_5$. What results may be expected from controlling the temperature at 273 K instead of at 298 K?

5.19. Pyrazole reacts with bromine in aqueous solutions to form 4-bromopyrazole.

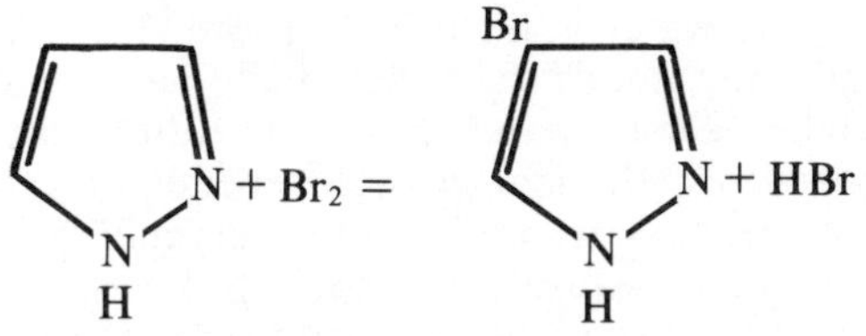

Pyrazole 4-bromopyrazole

(a) Reactions were studied in which the pyrazole concentration was approximately 2×10^{-4} mol dm^{-3} and the bromine concentration 1×10^{-5} mol dm^{-3}. Under these conditions it was found that

$$\ln(c_{Br_2,\,t=0}/c_{Br_2,\,t}) = k_{obs}t$$

with k_{obs} constant for constant pyrazole concentrations. What conclusion may be drawn about the rate law for this reaction?

(b) A series of rate measurements was made with varying concentrations of pyrazole and constant initial concentrations of bromine (1×10^{-5} mol dm^{-3}). At 298 K, the initial rates are:

$10^4\ c_{pyrazole}$ / mol dm^{-3}	$-10^5(dc_{Br_2}/dt)_{t=0}$ / mol $dm^{-3}\ s^{-1}$
2.3	6.2
1.8	5.0
1.5	4.3
1.1	3.0

What can be deduced about the rate law for this reaction?

(c) What can be deduced from (a) and (b) about the value of the rate coefficient at 298 K?

5.20. In a gas phase study of the reaction of chlorine atoms with chlorine dioxide

$$Cl + ClO_2 \rightarrow 2ClO$$

the following measurements were made.

At 298 K, and with initial chlorine atom concentrations of 2×10^{-9} mol dm^{-3}, the chlorine dioxide concentration is found to fall in the following way:

10^3 *time*/s	$10^{10}\ c_{ClO_2}$/mol dm^{-3}
0	1.60
7	0.90
12	0.71
16	0.50
20	0.36
30	0.20
40	0.10

(a) How does the rate of reaction depend on concentration of chlorine dioxide?

(b) The rate coefficients k found in experiments like that described increase with concentration of chlorine atoms according to the equation

$$k = 3.5 \times 10^{10}\ c_{Cl}$$

What is the rate law?

(c) Suggest a mechanism for the reaction.

6. Chemical Potentials

In this chapter we develop the alternative approach, promised in Section 4.2, to the analysis of the direction in which chemical reaction occurs. We begin by examining some of the general features of spontaneous changes and the conditions under which other processes, those which do not normally occur, may be carried out. The idea of potential to which this study will lead us will then be correlated with the ideas of chemical activity developed in Chapter 4, and used in some calculations on systems in equilibrium.

6.1 Driven or Spontaneous Reaction?

Spontaneous chemical processes make all sorts of things possible. They lead to the manufacture of desired materials and to their deterioration in use. They drive motor cars and power plants and human beings. They have key functions in the weather and the landscape, in printing and painting and, indeed, in all events. But there are also many changes which occur only because they are driven by coupling to some other, spontaneous, change. These changes occur in biological systems as well as in technology. One of the consequences of the method of examination used in this section is that we will learn how to find the least "driving" a process needs in order to make it happen.

6.1.1 Natural changes are spontaneous

There are many familiar changes which occur spontaneously (Fig. 6.1). When we raise a brick, it is likely to fall, when we puncture a tank full of water, water is likely to run out. Rain falls from the clouds; pricked balloons deflate; stretched elastic bands relax. The direction of spontaneous thermal change depends on relative hotness—hot bricks cool but cold bricks warm up. Natural electrical processes like lightning can be shown to shift charge in one direction. In certain conditions only setting off is required for fire crackers to explode and for bushfires to burn. These are examples of the many uncontrolled processes in nature. They do not spontaneously reverse themselves. The changes in the things described are coupled only with changes in their environment—the atmosphere and the earth.

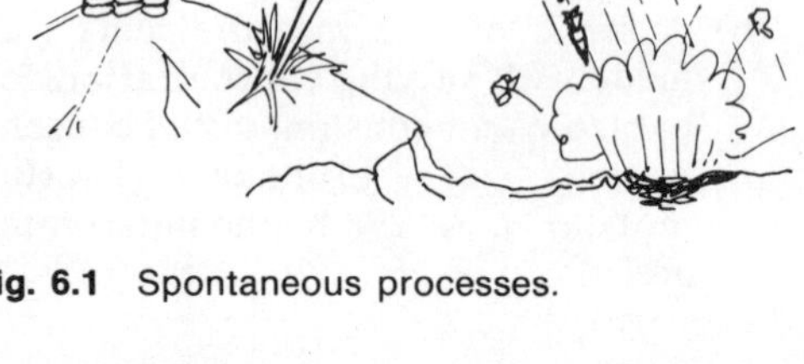

Fig. 6.1 Spontaneous processes.

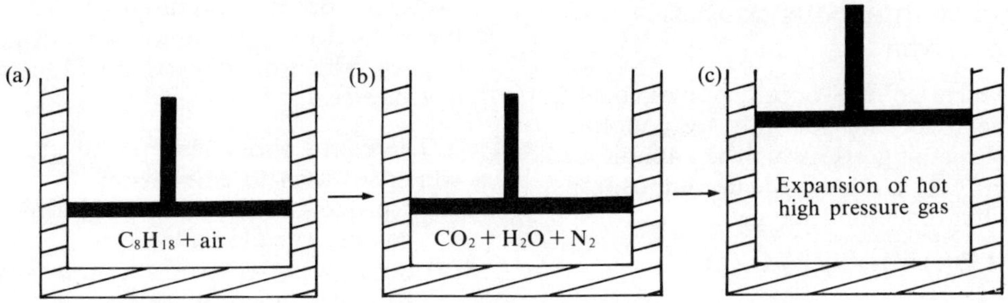

Fig. 6.2 Combustion of octane.

6.1.2 Spontaneous processes may drive things for us!

Any one of the foregoing spontaneous processes can be put to use if we can get it under control. Consider, for example, the burning of a hydrocarbon such as octane in air. The reaction occurs readily on the surface of a platinum catalyst (low temperature reaction) or on ignition (high temperature reaction). It may also be carried out in an internal combustion engine. In the cylinder there is a mixture of octane and oxygen at, perhaps, 80° C and $9P^\circ$ (Fig. 6.2a). The mixture is ignited and burns very quickly to form carbon dioxide and water at a very much higher temperature (perhaps 1600° C) and pressure (Fig. 6.2b).

$$C_8H_{18} + 12\tfrac{1}{2}O_2 = 8CO_2 + 9H_2O$$

These changes are followed by changes in volume and temperature (Fig. 6.2c) which result in the turning of the crankshaft and the driving of the machine.‡

Query What can we learn from this example?
Reply The important observation is that the transformation of a mixture of octane and oxygen to carbon dioxide and water occurs naturally and may be arranged to achieve useful work.

‡The actual changes which occur during the combustion and the expansion are more complex than has been implied. On combustion, considerable amounts of other species such as carbon monoxide are formed which undergo further reaction during the expansion. In addition, changes in the temperature of parts of the engine result in the heating of the circulating water. The details do not concern us at present.

Another example is the reaction of zinc with mercuric oxide —

$$Zn(s) + HgO(s) = ZnO(s) + Hg(l)$$

The Mallory cell has been designed to couple this spontaneous process with the transfer of electric charge. The voltage of 1.347 volt that is developed is used to drive electric current through such devices as hearing aids. Again a spontaneous chemical process has been harnessed to drive things for us.

Query Is it generally true that spontaneous processes can be made to do something useful?
Reply Yes. In principle every spontaneous change has associated with it an amount of "available work" — i.e. work which can be harnessed to drive machines. Whether or not work is obtained in a given practical situation depends on the mechanism by which the process is carried out.

Exercise 6.1
Give examples of situations where the following spontaneous processes are harnessed to perform useful work:
(a) the falling of water;
(b) the falling of masses;
(c) the expansion of compressed air;
(d) the expansion of steam from a region of high pressure to a region of lower pressure;
(e) the decomposition of trinitrotoluene;
(f) the oxidation of glucose to carbon dioxide and water;
(g) the reaction

$$PbO_2(s) + 4H_3O^+(aq) + 2SO_4^{2-}(aq) + Pb(s) = 2PbSO_4(s) + 6H_2O(l)$$

(h) the oxidation of hydrogen to water.

6.1.3 Non-spontaneous processes must be driven

Processes which do not occur spontaneously can be made to occur, but only by coupling them with something else which is capable of driving them. An example is the decomposition of water to hydrogen and oxygen:

$$H_2O(l) = H_2(g) + \tfrac{1}{2}O_2(g)$$

Figure 6.3 is a sketch of one possible arrangement for carrying out this transformation. In this method the electrolytic decomposition of water is driven by applying a voltage to the platinum electrodes immersed in the solution. This causes the passage of electric charge, which is accompanied by evolution of gaseous hydrogen and gaseous oxygen.

Delivery of the hydrogen generated by this simple process has been suggested as a practical stratagem for long distance transport of energy, instead of using high voltage electrical transmission lines.

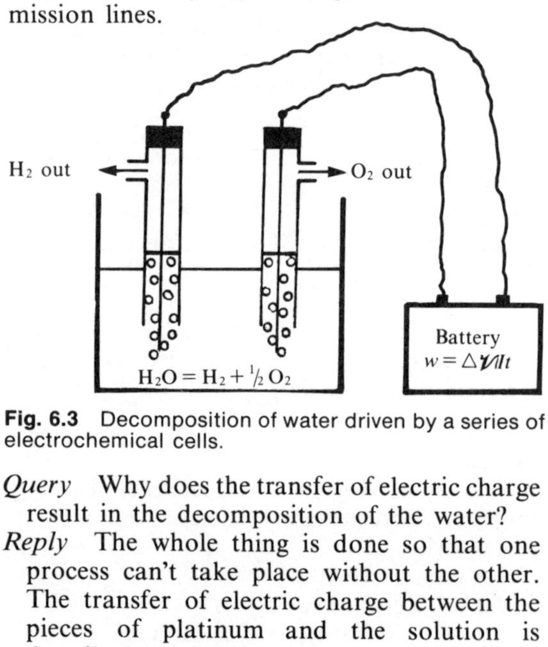

Fig. 6.3 Decomposition of water driven by a series of electrochemical cells.

Query Why does the transfer of electric charge result in the decomposition of the water?

Reply The whole thing is done so that one process can't take place without the other. The transfer of electric charge between the pieces of platinum and the solution is described by the equations

$$2H^+(aq) + 2e^-(Pt) = H_2$$

(on the surface of one piece of platinum)

$$2OH^-(aq) = \tfrac{1}{2}O_2 + H_2O + 2e^-(Pt)$$

(on the surface of the other piece of platinum)

When the battery drives electrons through the wires connected to the platinum plates, both processes occur continuously. They lead to the evolution of gaseous hydrogen and oxygen. This whole question will be taken up in Chapter 7.

There are three features of this process which we wish to emphasise:

(a) a process which does not normally occur has been made to happen;
(b) the process required action by something else—an "operator";
(c) the "operator" performed work.

Query What is this "operator"?

Reply The "operator" is the agent which drives the non-spontaneous process. In this case it was the battery. It might equally well have been a human operator turning the handle of an electric generator!

Query Do these statements apply to non-spontaneous processes in general?

Reply Yes. In general an "operator", by doing work, can drive a non-spontaneous process. It is only necessary to find a suitable mechanism for applying the work. The non-spontaneous transfer of water from a river to a water tower is achieved by turning the impeller of a pump. The transfer of electric charge from low voltage to high voltage may be driven by turning a generator. The extraction of drinkable water from a salty supply may be brought about by reverse osmosis, by the application of pressure to the salty water to drive the water but not the salt through a specially selected membrane (see Section 3.9.2).

Query Does this mean that non-spontaneous processes always require work by an "operator" in order to proceed?

Reply No, not always. Action by an "operator" is required but the action need not be the performance of *work*. An alternative is to drive the process by another process which can proceed spontaneously. For example, the photosynthesis of glucose (at its simplest level) can be thought of as the driving of the non-spontaneous process,

$$6CO_2(g) + 6H_2O(l) =$$

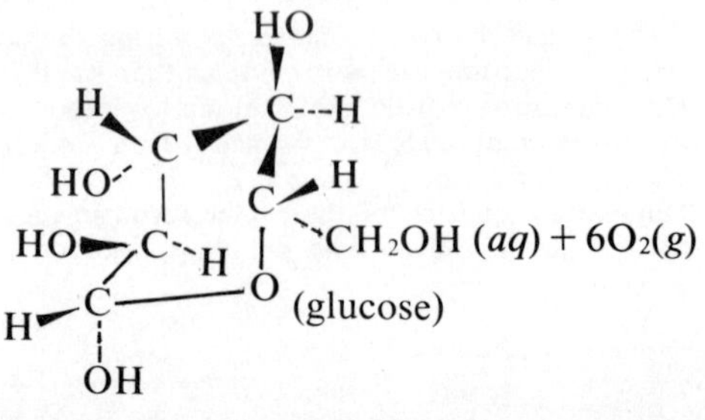

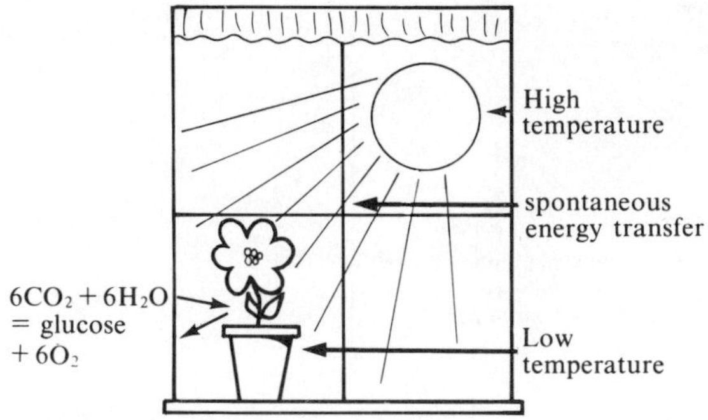

Fig. 6.4 Production of carbohydrate driven by solar radiation.

by the spontaneous process of energy transfer (radiation) from a hot body (the sun) to a cooler body (the plant) (Fig. 6.4). Another important example is the reduction of ferric oxide with carbon to produce pig iron:

$$Fe_2O_3(s) + 3C(s) = 2Fe(s) + 3CO(g)$$

In this case the non-spontaneous process

$$Fe_2O_3(s) = 2Fe(s) + \tfrac{3}{2}O_2(g)$$

is driven by the spontaneous process

$$3C(s) + \tfrac{3}{2}O_2(g) = 3CO(g)$$

Exercise 6.2
Suggest some of the methods or processes by which the following normally non-spontaneous processes may be driven.

(a) the formation of aluminium from alumina

$$Al_2O_3(s) = 2Al(s) + \tfrac{3}{2}O_2(g)$$

(b) the recovery of sodium chloride from sea water;
(c) the production of iodine from seaweed.

6.1.4 The amount of driving can be measured

The amount of driving done by an operator is measured by the amount of work done by the operator, $w_{\text{♀}}$. For a simple process such as the raising of a bag of flour by the mechanism illustrated in Fig. 6.5, the work done by the operator is simply calculated by multiplying the force applied by the operator ($F_{\text{♀}}$) by the distance he traverses ($s_{\text{♀}}$).

$$s_{\text{♀}} = F_{\text{♀}} \times s_{\text{♀}}$$

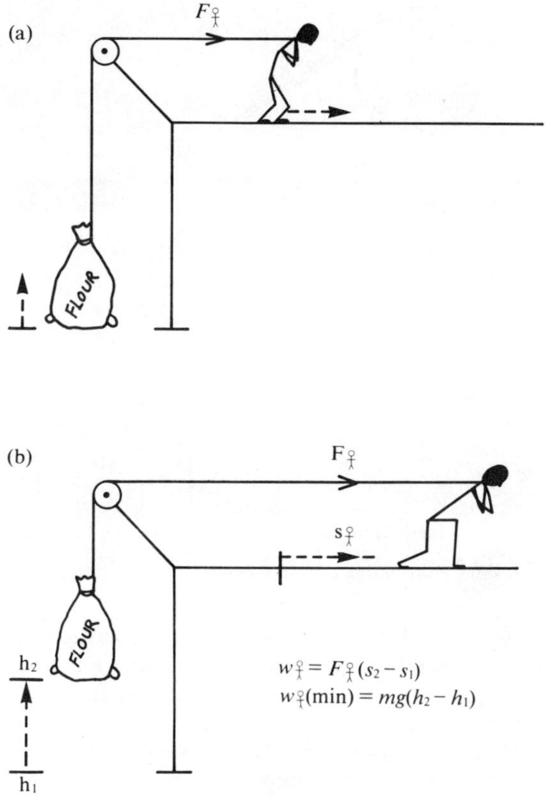

Fig. 6.5 Work done in lifting a mass.

It is clear that if the pulley in Fig. 6.5 is stiff and rusty, the operator will have to apply a much greater force (that is, do more work) than if the pulley is well oiled. Improvement of the pulley reduces the force (and hence the work) required, to the value just sufficient to overcome the earth's gravitational attraction (mass of flour × gravitational acceleration). That is,

$$F_{\text{♀}}(\text{minimum required}) = mg$$

and

$$w_{\text{♀}}(\text{minimum required}) = mg\, s_{\text{♀}}$$

Because the rope has a fixed length

$$s_{\text{♀}} = h_2 - h_1$$

hence, $w_{\text{♀}}(\textit{minimum}) = mg(h_2 - h_1)$

There is one very important feature of this result which applies to any process. The value of $w_{\text{♀}}$ (*minimum*) is determined only by what the change is and not by the mechanism or path by which it is carried out.

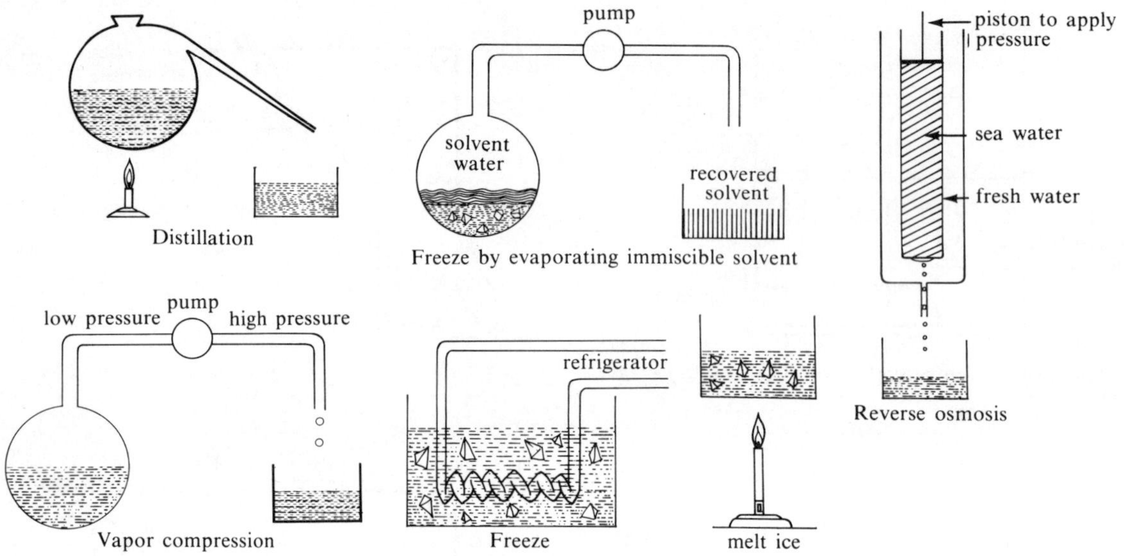

Fig. 6.6 Purification of sea water.

Comment I don't see what you are getting at.

Reply Another example might help. Take the preparation of pure water at 290 K from sea water at 290K,

$H_2O(sea, 290K) = H_2O(pure, 290K)$.

+ slightly concentrated sea water).

A number of ways by which this process might be carried out are illustrated schematically in Fig. 6.6. In each case it can be shown (Lane, J.E. and Mansfield, W.W. in *Salinity and Water Use*, edited by T. Talsma and J.R. Philip, Macmillan, 1971) that *w*♀ (*minimum*) is the same. The value is 41 J per mole of pure water produced. A spontaneous process can be treated in a similar manner. In this case *w*♀ (*minimum*) is a negative quantity.

Query What do negative values of *w*♀ (*minimum*) mean?

Reply Positive values of *w*♀ mean that work is being done on the system by an operator. Negative values of *w*♀ mean that the system is doing work on some other system. In fact the value of {− *w*♀ (*minimum*)} is the maximum amount of work which the system can do as a result of the change being considered. It is the maximum work available from the change.

Comment I don't get it!

Reply Consider the reaction between hydrogen and oxygen

$$H_2(g) + \tfrac{1}{2}O_2(g) = H_2O(l)$$

In Fig. 6.7a we have sketched diagramatically the value of *w*♀ (*minimum*) together with the actual values of *w*♀ when a particular fuel cell was used to run a transistor radio and when the hydrogen was burnt in a gas burner.

In Fig. 6.7b we have sketched the maximum amount of "usable" work obtainable together with the amount actually obtained for the same two processes.

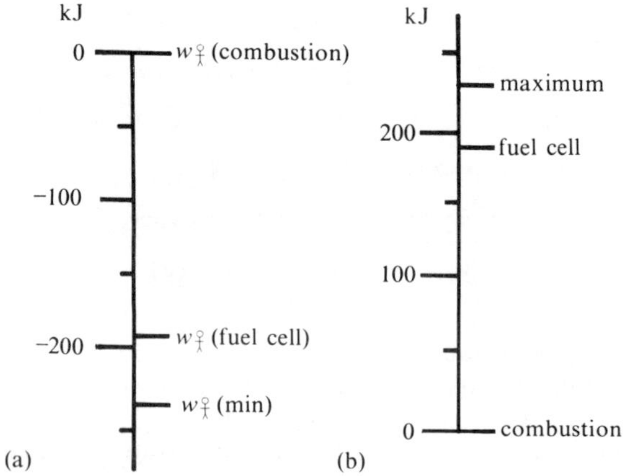

Fig. 6.7 Operator work and usable work in reaction of hydrogen and oxygen.

Query If *w*♀ is greater than *w*♀ (*minimum*) what happens to this extra work?

Reply Either the extra work goes into raising the temperature, or it is dissipated by heating the environment.

Query When might *w*♀ be greater than *w*♀ (*minimum*)?

Reply In mechanical situations the answer is simple. The work done by the operator ($w_♀$) is greater than the minimum value for the process whenever there are frictional forces to be overcome. In electrical examples the corresponding factor is the electrical resistance. Other situations, particularly those involving chemical or biological processes, are not so clear. As a general rule it is found that whenever a process is carried out so that it proceeds at a reasonable rate, $w_♀$ is greater than $w_♀$ (*minimum*).

6.2 The Idea of Potential

To describe in more detail the action required to carry out a non-spontaneous change or the action available from a spontaneous change, we introduce the concept of the "potential for change".

When something (or somebody) is said to "have potential" it just means that we recognise some latent power or capacity which can develop into activity. This idea can be used to determine whether or not a great variety of processes, such as the falling of a mass, the combustion of octane and the decomposition of water, can occur unaided or whether they need to be driven. The clue to the use of this idea is the relationship between the process and the action required in order to drive it.

We begin by looking at factors which need to be overcome to carry out a non-spontaneous change. Our first examples involve little chemistry, but set up the background required in order to tackle problems of chemical and biological interest.

6.2.1 Opposition to change

The factors opposing a non-spontaneous change are usually readily identified. In the example illustrated in Fig. 6.5, it is clearly the gravitational attraction between the flour and the earth which opposes the lifting of the flour. Let us examine other examples in somewhat more detail.

Consider a partly stretched spring (Fig. 6.8). In order to stretch the spring more, we must apply a force F sufficient to overcome the tension τ in the spring. If we extend the spring by a small amount dl we must do some work ($w_♀$). In practice

$$w_♀(minimum) = \tau \, dl$$

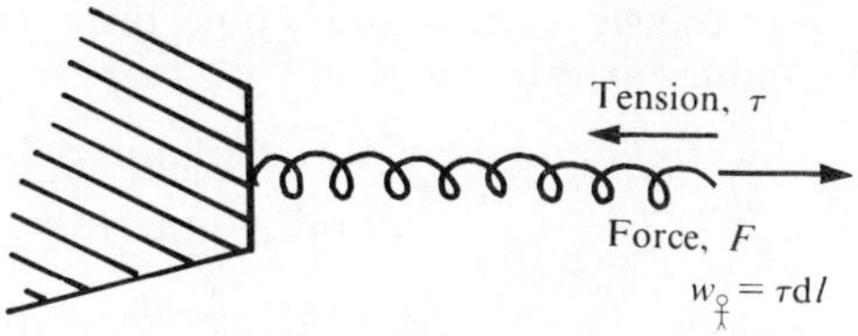

Fig. 6.8 Equilibrium in a partly stretched spring.

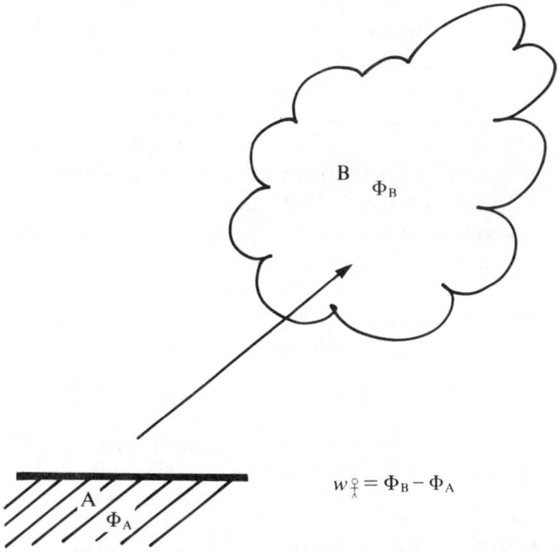

Fig. 6.9 Charge transfer from earth to a cloud.

Compare the stretching of the spring with the transfer of a small amount of electric charge (dQ) between the earth's surface, A, and a cloud, B (Fig. 6.9). In this case it is the electrical potential difference ($\Phi_B - \Phi_A$) between the earth and the cloud which must be overcome. In this example the minimum amount of work which must be done by an operator is given by

$$w_♀(minimum) = (\Phi_B - \Phi_A)dQ$$

The force acting on the charge is due to the variation of electrical potential with distance ($d\Phi/dl$). In fact, at any particular point in the transfer of charge, dQ,

$$F = dQ(d\Phi/dl)$$

and the corresponding element of operator work to transfer the charge through a distance dl is

$$dw_♀(minimum) = F\,dl = (d\Phi/dl)\,dl\,dQ$$

For transfer across the whole distance, the minimum operator work is

$$
\begin{aligned}
w_{♀}(\textit{minimum}) &= \int_A^B \mathrm{d}w_{♀}(\textit{minimum})\,\mathrm{d}l \\
&= \int_A^B (\mathrm{d}\Phi/\mathrm{d}l)\mathrm{d}l\mathrm{d}Q \\
&= \quad (\Phi_B - \Phi_A)\mathrm{d}Q
\end{aligned}
$$

Spring tension and electrical potential have three important properties in common.

(a) They both may cause spontaneous changes. It is the tension in a stretched spring which causes it to contract. It is the difference in electrical potential which causes electric charge to flow from one region to another.
(b) If the change is a non-spontaneous one, such as in the examples discussed, then it is the tension which must be overcome in order to stretch the spring, and the electrical potential difference which must be overcome in order to transfer the electric charge.
(c) The third property in common can be seen from the equations

$$\tau = \frac{\text{minimum operator work required to stretch spring}}{\text{increase in length}}$$

$$\Phi_B - \Phi_A = \frac{\text{minimum operator work required to transfer a quantity of charge from A to B}}{\text{amount of charge transferred}}$$

6.2.2 The definition of potential

Both τ and Φ are *potentials*, where "potential" is defined in terms of the operator work needed for an infinitesimal amount of change.

$$\text{itial for change} = \frac{\text{minimum operator work required for infinitesimal change}}{\text{amount of such change}}$$

Query What is the most important difference between spring tension and electrical potential?

Reply The potentials are different because the types of change are different. The total charge in a system is constant apart from charge entering or leaving the system; that is, charge is a "conserved quantity". Consequently the change which is associated with an electrical potential difference is a *transfer* of charge from one region to another (and/or from one chemical form to another).

Length, on the other hand, is not a "conserved" quantity. Change of length may alter the properties of the system very much but transfer is not involved, only change. Because charge is conserved, all changes involving electric charges merely transfer charge from one place to another or from one chemical species to another. For this reason the electrical potential difference is called a "transfer potential".

Query What about gravitational potentials?

Reply Gravitational potentials govern the transfer of mass from one place to another. Take the problem of moving an infinitesimal mass dm from a height (h_1) above the earth's surface to a new height (h_2). Provided that the change is small enough so that the earth's gravitational field (g) may be taken as constant over the change, then

$$w_{♀}(\textit{minimum}) = g(h_2 - h_1)\mathrm{d}m$$

we can rewrite the potential for transfer of mass as

$$(gh_2 - gh_1) = \frac{w_{♀}(\textit{minimum})}{\mathrm{d}m}$$

which is the difference in gravitational potential. By choosing one particular height to have the value "zero", we can give the gravitational potentials at other heights as values of gh.

Note that if the gravitational field (g) is not constant over the change in height then the difference in gravitational potential is not given by the simple formula ($gh_2 - gh_1$). In all terrestrial examples of the effect of gravitational fields on chemical composition the simple formula is adequate.

Query Then this is another "transfer potential"?

Reply Yes. Mass is another "conserved quantity" and hence the only changes which can occur are transfer from one place to another or transformation into different chemical forms.

Query Are transfer potentials always differences between potentials?

Reply Yes. One consequence is that we can always assign a numerical value to one of the potentials and hence fit numerical values for the others. For example, we might give the electrical potential of the earth the value zero and quote the values of other electrical potentials as ($\Phi - \Phi_{\text{earth}}$).

Query How can we use the potential for a process to determine whether it is spontaneous or needs to be driven?

Reply From the equation

$$\text{potential for change} = \frac{w_{\text{♀}}(\textit{minimum})\text{ for an infinitesimal amount of change}}{\text{amount of change}}$$

we get the following results:

(a) if the potential for change (for a transfer process, the difference in potential) is greater than zero, $w_{\text{♀}}$ (*minimum*) is positive and, under the present conditions the process must be driven;

(b) if the potential for change is less than zero, $w_{\text{♀}}$ (*minimum*) is negative and the process envisaged is spontaneous;

(c) if the potential for change is zero, $w_{\text{♀}}$ (*minimum*) is zero and the process is in equilibrium.

6.2.3 Potential energy (available work)

Query Isn't this work $w_{\text{♀}}$ (*minimum*) just the change in potential energy?

Reply The value of $\{-w_{\text{♀}}\ (\textit{minimum})\}$ is a measure of the change in the amount of work which can be obtained from the system.

If we have carried out a process which is not spontaneous, $w_{\text{♀}}$ (*minimum*) is positive and we end up with a system which can do more work. If the process was spontaneous, $w_{\text{♀}}$ (*minimum*) is negative and after it the system will be able to do less work than at the beginning.

Comment You haven't answered the question.

Reply The answer is yes. We want to define the potential energy of the system so that it is the amount of usable work which is available.

Query What has all this to do with chemistry?

6.3 Chemical Potential

The ideas of "potential" developed in the previous section can be applied directly to processes which involve changes in chemical composition. In fact many of our examples have been just of this type.

The transformations with which chemistry is concerned can be separated into two different groups. One type involves the movement of a substance from one phase to another, such as when sugar dissolves in a cup of tea, or from one region to another, as when river water runs into the sea. This type of transformation we can sketch as in Fig. 6.10. The potential associated

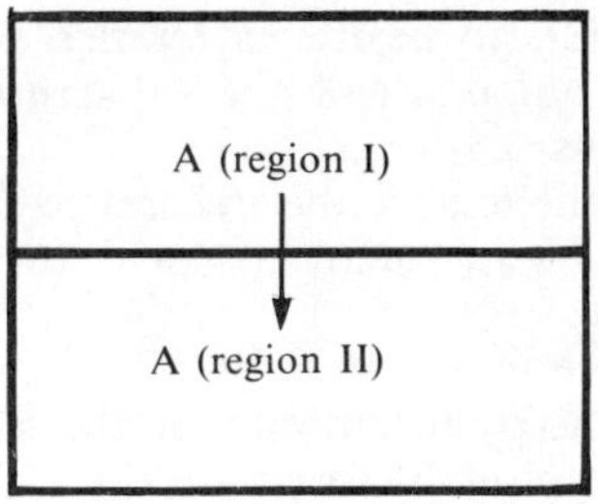

Fig. 6.10 Transfer of a substance from one phase to another.

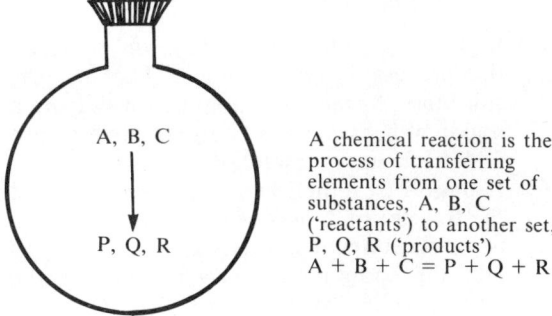

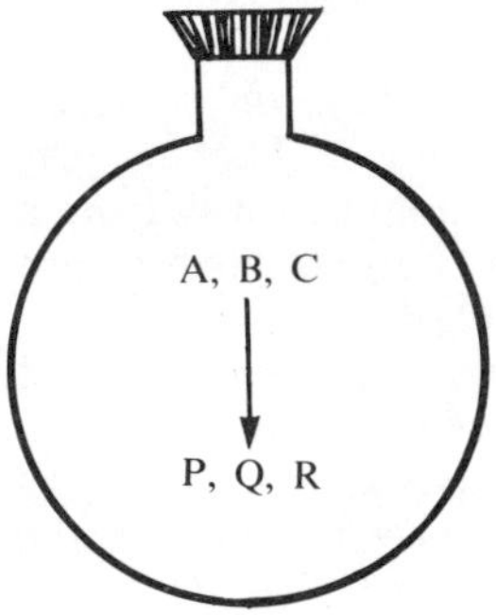

A chemical reaction is the process of transferring elements from one set of substances, A, B, C ('reactants') to another set, P, Q, R ('products') A + B + C = P + Q + R

Fig. 6.11 A chemical reaction.

with this change is clearly a transfer potential. The other type of transformation includes all cases where a set of "products" is formed from a set of "reactants" within some particular region (Fig. 6.11). The "transformation" potential is like a "transfer" potential. Direction of reaction is governed by the difference in potential between reactants and products.

Both types of transformation can be described by chemical equations (Section 1.4.2). For example, that of Fig. 6.10 is described by an equation like:

$$A(1) = A(2)$$

and that of Fig. 6.11 by an equation like:

$$C_2H_6 + 3\tfrac{1}{2}O_2 = 2CO_2 + 3H_2O$$

6.3.1 Definition of chemical potential

We define the chemical potential, μ, of a substance as the potential which governs transfer or transformation of that substance in either type of chemical change. By analogy with other transfer potentials (for example electrical potential), we define the potential for the transfer of the substance A from region (1) to region (2) (Fig. 6.10), that is, for the reaction

A(1) = A(2), to be the *difference between the chemical potentials* of the substance A in the two regions, $\mu_{A(2)} - \mu_{A(1)}$.

We might expect the difference in chemical potentials $\mu_{A(2)} - \mu_{A(1)}$ to have the following properties.

(a) The minimum amount of driving required to transfer an amount of the substance A from region (1) to region (2) is given by the product of the (amount of change) and (the difference in potential). That is

$$w_{♀}(minimum) = (\mu_{A(2)} - \mu_{A(1)})\,dn$$

(b) If $\mu_{A(1)} > \mu_{A(2)}$ then $w_{♀}(minimum) < 0$ and the process is spontaneous. That is, the substance A transfers spontaneously from a region of high chemical potential to a region of low chemical potential.

(c) If $\mu_{A(2)} > \mu_{A(1)}$ then $w_{♀}(minimum) > 0$ and the transfer of A between regions (1) and (2) is not spontaneous; it needs to be driven. In fact the spontaneous process will be the transfer of A from region (2) to (1).

(d) If $\mu_{A(2)} = \mu_{A(1)}$, then $w_{♀}$ $(minimum) = 0$. That is, there is no tendency for A to transfer between the two regions. The system is in equilibrium.

(e) The units of chemical potential can be determined by examining (a) above: $w_{♀}$ (*minimum*) has units of J, d*n* has units of mol, hence the units of chemical potential are J mol^{-1}.

Comment You are simply *assuming* that it is possible to use chemical potentials.

Reply Yes. That is why we delayed sorting out the meaning of the word "potential". Before chemical potentials can be used, we need to find ways of estimating values of chemical potential differences.

The potential for the transformation of reactants into products can be treated in a similar way. For example we can define the potential for the transformation of methyl cyanide into methyl isocyanide, that is

$$CH_3-CN = CH_3-NC$$

to be the *difference between* the *chemical potentials* of the *reactants* and the *products.*

$$\text{potential for reaction} = \mu_{CH_3-NC} - \mu_{CH_3-CN}$$

From the idea of potential we would expect that:

(a) the minimum amount of driving required to transform an amount (d*n*) of methyl cyanide into methyl isocyanide would be given by

$$w_{♀}(minimum) = (\mu_{CH_3-NC} - \mu_{CH_3-CN})\,dn$$

(b) if $\mu_{CH_3-CN} > \mu_{CH_3-NC}$

then $w_{♀}(minimum) < 0$

then the reaction would be spontaneous. That is, a substance of high potential converts to one of lower potential;

(c) if $\mu_{CH_3-NC} > \mu_{CH_3-CN}$

then $w_{♀}(minimum) > 0$

then the conversion of methyl cyanide to methyl isocyanide under such conditions would not be spontaneous but would need to be driven. The spontaneous process would be the conversion of methyl isocyanide to methyl cyanide;

(d) under conditions such that

$$\mu_{CH_3-NC} = \mu_{CH_3-CN}$$

$$w_{♀}(minimum) = 0$$

there would be no tendency for any reaction to take place. It is in equilibrium.

Query Do the same values of chemical potential govern both the transfer of a compound from one region to another and its tendency to take part in chemical reactions?

Reply Yes. The significant feature of both types of process is the change in the amount of a substance present in a particular region.

Comment You haven't really defined the chemical potential of a substance. You have only defined some differences in chemical potential!

Reply True. Like electrical and gravitational potentials, only *differences* in chemical potentials can be measured. However it is useful to think about the chemical potential of a substance.

We can define such a chemical potential in terms of the minimum amount of work which must be done by an operator to increase infinitesimally (d*n* mol) the amount of the substance present in a particular region. That this definition is compatible with the definitions of differences in chemical potential can be seen by calculating the value of $w_{♀}(minimum)$ required to convert a small amount (d*n* mol) of methyl cyanide to methyl isocyanide in a mixture of the two compounds.

$w_{♀}$(*minimum*) to change the amount of methyl cyanide $= \mu_{CH_3\text{-}CN}\, dn_{CH_3\text{-}CN}$

and, as $\quad dn_{CH_3\text{-}CN} = -dn$

$$w_{♀}(minimum) = -\mu_{CH_3\text{-}CN}dn$$

Similarly

$w_{♀}$(*minimum*) to change the amount of methyl isocyanide $= \mu_{CH_3\text{-}NC}\, dn_{CH_3\text{-}NC}$

or, as $\quad dn_{CH_3\text{-}NC} = dn$

$$w_{♀}(minimum) = \mu_{CH_3\text{-}CN}dn$$

Of course the two processes are coupled so that one can't take place without the other. Hence the driving required for the reaction is the sum of that for the two processes. That is,

$$w_{♀}(minimum,\text{reaction}) = (\mu_{CH_3\text{-}NC} - \mu_{CH_3\text{-}CN})\, dn$$

as before.

Example 6.1

(a) What is the potential which must be overcome in order to extract pure water from sea water?

Answer

The process is

$$H_2O(sea) = H_2O(pure)$$

The potential is

$$\mu(H_2O,pure) - \mu(H_2O,sea)$$

(b) How much work must be done by an operator in order to convert n mol of graphite into diamonds?

Answer

The process is

$$\text{C(graphite)} = \text{C(diamond)}$$

The potential for the process is

$$\mu(\text{diamond}) - \mu(\text{graphite})$$

The amount of reaction is n mol

$$\therefore \quad w_{♀}(minimum) = (\mu_{\text{diamond}} - \mu_{\text{graphite}})\, n$$

PROBLEM

6.1. (a) The chemical potential of the water in the sample of sea water is 237 159 J mol^{-1}. What is the minimum amount of driving required to prepare 1 tonne (5.5×10^4 mol) of fresh water with a value of chemical potential equal to 237 200 J mol^{-1}?

(b) In a solution of *o*-dinitrobenzene

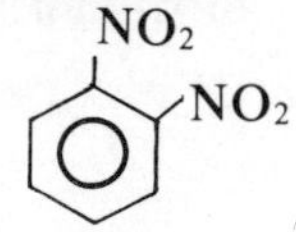

and *p*-dinitrobenzene

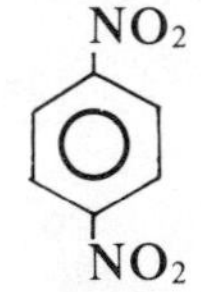

μ(*o*-dinitrobenzene) = 11.3 kJ mol^{-1}
μ(*p*-dinitrobenzene) = −39.7 kJ mol^{-1}

(i) In which direction will the reaction proceed?

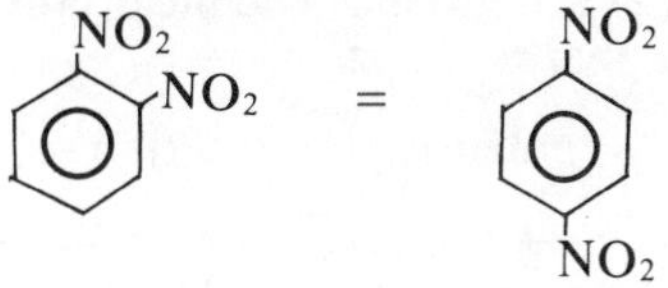

(ii) The reaction proceeds until equilibrium is reached. If at equilibrium μ(*o*-dinitrobenzene) = −10.0 kJ mol^{-1}, what is the value of μ(*p*-dinitrobenzene)?

(c) The chemical potential of solid iodine has the value $\mu_{I_2}(solid) = 0$. What is the value of the chemical potential of iodine in a saturated solution of iodine in cyclohexane?

(d) Describe a supersaturated solution in terms of the chemical potential of the solute.

6.3.2 The potential for chemical change

The potentials which govern more complex reactions are obtained in the same way as those for reactions involving a single reactant and single product. Consider a particular example, the burning of ethane:

$$C_2H_6 + 3\tfrac{1}{2}O_2 = 2CO_2 + 3H_2O$$

If an amount of reaction ($d\xi$ mol) occurs, this *removes* $d\xi$ mol of ethane and $3\tfrac{1}{2}\,d\xi$ mol of oxygen *and forms* $2d\xi$ mol of carbon dioxide and $3d\xi$ mol of water. So that $w_{♀}$(*minimum*) due to the change in the amounts of reactants is

$$\mu_{C_2H_6}\, dn_{C_2H_6} + \mu_{O_2}\, dn_{O_2} = -(\mu_{C_2H_6} + 3\tfrac{1}{2}\mu_{O_2})\, d\xi$$

and $w_{♀}$(*minimum*) due to the change in the amounts of products is

$$\mu_{CO_2}\,dn_{CO_2} + \mu_{H_2O}\,dn_{H_2O} = +(2\mu_{CO_2} + 3\mu_{H_2O})\,d\xi$$

For $d\xi$ mol of reaction

$$w_{♀}(minimum) = \{(2\mu_{CO_2} + 3\mu_{H_2O}) - (\mu_{C_2H_6} + 3\tfrac{1}{2}\mu_{O_2})\}d\xi$$

The potential for reaction

$$= \frac{w_{♀}(minimum)}{\text{amount of reaction}}$$

$$= (2\mu_{CO_2} + 3\mu_{H_2O}) - (\mu_{C_2H_6} + 3\tfrac{1}{2}\mu_{O_2})$$

Thus the *difference in potential* between reactants and products is

$$2\mu_{CO_2} + 3\mu_{H_2O} - \mu_{C_2H_6} - 3\tfrac{1}{2}\mu_{O_2}$$

For any reaction (or transfer process) for which we have written a chemical equation, we can write this result as

$$\text{potential for reaction} = \Sigma\nu_B\,\mu_B$$

where ν_B is the coefficient of the substance B in the chemical equation for the reaction. It is positive for products and negative for reactants.

The expressions used in this section to describe the potentials which govern changes in chemical composition need to be used rather often. As they are rather clumsy, some new names are appropriate. The names chosen honour J. Willard Gibbs, (1839–1903) Professor of Mathematical Physics at Yale College (1871–1903), who pioneered the application of the ideas of potential to changes of chemical composition.

6.3.3 $\Delta\bar{G}$, the potential for chemical change

$\Delta\bar{G}$ (pronounced "delta-G-bar") is defined as *the potential for chemical change*. That is,

$$\Delta\bar{G} \equiv \Sigma\nu_B\mu_B$$

This use is quite similar to that for $\Delta\bar{V}$ (Section 2.9)

$$\Delta\bar{V} \equiv \Sigma\nu_B\bar{V}_B$$

so the chemical potential of each substance (μ) is a "molar Gibbs energy". The value of $\Delta\bar{G}$ is characteristic of the *system* (reacting mixture) for which it is determined, *in relation to* a particular reaction. $\Delta\bar{G}$ is often called the Gibbs energy change per mole of reaction.

Query How do we determine the value of $\Delta\bar{G}$ for the reaction

$$C_2H_6 + 3\tfrac{1}{2}O_2 = 2CO_2 + 3H_2O$$

Reply From its definition

$$\Delta\bar{G} \equiv \Sigma\nu_B\mu_B$$

$$= 2\mu_{CO_2} + 3\mu_{H_2O} - 3\tfrac{1}{2}\mu_{O_2} - \mu_{C_2H_6}$$

Query Does this mean that the value of $\Delta\bar{G}$ depends on how we write the equation?

Reply Yes. For the reaction

$$2C_2H_6 + 7O_2 = 4CO_2 + 6H_2O$$

$$\Delta\bar{G} = 4\mu_{CO_2} + 6\mu_{H_2O} - 7\mu_{O_2} - 2\mu_{C_2H_6}$$

And $\Delta\bar{G}$ may also change in value as the reaction progresses.

Query What are the properties of $\Delta\bar{G}$?

Reply The value of $w_{♀}$ (*minimum*) for a process involving an infinitesimal amount of transfer or transformation of chemical substances is

$$w_{♀}(minimum) = \Delta\bar{G}\,d\xi$$

where $d\xi$ is the amount of transfer or transformation which occurs.

There are three possible results:

(a) if $\Delta\bar{G} < 0$ the process as written is spontaneous;

(b) if $\Delta\bar{G} > 0$ the process as written is non-spontaneous;

(c) if $\Delta\bar{G} = 0$ the reaction is in equilibrium.

6.3.4 The Gibbs energy (G)

The Gibbs energy of a system is sometimes known as the "chemical potential energy". Different texts define it in different ways. For our purposes, the most useful definition is in terms of the amounts of chemical substances present and their chemical potentials.

For any phase (α), we can evaluate the product of the chemical potential of each substance and the amount of that substance present in the phase ($\mu_{B,\alpha}n_{B,\alpha}$). The sum of these products is equal to the Gibbs energy of the phase.

$$G(phase\ \alpha) = \sum_{\substack{\text{all substances} \\ \text{present (B)}}} \mu_{B,\alpha}\,n_{B\alpha}$$

The Gibbs energy of the whole system is the sum of the values for each of the phases.

$$G(system) = \sum_{\substack{\text{all phases} \\ \alpha}} G(\alpha)$$

That is, $G\,(system) \equiv$ sum for all phases (α) and all substances (B) in each phase of the values of $\mu_{B,\alpha} n_{B,\alpha}$

$$\equiv \sum_{\alpha} \sum_{B} \mu_{B,\alpha} n_{B,\alpha}$$

Example 6.2
(a) What is the value of the Gibbs energy of a gaseous mixture containing 0.02 mol of oxygen, 0.1 mol of carbon dioxide and 0.01 mol of carbon monoxide, under conditions such that $\mu_{O_2} = 0$, $\mu_{CO_2} = -250$ kJ mol^{-1} and $\mu_{CO} = -250$ kJ mol^{-1}.

Answer

$$G = n_{O_2}\,\mu_{O_2} + n_{CO}\,\mu_{CO} + n_{CO_2}\,\mu_{CO_2}$$
$$= 0.02 \times 0 + 0.01 \times (-250 \times 10^3) + 0.1 \times (-250 \times 10^3)\,\text{J}$$
$$= -27.5 \times 10^3\ \text{J}.$$

PROBLEM

6.2. What is the value of the Gibbs energy of a system containing 36 g (2 mol) of ice in equilibrium with a solution of 0.3 mol of sodium sulphate in 270 g (15 mol) of water? Under these conditions

$$\mu_{H_2O}\,(ice) = -241\ \text{kJ mol}^{-1}$$
$$\mu_{H_2O}\,(solution) = -241\ \text{kJ mol}^{-1}$$
$$\mu_{Na_2SO_4}\,(solution) = -1290\ \text{kJ mol}^{-1}$$

Answer $-$4484 kJ

6.3.5 The change in Gibbs energy (ΔG).

Any reaction or process where there is a change in either the amount or the chemical potential of any of the substances present results in a change in the value of the Gibbs energy.

$$\Delta G = G\,(\text{after reaction}) - G\,(\text{before reaction})$$

Example 6.3
Calculate the change in the Gibbs energy when 1 dm^3 of a solution of sodium chloride (1 mol dm^{-3}) is evaporated to a concentration of (2 mol dm^{-3}) and the water collected and condensed. The conditions are such that

$$\mu_{NaCl}\,(1\ \text{mol dm}^{-3}) = -395.0\ \text{kJ mol}^{-1}$$
$$\mu_{H_2O}\,(1\ \text{mol dm}^{-3}\,\text{NaCl}) = -237.1\ \text{kJ mol}^{-1}$$
$$\mu_{NaCl}\,(2\ \text{mol dm}^{-3}) = -391.6\ \text{kJ mol}^{-1}$$
$$\mu_{H_2O}\,(2\ \text{mol dm}^{-3}\ \text{NaCl}) = -237.2\ \text{kJ mol}^{-1}$$
$$\mu_{H_2O}\,(pure) = -237.0\ \text{kJ mol}^{-1}$$

Answer
G(after reaction)

$$= n_{NaCl}\mu_{NaCl} + n_{H_2O}(solution)\mu_{H_2O}(solution) + n_{H_2O}(pure)\mu_{H_2O}(pure)$$
$$= 1(-391.6 \times 10^3) + 27.8(-237.2 \times 10^3) + 27.8(-237.0 \times 10^3)\ \text{J}$$

G(before reaction)

$$= n_{NaCl}\mu_{NaCl} + n_{H_2O}(solution)\mu_{H_2O}(solution)$$
$$= 1(-395.0 \times 10^3) + 55.6(-237.1 \times 10^3)\ \text{J}$$

$$\Delta G = G\,(\text{after}) - G\,(\text{before})$$
$$= 3.4 \times 10^3\ \text{J}$$

Query Is there any relationship between ΔG and $\Delta \bar{G}$?

Reply In general, no. $\Delta \bar{G}$ is evaluated for a particular mixture. In most reactions the value of $\Delta \bar{G}$ changes as the reaction proceeds. ΔG on the other hand depends only on the differences in the properties before and after reaction. There is one type of process for which they are related. When the value of $\Delta \bar{G}$ remains constant throughout the reaction then, and only then, we can write

$$\Delta G = \Delta \bar{G} \Delta \xi$$

where $\Delta \xi$ is the amount of reaction which has taken place. An example is the extraction of magnesium and chlorine from sea water. The process is

$$Mg^{2+}(sea) + 2Cl^-(sea) = Mg(s) + Cl_2(l)$$

The chemical potentials of $Mg(s)$ and $Cl_2(l)$ remain constant throughout and provided that only a small proportion of the magnesium in the sea is extracted, $\mu_{Mg^{2+}}$ and μ_{Cl^-} will also remain constant. (In practice this requires a large throughput of sea water.)

That is, $\Delta \bar{G}$ will be constant and

$$\Delta G = \Delta \bar{G} \Delta \xi$$

In practice $\Delta \bar{G} = 764$ kJ mol^{-1} so that ΔG to produce 1 tonne (4.11×10^4 mol) of magnesium ($\Delta \xi = 4.11 \times 10^4$ mol) is given by

$$\Delta G = 764 \times 10^3 \times 4.11 \times 10^4$$
$$= 31.5 \times 10^9 \text{ J}$$
$$= (8.7 \times 10^3 \text{ kW hr})$$

6.3.6 Relationship between ΔG and $w_{♀}$ *(minimum)*

ΔG is the contribution to $w_{♀}$*(minimum)* due to the change in chemical composition. Other factors, such as expansion of the system or heat transfer, may also contribute to $w_{♀}$*(minimum)*. Hence, in general there is no simple relation between ΔG for the change and $w_{♀}$*(minimum)*. If, however, a reaction is carried out so that before the change and after the change is completed, the temperature and pressure of the system are equal to the temperature and pressure of the environment, then ΔG is the only contribution to $w_{♀}$ *(minimum)*. For this type of change $w_{♀}$*(minimum)* $= \Delta G$.

Query Does this mean that the reaction would have to be carried out at constant temperature and pressure?

Reply No. Reactions carried out at constant temperature and pressure clearly satisfy the requirements so that $w_{♀}$ *(minimum)* $= \Delta G$. However, it is only the temperatures and pressures *before* and *after* reaction which matter. The temperatures and pressures during reaction can have any value.

Query This seems complicated. What use is it?

Reply In most industrial processes the raw materials and the products are at the temperature and pressure of the environment. Hence for the process

$$w_{♀}(minimum) = \Delta G$$

and the efficiency of the process may be measured by quantities such as

$$\Delta G \,/\, w_{♀}$$

6.3.7 Range of application

The relationships defining $\Delta \bar{G}$, G and ΔG are completely general. In treatments in textbooks these relationships often emerge after arguments which include restriction to constant temperature and pressure. The only necessary restriction is that placed on the relationship between $w_{♀}$ *(minimum)* and ΔG.

6.4 Chemical Potential and Activity

6.4.1 Absolute activity

We have developed two measures of the possibility, or "tendency for advancement", of chemical change: namely the chemical potential, and the absolute activity. So they must be related to one another.

As an example, take the process:

$$2NO(g) + O_2(g) = 2NO_2(g)$$

The two criteria for determining the direction in which change occurs are

$$\lambda_{NO_2}^{2}\,\lambda_{O_2}^{-1}\,\lambda_{NO}^{-2} \begin{cases} < 1 & \text{forward reaction} \\ > 1 & \text{backward reaction} \\ = 1 & \text{equilibrium} \end{cases}$$

and

$$2\mu_{NO_2} - \mu_{O_2} - 2\mu_{NO} \begin{cases} < 0 & \text{forward reaction} \\ > 0 & \text{backward reaction} \\ = 0 & \text{equilibrium} \end{cases}$$

The nature of the relationship can be seen from the facts that the logarithm of unity $= 0$, and that

$$\ln(AB) = \ln A + \ln B$$

So the required relation is of the form:

$$\mu \text{ proportional to } \ln \lambda$$

The proportionality constant is usually obtained by examining the properties of gases at low pressures. The proportionality constant turns out to be the product of the gas constant (R) and the absolute temperature (T). Hence the relation between the chemical potential and the absolute activity is:

$$\mu \equiv RT \ln \lambda$$

6.5 The Evaluation of Chemical Potentials

In the following sections we will be concerned with the evaluation of the chemical potential of a substance. At each stage the relationships between the potentials and activities will be pointed out.

6.5.1 Zero of the chemical potential scale

With gravitational and electrical potentials we have to choose reference states of zero potential in order to give a value to the potential. The choices are often sea level and earth. Thus gravitational potential $= gh$, where $h =$ altitude above sea level, and electrical potential $= \Phi$, relative to earth potential. A convention is also needed to fix the zero of the scale of chemical potential. We observe that any chemical substance can, in principle, be synthesised from the appropriate elementary substances. Thus the state of "zero" potential for a chemical substance at any temperature can conveniently be taken as the elementary substances from which it could be formed. We still need a specification for the state of the elementary substances, since this is the point of reference. The choice made is that each elementary substance shall be pure, in its stable form at standard atmospheric pressure ($P^\circ = 101.3$ kPa), and at the experimental temperature. The convention for elements is thus

μ (elementary substance, pure, stable form, standard atmospheric pressure, any given T)
$\equiv \mathring{\mu}_{\text{ref}}$ (element, any T)

$\equiv$ zero

6.5.2 The chemical potential of a substance

The potential of any substance is given a value equal to the potential for its formation from its elements in their reference states at the same temperature. That is,

$\mu_{\text{B}}(\beta) = \Delta \bar{G}$(formation of substance B in phase β from its elements in their reference states)

These conventions agree with the identity

$$\mu \equiv RT \ln \lambda$$

since they assign to the same state of the elements

$$\mathring{\mu}_{\text{ref}}(\text{element}) \equiv 0$$

and $$\mathring{\lambda}_{\text{ref}}(\text{element}) \equiv 1$$

Example 6.4

(a) What are the stable forms of the following elements at atmospheric pressure and 298 K?
(i) hydrogen;
(ii) carbon;
(iii) sodium;
(iv) chlorine;
(v) bromine;
(vi) iodine.

Answer

(i) gaseous;
(ii) solid, graphite;
(iii) solid, metallic;
(iv) gaseous;
(v) liquid;
(vi) solid.

(b) What can we say about the potentials of carbon (diamond) and bromine (solid) at P° and 298 K?

Answer

They must both be higher than zero, because these are unstable forms of the elements under these conditions.

6.5.3 Standard chemical potential

Under the conditions in which they are used, the chemical potentials of substances depend on the *type of phase* in which the substance is present, on the local *pressure*, the local *temperature* and on the local *concentration* of the substance. Occasionally the electrical potential and/or the gravitational potential will also be important.

As with absolute activities, it is convenient to be able to relate the chemical potential of a substance in a reacting mixture to other properties (particularly those which specify composition) which are easily determined, easily controlled or which we wish to know about. In analysing absolute activities we dealt with this problem by picking a standard or reference state which was convenient to use with a particular concentration scale. We then determined the activity of the substance for the experimental conditions relative to the value for the reference conditions. We do the same sort of thing with chemical potentials.

The value of the chemical potential of the substance in its standard state is called the *standard chemical potential* of the substance and is given the symbol μ°, that is,

$\mu_B^\ominus(\beta), T \equiv \mu$(substance B, in phase type β, under the standard conditions at standard pressure and given T)

$\mu_B^\ominus(\beta), T \equiv \Delta G^\circ$(formation of B($\beta$) in its standard state from the elements in their reference states)

Query What are the standard states for which $\mu_B^\ominus$ is the chemical potential?

Reply The same standard states as those used for defining the standard absolute activity (Section 4.4), namely:

nature of substance	*standard state*
component of a gaseous mixture	gas (pure, standard pressure, P°, temperature, T)
pure solid	solid (pure, standard pressure, P°, temperature, T)
pure liquid	liquid (pure, standard pressure, P°, temperature, T)
solvent in solution	solvent (pure, standard pressure, P°, temperature, T)
solute in solution	the state which would exist at the standard concentration and standard pressure if the solute behaved "ideally" (that is, according to some model) at all concentrations up to the standard value. Usually this state exists only in the imagination.

Query What is the relationship between the standard chemical potential ($\mu_B^\ominus$) and the standard absolute activity ($\lambda_B^\ominus$) of a substance?

Reply In a phase β and at a temperature T

$$\mu_B^\ominus(\beta, T) \equiv RT \ln \lambda^\circ(\beta, T)$$

For a pure substance, the Gibbs energy is given by $G = n\mu$ and so $\mu = G/n \equiv \bar{G}$, the molar Gibbs energy of the pure substance. So, using the same symbols the standard chemical potential is the standard molar Gibbs energy $\bar{G}^\circ$. For a substance B in phase β and at temperature T, we therefore have $\mu_B^\ominus(\beta, T) \equiv \bar{G}_B^\ominus(\beta, T)$.

Tables of standard chemical potentials are available in many books. The symbols they use are most often

$$G^\circ \text{ or } \bar{G}^\circ \text{ or } \Delta G_f^\ominus \text{ or } \Delta \bar{G}_f^\ominus$$

In older books the symbol F is used for the Gibbs energy (free energy), so similar tables will use

$$F^\circ \text{ or } F^\circ \text{ or } \Delta \bar{F}_f^\ominus \text{ or } \Delta F_f^\ominus$$

These are all equivalent.

Example 6.5

Use Table 6.1 to find the standard chemical potential of solid benzoic acid (C_6H_5COOH) at 298 K.

Table 6.1

Some Standard Enthalpies, $\Delta H_f^\ominus$, and Gibbs Energies, $\Delta G_f^\ominus$, of Formation at 298.15 K.

Substance	$\Delta \bar{H}_f^\ominus$/kJ mol^{-1}	$\Delta \bar{G}_f^\ominus$/kJ mol^{-1}
Al^{3+} (*aq*)	− 531	− 485
Al_2O_3 (*s*)	−1676	−1582
$Al(OH)_4^-$ (*aq*)	−1490	−1297
Br^- (*aq*)	− 121	− 104
Cl^- (*aq*)	− 167	− 131
C (graphite)	0.00	0.00
C (diamond)	1.88	2.89
CH_4 (*g*)	− 75	− 51
C_2H_6 (*g*)	− 85	− 33
C_2H_4 (*g*)	52	68
C_2H_2 (*g*)	227	209
C_6H_6 (*g*)	83	130
C_6H_{14} (*g*)	− 167	− 0.3
C_8H_8 (*g*) (styrene)	147	214
C_8H_8 (*s*) (styrene)	104	202
$C_{11}H_{24}$ (*g*)	− 270	43
$C_{11}H_{22}$ (*g*)	− 145	130
C_6H_5COOH (*s*)	− 385.1	− 245.2
CO (*g*)	− 110	− 137
CO_2 (*g*)	− 394	− 394
Ca^{2+} (*aq*)	− 543	− 553
CaF_2 (*s*)	−1215	−1162
$CaSO_4 \cdot \frac{1}{2}H_2O$ (α,*s*)	−1575	−1435
$CaSO_4 \cdot 2H_2O$ (*s*)	−2021	−1796
F^- (*aq*)	− 333	− 279
Fe(II) (*aq*)	− 88	− 85
Fe(III) (*aq*)	− 48	− 10
H_2 (*g*)	0.00	0.00
HBr (*g*)	− 36	− 53
HCl (*g*)	− 92	− 95
HI (*g*)	26	2
H_2O (*g*)	− 242	− 229
H_2O (*l*)	− 285	− 237
H^+ (*aq*)	0.00	0.00
H_2O_2 (*aq*)	− 188	− 120
I^- (*aq*)	− 56	− 52
NH_3 (*g*)	− 46	− 16
N_2H_4 (*l*)		140?
NO (*g*)	90	87
NO_2 (*g*)	33	51
N_2O_4 (*g*)	10	98
N_2O_5 (*g*)	11	115
$^+NH_3CH_2CO_2^-$ (*s*)	− 528	− 369
O_3 (*g*)	143	163
OH^- (*aq*)	− 230	− 157

Answer

$$\mu^{\circ}_{\text{benzoic acid}}(s,298\text{ K}) \equiv \Delta G^{\circ}_f(s,298\text{ K})$$
$$= -245.2\text{ kJ mol}^{-1}$$

Exercise 6.3

(a) Use Table 6.1 to find the standard chemical potential of gaseous hydrogen chloride.

(b) From the tables in a collection of data such as Aylward, G. and Findlay, T., *S.I. Chemical Data*, Wiley, Sydney, 1971 or Chemical Rubber Co., *Handbook of Chemistry and Physics*, find the values of the standard chemical potentials at 298 K of
   (i) gaseous methylamine;
   (ii) gaseous uranium hexafluoride;
   (iii) solid copper (I) oxide;
   (iv) aqueous potassium thiosulphate

$$K^+(aq)_2\, S_2O_3^{2-}(aq)$$

6.5.4 Relative activity and chemical potentials

In most cases the value of the chemical potential of a substance (μ_B) will be different from its standard value (μ°_B). The difference may be called "the relative chemical potential", $\mu_B(rel)$, defined as

$$\mu_B(rel) \equiv \mu_B(\textit{particular conditions}) - \mu_B(\text{reference conditions})$$

that is, $\mu_B \equiv \mu^{\circ}_B + \mu_B(rel)$

The more usual practice is to write $\mu_B(rel)$ in terms of the relative activity. This follows directly from the equations already written

$$\mu_B \equiv RT\ln\lambda_B$$
$$\mu^{\circ}_B \equiv RT\ln\lambda^{\circ}_B$$
$$\therefore \quad \mu_B - \mu^{\circ}_B \equiv RT\ln(\lambda_B/\lambda^{\circ}_B)$$
$$= RT\ln a_B$$

that is, $\mu_B \equiv \mu^{\circ}_B + RT\ln a_B$

Example 6.6

Estimate the chemical potential of $Cd^{2+}(aq)$ in a 0.1 mol dm^{-3} solution of $CdCl_2$ in water at 300 K and 100 kPa.

$$(y_{Cd^{2+}} = 0.2,\ \mu^{\circ}_{Cd^{2+}}(aq) = -78\text{ kJ mol}^{-1})$$
$$\mu_{Cd^{2+}} = \mu^{\circ}_{Cd^{2+}} + RT\ln a_{Cd^{2+}}$$
$$= -78\times10^3\text{ J mol}^{-1} + 8.314\text{ J mol}^{-1}\text{ K}^{-1}\times 300\text{ K} \times \ln(0.1\times0.2)$$
$$= -78\times10^3\text{ J mol}^{-1} + 2.49\times10^3\text{ J mol}^{-1}\times(-3.912)$$
$$= -88\times10^3\text{ J mol}^{-1}$$

Query Is there any other way of estimating numerical values of chemical potentials other than by using the "standard rules" for the estimation of relative activities?

Reply Yes, there is another way which we will explore in the next section.

6.6 Measurement of Chemical Potential

There are three main ways of approaching the measurement of the way chemical potential is affected by experimental parameters such as the pressure and the chemical composition.

The first method depends on comparison of states for which the substance, A, is at equilibrium. So, if there is a process

$$A(\text{state }\alpha) = A(\text{state }\beta)$$

it will reach equilibrium when

$$\mu_A(\text{state }\alpha) = \mu_A(\text{state }\beta)$$

Examination of the actual experimental behaviour of a large number of such processes has shown that there are general patterns of behaviour—and these are summarised in the "standard rules" for the determination of relative activities (see Section 4.4) and hence of chemical potentials.

The second method simulates the experimental system by a suitable "molecular model". Calculation of the properties of such models lead to the "standard rules" for the determination of relative activities

(a) for gases when the interactions between molecules can be neglected (see Section 4.4.5) and

(b) for the components of solutions when the interactions between solute molecules can be neglected (see Section 4.4.5).

Modification of the model used to represent ionic solutions to include electrostatic interactions between ions leads to calculated activity coefficients of sufficient accuracy for use for estimating the properties of dilute solutions (see Section 4.4.6).

The third procedure is based on the relationship between the chemical potentials of substances and the work required to carry out processes. We will discuss this method in more detail.

Consider a simple transfer of a substance, A, from region (1) to region (2) (Fig. 6.12).

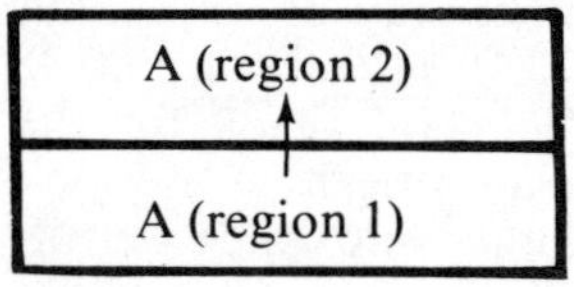

Fig. 6.12 Simple transfer of a substance.

The potential for the transfer of A (the difference in the chemical potentials) is given by

$$\text{potential for transfer} = \frac{w_{♀}(\textit{minimum}) \text{ for the transfer of an infinitesimal amount}}{\text{amount transferred}}$$

or

$$\mu_A(\text{region 2}) - \mu_A(\text{region 1}) \equiv \frac{w_{♀}(\textit{minimum}) \text{ for transfer of } dn_A \text{mol A from (1) to (2)}}{dn_A}$$

For many types of situation, the right hand side of this equation can be evaluated exactly, and hence the differences in the chemical potential can be determined even though the two regions are not in equilibrium. In the following discussion we will examine the factors which influence the transfer of a substance from one region to another. Hence we can determine the variation of the chemical potential of a substance with varying experimental conditions.

The main factors which cause substances to transfer between regions are:

(a) a difference in pressure;
(b) a difference in concentration;
(c) a difference in electrical potential (if the substance is charged); and
(d) a difference in gravitational potential (if the height differs).

So the factors which contribute to the value of the chemical potential of a substance in a particular region (apart from the nature of the substance itself) are the pressure, the concentration of the substance, the electrical potential and the gravitational potential. The chemical potential of a substance, A, can be considered as

$\mu_A \equiv$ (contribution to μ due to the nature of the substance and the phase β in which it is present) + (contribution to μ due to the pressure and the concentration) + (contribution to μ due to the electrical potential) + (contribution to μ due to the gravitational potential)

$$= \mu^{\circ}_{A,\beta,T} + RT\ln a + \mu_{\text{electrical}} + \mu_{\text{gravitational}}$$

We will examine each of these effects separately.

6.6.1 The effect of gravitational potential (gh)

The minimum work which must be done by an operator ($w_{♀}$,*minimum*) in order to transfer an amount of *matter* (dn mol) through a small change in height is given by

$$w_{♀}(\textit{minimum,grav}) = (gh_2 - gh_1)m$$

where m is the mass transferred.

If the molar mass of the material is $\bar{M}$ then the mass of the material transferred $m = \bar{M}dn$, that is

$$w_{♀}(\textit{minimum,grav}) = (gh_2 - gh_1)\bar{M}dn$$

Now

$$\mu(\text{region 2}) - \mu(\text{region 1}) = \frac{w_{♀}(\textit{minimum})}{dn}$$

that is

$$\mu_{\text{grav}}(\text{region 2}) - \mu_{\text{grav}}(\text{region 1}) = (gh_2 - gh_1)\bar{M}$$

6.6.2 The effect of electrical potential (Φ)

This time we require the minimum work which must be done by an operator in order to transfer an amount of charge (dn mol) from one region of electrical potential, Φ_1, to another region of different electrical potential, Φ_2. That is,

$$w_{♀}(\textit{minimum,elec}) = (\Phi_2 - \Phi_1)dQ$$

where dQ is the amount of charge transferred.

If the charge per mole of the substance is $\bar{Q}$ then $\bar{Q} = zF$, where z is the ionic charge number and F is the charge of 1 mol of protons, so $dQ = \bar{Q}dn = zFdn$. That is,

$$w_{♀}(\textit{minimum,elec}) = (\Phi_2 - \Phi_1)\bar{Q}dn$$

now

$$\mu_2(\text{region 2}) - \mu(\text{region 1}) = \frac{w_{♀}(\textit{minimum})}{dn}$$

that is,

$$\mu_{\text{elec}}(\text{region 2}) - \mu_{\text{elec}}(\text{region 1}) = (\Phi_2 - \Phi_1)\bar{Q} = (\Phi_2 - \Phi_1)zF$$

Example 6.7
What is the molar charge of an SO_4^{2-} ion?

Answer

$\bar{Q} = zF$, $z_{SO_4^{2-}} = -2$, $F = 96\,487$ C mol^{-1}

$\bar{Q}_{SO_4} = -192\,974$ C mol^{-1}

Exercise 6.4
Write down the values of z for
(a) electron;
(b) potassium ion;
(c) proton;
(d) methane molecule;
(e) phosphate ion (PO_4^{3-})

Answer
(a) −1;
(b) +1;
(c) +1;
(d) 0;
(e) −3.

6.6.3 The effect of pressure

As in the two previous examples, we need to determine the minimum work which must be done by an operator in order to transfer an amount of a substance (dn mol) from one region to another region at a different pressure. Figure 6.13 shows the sort of transfer we are considering. In its operation the source (1) is allowed to push a *small* amount of the substance (δn mol) volume, V, into the engine (3). The operator then compresses this fluid from pressure P_1 to pressure P_2, opens the valves and pushes the compressed gas into (2). The sequence of values of the volume and pressure of the gas in the engine are shown in Fig. 6.14. The minimum operator work involved in each step is shown in Fig. 6.15. The value of $w_{♀}$ *(minimum)* for the transfer is given by the area enclosed by the heavy lines.

The relation between the areas in Fig. 6.15 and $w_{♀}$ *(minimum)* can be seen by looking at each stage in turn. Consider the compression stage. The force opposing compression is that due to the pressure of the gas while the forces causing compression are due to the pressure of the environment plus the force applied by the operator. The minimum force which must be applied by the operator is given by

$$F_{♀}(minimum) = F_{gas} - F_e$$
$$= (P - P_e)\,A$$

where A is the area of the piston and P_e is the pressure exerted by the surrounding atmospheric environment.

If, as a result, the piston moves a distance δx, then

$$w_{♀}(minimum) = F_{♀}(minimum)\,\delta x$$
$$= (P - P_e)\,A\delta x$$

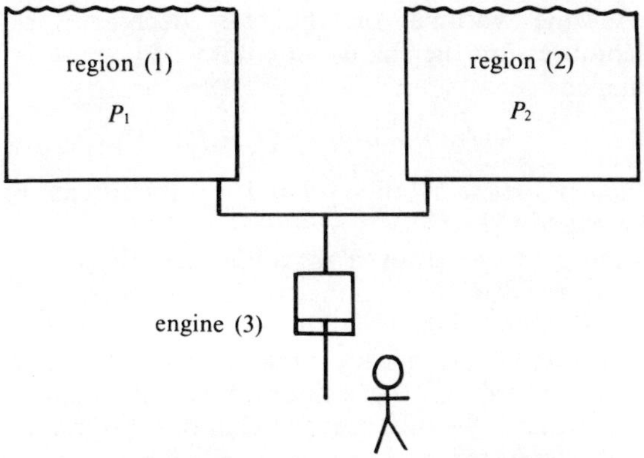

Fig. 6.13 Engine (pump) for transfer of fluid.

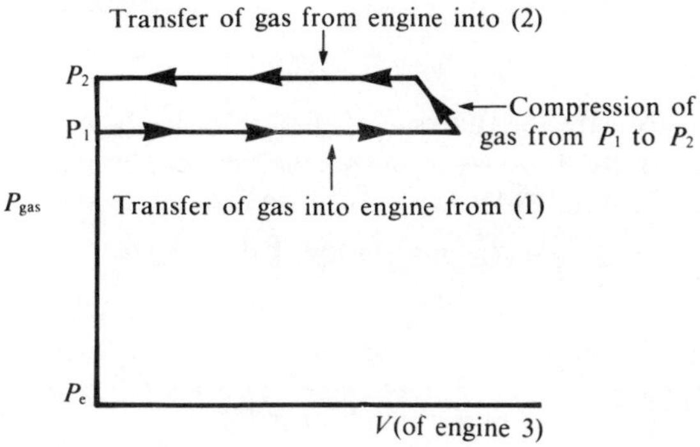

Fig. 6.14 Stages in transfer of fluid.

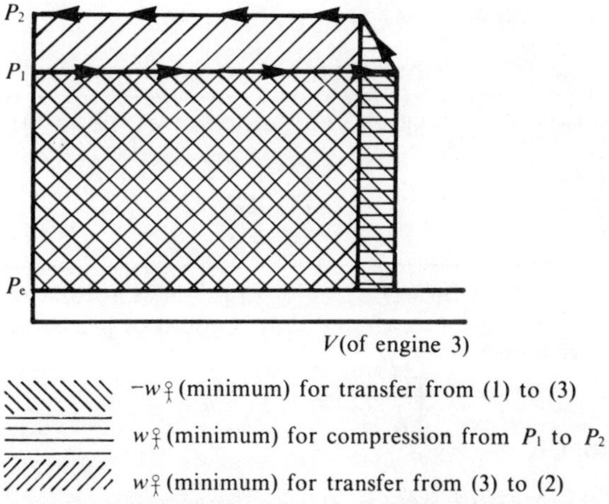

Fig. 6.15 Work in transfer of a fluid.

As the volume of the gas decreases on compression the change of volume, $\delta V = -A\,\delta x$, hence

$$w_{♀}(minimum) = -(P - P_e)\,\delta V$$

Query How is this related to the areas in Fig. 6.15?

Reply If we draw a rectangle, length $(P - P_e)$ and width δV, then the area of the rectangle is $(P - P_e)\delta V$ which is the value of $w_{♀}$ *(minimum)*. Therefore $w_{♀}$ *(minimum)* is the area of the graph of the values of $(P - P_e)$ against volume. The difference in chemical potentials between the two regions is given by

$$\mu(\text{region 2}) - \mu(\text{region 1}) = \frac{w_{♀}(minimum)}{dn} = \frac{\text{area enclosed by the heavy lines}}{dn}$$

A careful analysis of Fig. 6.15 leads to the result that $w_{♀}$ *(minimum)*, the area enclosed by the heavy lines, is given by

$$w_{♀}(minimum) = \int_{P_1}^{P_2} V_{\text{engine}}\,dP = \int_{P_1}^{P_2} \delta n\,\bar{V}\,dP$$

and that, if the molar volume of the substance at any time is $\bar{V}$ then, for a transfer of δn mol,

$$w_{♀}(minimum) = \delta n \int_{P_1}^{P_2} \bar{V}\,dP$$

That is

$$d\mu_B/dP \equiv \bar{V}_B$$

at fixed T and composition, or, in general,

$$\mu(\text{pressure (region 2)}) - \mu(\text{pressure (region 1)}) = \int_{P_1}^{P_2} \bar{V}\,dP$$

Liquids or Solids For a liquid or a solid the molar volume is almost independent of pressure and hence

$$\int_{P_1}^{P_2} \bar{V}\,dP \simeq \bar{V}(P_2 - P_1)$$

that is

$$\mu_A(\text{pressure (region 2)}) - \mu_A(\text{pressure (region-1)}) \simeq \bar{V}_A(P_2 - P_1)$$

Query How large is this quantity?

Reply Compare liquid cyclohexane (C_6H_{12}) at a pressure of 11 $P°$ with its value at a pressure $P°$.

$$\bar{V}_A \simeq 100\ \text{cm}^3\ \text{mol}^{-1} = 100 \times 10^{-6}\ \text{m}^3\ \text{mol}^{-1}$$

$$P° \simeq 10^5\ \text{Pa}$$

$\therefore$

$$\mu_{C_6H_{12}}(11\ P°) - \mu_{C_6H_{12}}(P°) \simeq 10^{-4}\ \text{m}^3\ \text{mol}^{-1} \times 10^6\ \text{Pa} = 100\ \text{J mol}^{-1}$$

Query What effect would pressure have on the relative activity of cyclohexane at 11 $P°$ and 300 K?

Reply

$$a = \exp\frac{\mu - \mu°}{RT} = \exp\frac{100\ \text{J}}{8 \times 300\ \text{J}} = 1.04$$

Gases. Treatment of the gases (or the components of a gas mixture) is a little less straightforward because the molar volume changes with change of pressure. The molar volume (see Section 2.7) may be written

$$\bar{V} \equiv \frac{ZRT}{P} \quad (Z \text{ is the compressibility factor})$$

and hence

$$\mu(P_2) - \mu(P_1) = \int_{P_1}^{P_2} \frac{ZRT}{P}\,dP$$

For a gas at low pressure it is quite adequate to assume that Z approaches unity. In that case the expression on the right has a general solution, that is

$$\mu(P_2) - \mu(P_1) \simeq \int_{P_1}^{P_2} \frac{RT}{P}\,dP = RT\int_{P_1}^{P_2} \frac{1}{P}\,dP = RT\ln\frac{P_2}{P_1}$$

Thus for gases at low pressures

$$\mu_A(\text{pressure}(2)) - \mu_A(\text{pressure}(1)) \simeq RT\ln\frac{P_2}{P_1}$$

Query How large is this effect?
Reply Take, for example, a gas at a pressure $10\ P^\circ$

$$\mu_A(10\ P^\circ) - \mu_A(P^\circ) \approx RT\ln \frac{10\ P^\circ}{P^\circ}$$
$$= RT\ln 10$$
$$= 8.3\ \mathrm{J\ mol^{-1}\ K^{-1}} \times 300\ \mathrm{K} \times 2.3$$
$$= 5.7\ \mathrm{kJ\ mol^{-1}}$$

This is a significant effect.
Query How much error is introduced by approximating Z by unity?
Reply Usually the error is small. Up to medium pressures,

$$Z = 1 + \frac{BP}{RT}$$ (B is the second virial coefficient of the gas)

A better approximation is therefore

$$\mu_A(P_2) - \mu_A(P_1) = \int_{P_1}^{P_2} \frac{RT}{P}\left(1 + \frac{BP}{RT}\right) \mathrm{d}P$$
$$= RT\ln \frac{P_2}{P_1} + B(P_2 - P_1)$$

For example, for cyanogen $(CN)_2$ at 400 K

$$B(10P^\circ - P^\circ) = -174.9 \times 10^{-6} \times 9 \times 10^5$$
$$= 157\ \mathrm{J}$$

which is 3% of the difference $\mu_A(10P^\circ) - \mu_A(P^\circ)$.
Query Does this mean that the chemical potential of a gaseous substance is related to its reference value by the relation

$$\mu(P) = \mu(P^\circ) + RT\ln(P/P^\circ) + B(P - P^\circ)$$

Reply No. the relation is

$$\mu(P) = \mu^\circ + RT\ln(P/P^\circ) + BP$$

Query Why?
Reply In Section 6.5.3 we said that for a gaseous substance μ° is the chemical potential of the pure gas at the reference pressure P°. If $P^\circ = 101$ kPa then this statement is sufficiently accurate for most applications. However, μ° is taken to be the value that the chemical potential of the pure gas would be at P° *if* the gas behaved ideally.

$$\mu(P^\circ) \approx \mu^\circ + BP^\circ$$

Query Does this mean that for a pure gas at P°, μ is not equal to μ°?
Reply Yes. $\mu \simeq \mu^\circ + BP^\circ$ and the relative activity of the pure gas at P° is not equal to unity but is given by the expression

$$a(P^\circ) = \exp(BP^\circ/RT)$$

This value is rarely smaller than 0.98 and rarely larger than 1.02.

Example 6.8
Consider the changes in chemical potential for a pure substance over a region close to the phase change from liquid to gas. The changes in molar volume, $\bar{V}$, with changing pressure, and at some constant temperature may be graphed as in Fig. 6.16. In consequence there are changes in chemical potential, as depicted in Fig. 6.17. This type of graph shows very clearly the range of pressure for which the liquid is the stable phase, and the range for which the gas is the stable phase. The pressure at which $\mu(liquid) = \mu(gas)$ is the pressure at which the two phases are in equilibrium (that is, the vapor pressure).

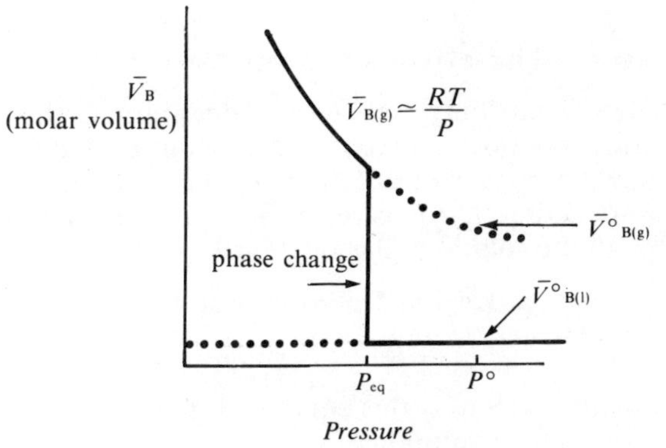

Fig. 6.16 Molar volumes and pressures in the vicinity of a change of phase.

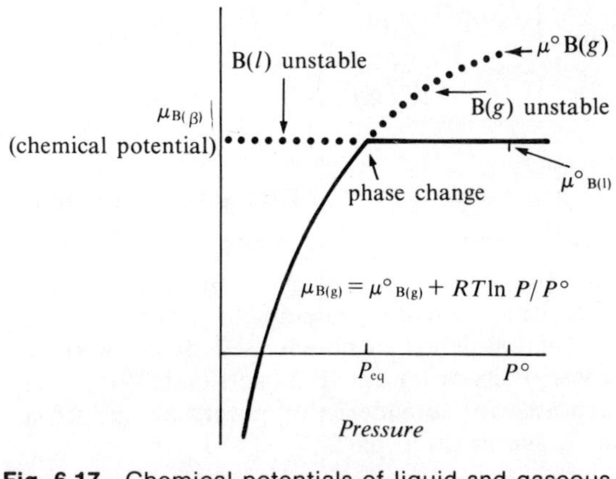

Fig. 6.17 Chemical potentials of liquid and gaseous forms of a substance in the vicinity of a phase change.

Query Do these results apply equally to a component of a gaseous mixture or of a solution?

Reply Yes.

Query In that case, what is meant by the molar volume of a component of a solution (for example, of urea in water)?

Reply The quantity required is the partial molar volume of the substance in the solution. This is determined by the effect on the volume of a solution of the addition of a small amount of the substance.

$$\bar{V}_A = \frac{\text{change in volume on adding an infinitesimal amount } (dn) \text{ of A}}{\text{amount } (dn) \text{ of A added}}$$

For dilute solutions and gases at low pressures

$$\bar{V}_A \text{ (solvent in solution)} \simeq \bar{V}_A^\circ \text{ (pure solvent)}$$

$$\bar{V}_B \text{ (solute in solution)} \simeq \bar{V}_B^\circ \text{ (in solvent)}$$

$$\bar{V}_A \text{ (gas at low pressure)} \simeq \frac{RT}{P}$$

6.6.4 The effect of composition

There is no "universal" way of determining the amount of work required to transfer a substance between regions of different chemical composition. However, for gases at low pressures and for dilute solutions the relationship is:

$$\mu_A \text{ (region 2)} - \mu_A \text{ (region 1)} = RT\ln\frac{x_A(2)}{x_A(1)}$$

Comment Surely this equation applies only to very dilute solutions?

Reply Yes, for more concentrated solutions we need to introduce the idea of activity coefficient, in one of the forms

$$\mu(\text{region 2}) - \mu(\text{region 1}) = RT\ln\frac{(fx)_2}{(fx)_1}$$

$$= RT\ln\frac{(yc)_2}{(yc)_1} = RT\ln\frac{(\gamma m)_2}{(\gamma m)_1}$$

6.6.5 Summary of effects on chemical potential

The possibility of transfer of a chemical substance from one region to another in a chemical system can be assessed by comparing values of its potential. It is sufficient for most purposes to consider the potential, μ, of a substance in three parts:

$$\mu = \mu(\text{chemical}) + \mu(\text{electrical}) + \mu(\text{gravitational})$$

$$= \mu^\circ + RT\ln a + \bar{Q}(\Phi - \Phi^\circ) + \bar{M}(gh - gh^\circ)$$

Movement of the substance is possible from regions or forms of high potential to those of lower potential. *Equilibrium* exists *when the potential, μ, has the same value wherever the substance is present in the system.*

The *chemical part* of the potential may be split up into

$$\mu(\text{chemical}) = \mu(\text{standard } P, x) + \mu(\text{relative } P, x)$$

$$= \mu^\circ + RT\ln a$$

where μ° is contained in tables under the heading ΔG_f° for a given temperature for the substance B in a phase type (β). The relative chemical potential, $RT\ln a$, is usefully expressed in the form

$$RT\ln a = RT\ln fxP/P^\circ = RT\ln fp/P^\circ$$

for gaseous substances, and

$$RT\ln a = RT\ln fx$$

for substances in liquid or solid phases, where the activity coefficient has contributions from both pressure and composition. For substances in liquid or solid phases, for example,

$$RT\ln f_{\text{pressure}} = (P - P^\circ)\bar{V}^\circ$$

The *electrical part* of the potential comes in when ionic substances are involved.

$$\mu(\text{electrical}) = \bar{Q}(\Phi - \Phi^\circ)$$

where the ionic substance B^{z+} has molar charge $\bar{Q} = +zF$. Φ is the electric potential in the region where the ionic substance is situated.

The *gravitational term* is calculated from the molar mass $\bar{M}$ and the height, h, above a standard level, h°.

$$\mu(\text{gravitational}) = \bar{M}(gh - gh^\circ)$$

where $g = 9.81 \text{ J m}^{-1} \text{ kg}^{-1}$

(This is an approximation which needs modification when great heights are involved such that g changes significantly with altitude).

Some *other terms*, which we will not consider here, may arise in particular cases. They include effects at surfaces or at boundaries between phases, effects due to magnetic fields and the situation in a revolving centrifuge. Only the *chemical part* of the potential needs to be considered for most chemical processes. Where ionic substances and electrochemical cells are being considered we have to think of the *electrochemical potential*:

$$\mu(\text{electrochemical}) = \mu(\text{chemical}) + \mu(\text{electrical})$$
$$= \mu^\circ + RT\ln a + \bar{Q}(\Phi - \Phi^\circ)$$

The *distribution of phases and substances with height* in a container is affected by the combination of pressure and gravitational potential

$$\mu(\text{gravichemical}) = \mu(\text{chemical} + \mu(\text{gravitational})$$
$$= \mu^\circ + RT\ln a + \bar{M}(gh - gh^\circ)$$

When *chemical reactions* and *phase transfers* are *not involved* it is possible to ignore μ° in this equation.

6.7 Chemical Potentials and Chemical Equilibrium

6.7.1 Equilibrium constant

The relationship between the equilibrium constant for a chemical reaction and the chemical potentials is simply

$$RT\ln K_{eq} \equiv -\Sigma\nu_B\mu_B^\circ \equiv -\Delta\bar{G}^\circ$$

This relation follows directly from the definitions.

$$K_{eq} \equiv \Pi_B(\lambda^\circ{}_B{}^{\nu_B})^{-1}$$

$\therefore$ $$\ln K_{eq} = -\Sigma\nu_B\ln\lambda_B^\circ$$

$\therefore$ $$RT\ln K_{eq} = -\Sigma\nu_B RT\ln\lambda_B^\circ$$

As $$\mu_B^\circ \equiv RT\ln\lambda_B^\circ$$

$$RT\ln K_{eq} \equiv -\Sigma\nu_B\mu_B^\circ$$

Example 6.9
What is the value of the equilibrium constant at 298 K for the reaction

$$C_6H_5CH = CH_2(g) + H_2(g) = C_6H_5CH_2CH_3(g)$$

styrene ethyl benzene

$\Delta\bar{G}^\circ_{f,298K}$ + 213.8 + 130.5 kJ mol^{-1}

Answer

$$\Delta\bar{G}^\circ = -213.8 + 130.5 - 0 \text{ kJ mol}^{-1}$$

$\therefore$ $$RT\ln K_{eq} = 83.3 \text{ kJ mol}^{-1}$$

$$\ln K_{eq} = 33.6, \log K_{eq} = 14.59$$

$$K_{eq} = 3.9 \times 10^{14}$$

PROBLEM

6.3. What is the value of the equilibrium constant at 1000 K for the conversion of acetylene to benzene

$$3C_2H_2(g) = C_6H_6(g)$$

if, at 1000 K,

$$\Delta\bar{G}^\circ = -249.2 \text{ kJ mol}^{-1}$$

6.7.2 Possible direction of reaction

The criterion for determining the possible direction of chemical change is the sign of $\Delta\bar{G}$. That is,

$$\Delta\bar{G} > 0, \Delta\bar{G} < 0 \text{ or } \Delta\bar{G} = 0$$
$$\Delta\bar{G} = \Sigma\nu_B\mu_B = \Sigma\nu_B\mu_B^\circ + \Sigma\nu_B RT\ln a_B$$

so we can determine the direction of spontaneous change

$$\Sigma RT\ln a_B{}^{\nu_B} < -\Delta\bar{G}^\circ \quad \text{forward reaction}$$
$$\Sigma RT\ln a_B{}^{\nu_B} > -\Delta\bar{G}^\circ \quad \text{backward reaction}$$
$$\Sigma RT\ln a_B{}^{\nu_B} = -\Delta\bar{G}^\circ \quad \text{equilibrium}$$

Query Why aren't gravitational and electrical effects included when determining the possible course of a chemical reaction?

Reply In a chemical reaction both charge and mass are conserved. That is, if the equation for the reaction is properly balanced, then

$$\Sigma\nu_i\bar{Q}_i = 0$$

and $$\Sigma\nu_i\bar{M}_i = 0$$

Chemical reactions take place either at a "point" or across a surface. In either case the gravitational potential at the reaction site is uniform and so the gravitational term is zero.

$$\Sigma\nu_B\bar{M}_B gh = gh\Sigma\nu_B\bar{M} = 0$$

In a homogeneous fluid, and across most phase boundaries, the electrical potential is constant and hence the electrical term is also zero. However reaction of charged species in different phases which have different electrical potentials is common (*all* electrodes!). In these cases electrical effects are very important, and they will be considered in the next chapter.

PROBLEMS AND APPLICATIONS

6.4. *Direction of change*

(a) In what direction will the reaction $N_2 + 3H_2 = 2NH_3$ proceed if

(i) $2\mu_{NH_3} - \mu_{N_2} - 3\mu_{H_2} > 0$

(ii) $2\mu_{NH_3} - \mu_{N_2} - 3\mu_{H_2} = 0$

(b) On the sea floor on the continental shelf oil drums are observed to undergo the reaction

$$3Fe(s) + 2O_2(aq) = Fe_3O_4(s)$$

What can be said about the chemical potentials of the substances involved when the rusting process is advancing?

(c) When a spoonful of NH_4Cl crystals is poured into a glass of water, it slowly dissolves. What is the sign of $\Delta\bar{G}$?

(d) Will calcium oxide adsorb carbon dioxide from air

$$CaO(s) + CO_2(g) = CaCO_3(s)$$

if

$$\mu_{CaCO_3(s)} = -1129\ \text{kJ mol}^{-1}$$

$$\mu_{CaO(s)} = -604\ \text{kJ mol}^{-1}$$

$$\mu_{CO_2(g)} = -394\ \text{kJ mol}^{-1}$$

6.5. *Evaluation of chemical potentials*

(a) What is the value given to the standard chemical potential of gaseous chlorine at 25°C?

(b) Estimate the chemical potential of the gaseous chlorine at 25°C with $P = 8 \times 101.3\ \text{kN m}^{-2}$.

(c) (i) Use Table 6.1 to determine the standard chemical potential of water at 25°C.

The total concentration of dissolved substances in sea water is close to $1.0\ \text{mol dm}^{-3}$.

(ii) What is the mole fraction of water in the sea?

(iii) What is the approximate value of the chemical potential of water in warm (25°C) sea water?

(d) By how much would an electrical potential of 1 volt increase the electro-chemical potential of ionic calcium, Ca^{2+}?
(1 Faraday = $96\,487\ \text{C mol}^{-1}$,
1 volt = $1\ \text{JC}^{-1}$)

(e) (i) What is the chemical potential of iodine in a saturated solution of iodine in n-C_7H_{16} at 300 K?

(ii) The mole fraction of iodine in such a solution is 0.00679 at 300 K. Estimate the relative activity of iodine in the solution and the standard chemical potential at 300 K.

(f) How are the following pairs of chemical potentials related?

potential of material in (1)	*potential of material in* (2)
(i) $\mu_{air}(25°C, 10P°, h=0,)$	(i) $\mu_{air}(25°C, P°, h=0,)$
(ii) $\mu_{H_2O}(l, 25°C, P°, h=10\text{m})$	(ii) $\mu_{H_2O}(l, 25°C, P°, h=0\text{m})$
(iii) $\mu_{H_2O}(g, 110°C, P°)$	(iii) $\mu_{H_2O}(l, 110°C, P°)$
(iv) $\mu_{H_2O}(sea, P°, T)$	(iv) $\mu_{H_2O}(l, pure, P°, T)$
(v) $\mu_{O_2}(air, P°, T)$	(v) $\mu_{O_2}(\text{g, pure}, P°, T)$
(vi) $\mu_{H_2O}(l, P°, 0°C)$	(vi) $\mu_{H_2O}(s, P°, 0°C)$
(vii) $\mu_{Na^+}(aq, 1\ \text{mol dm}^{-3}, P°, T, +1\ \text{volt})$	(vii) $\mu_{Na^+}(aq, 1\ \text{mol dm}^{-3}, P°, T, 0\ \text{volt})$
(viii) $\mu_{e^-}(\text{in Cu metal}, +1\ \text{volt})$	(viii) $\mu_{e^-}(\text{in Cu metal}, +0\ \text{volt})$

6.6. *Examples using tabulated values of* $\Delta\bar{G}^\circ_f$.

(a) Determine the value of $\Delta\bar{G}^\circ$ (298K) for the reaction

$$CH_4(g) + 2O_2(g) = CO_2(g) + 2H_2O(l)$$

(b) Determine *the equilibrium constant* for the reaction

$$C(s, \text{graphite}) + O_2(g) = CO_2(g) \text{ at } 25°C$$

(c) Determine the values of $\Delta\bar{G}^\circ$ (298K) and the equilibrium constant at 298K for the reaction

$$Al^{3+}(aq) + 4OH^-(aq) = Al(OH)^-_4(aq)$$

(d) Determine the minimum electrical work required to make 1 k mol of aluminium using the electrolytic process

$$Al_2O_3(s, 298K) = 2Al\,(s, 298K) + 1.5\,O_2\,(g, \mathit{pure}, 101kPa, 298K)$$

(e) Is it likely that the processes

(i) $2NO + 2CO = N_2 + 2CO_2$

(ii) $2NO + 2O_2 = N_2 + 2O_3$

could be used for removal of nitrogen oxides from exhaust gases if a suitable catalyst could be found?

$(p_{NO} \simeq 1Pa, p_{CO} \simeq 0.1kPa,$
$p_{CO_2} \simeq 5kPa, p_{O_2} \simeq 15kPa,$
$p_{N_2} \simeq 80kPa, p_{O_3} \simeq 10^{-4}Pa)$

(f) The decomposition of hydrogen peroxide into oxygen and water

$$H_2O_2 = H_2O + \tfrac{1}{2}O_2$$

is greatly accelerated by the presence of trace quantities of iron(II) salts. One explanation might be that there are two simple fast reactions, (1) and (2),

$$H_2O_2 + 2H_3O^+ + 2Fe^{II} = 4H_2O + 2Fe^{III} \quad (1)$$

and

$$H_2O_2 + 2H_2O + 2Fe^{III} = O_2 + 2H_3O^+ + 2Fe^{II} \quad (2)$$

the first consuming Fe^{II} and the second producing Fe^{II}. Use the chemical potential data to determine if this reaction sequence is likely.

6.7. *Some other applications*

(a) The membrane of the glass electrode used with a pH-meter behaves as if it is permeable only to hydrogen ions. If the relative activities of H_3O^+ are a_1 in the test solution and a_2 inside the membrane:

(i) What is the equation for electric potential difference, $\Phi_2 - \Phi_1$, when the electro-chemical potentials μ_2 and μ_1 become equal? (This is a simple case of the Nernst equation.)

(ii) Using the equation obtained in problem (i), evaluate the approximate electric potential difference for a unit pH difference, $pH_2 - pH_1$, at 25°C ($\ln x = 2.303 \log_{10} x$).
What would be expected at 0°C?

(b) A nerve fibre consists of a cylindrical membrane which in the resting state is permeable to K^+ ions but almost impermeable to Na^+ ions.

(i) What is the condition that K^+ ions in the internal solution (the axoplasm) are in equilibrium with the K^+ ions in the external solution (the exoplasm)?

(ii) Experimental measurement indicates that the potential difference between the exoplasm and the axoplasm is 70 mV. What can be deduced about the amounts of K^+ in the two solutions?
(y_{K^+} internal $\simeq y_{K^+}$ external)

(c) Destruction of industrial cyanide wastes by chlorination proceeds by way of the reaction

$$Cl_2(g) + CN^-(aq) = ClCN(aq) + Cl^-(aq)$$

(i) Write down the condition for reaction to occur, in terms of the chemical potentials of the substances involved.

(ii) What is the value assigned to the standard chemical potential of gaseous chlorine?

(iii) The chlorine emerges from the solution at a partial pressure equal to the pressure of the atmosphere. Discuss whether the chemical potential of the chlorine on a day when the barometric pressure is 97 kPa will be greater than, less than, or equal to $\mu^\circ_{Cl_2(g)}$.

(iv) The solution rapidly becomes saturated with chlorine. What is the value of the chemical potential of the chlorine in the solution if the partial pressure of the chlorine gas is 101 kPa?

(v) $\Delta\bar{G}_f^\circ(Cl_2, aq, 298K) = -7 kJ\,mol^{-1}$. What is the relative activity of the chlorine in the solution?

(vi) Estimate the concentration of "free" chlorine in the solution.

(d) Calculate the solubility of calcium fluoride in water.
(*Hint*: write down the equation describing the solution of CaF_2 in water, calculate $\Delta\bar{G}^\circ$, then the solubility.)

(e) It has been suggested that a hydrazine (N_2H_4) fuel cell may make a suitable power source for operating an automobile.

$$N_2H_4(l) + O_2(g) = N_2(g) + 2H_2O(l)$$

Estimate the maximum amount of work available to drive an automobile from such a power source.

7.

Electrochemical Processes and Potentials

7.1 The Nature and Importance of Electrochemical Processes

The term "electrochemical" is used for processes for which it is necessary to take into account not only the amount of each chemical substance involved, but also its electric charge. There are many types of electrochemical processes. They include simple chemical reactions between charged species contained in a single, electrically neutral phase, such as the reaction

$$OH^-(aq) + H_3O^+(aq) = 2H_2O(l)$$

Electrochemical processes include the transfer of charged species from one region to another—the conduction of electricity—such as by the transfer of electrons along a wire or of ions through a solution.

Electrochemical processes find many applications. Batteries, accumulators and fuel cells harness spontaneous chemical reactions to drive hearing aids, start automobiles and drive electric motors. Non-spontaneous processes may be driven as in the commercial production of aluminium, sodium, chlorine and the electroplating of metals. Electrochemical processes are involved in corrosion and in many biological processes. They are also the basis of many methods of determining extents and rates of chemical reactions, and of determining the relative activities of the species involved (as in the use of pH meters, ion selective "electrodes", oxygen electrodes, polarographs, etc.).

Our study of electrochemical systems will be centered on electrochemical cells and on the combinations of phases which form the electrodes and ionic transfer junctions within the cell. In particular, we will be interested in the processes by which electric charge is transferred and the accompanying changes in chemical composition of the various phases. We will be interested in the potentials for these processes and the rates at which they occur. A result of particular importance will be the use of measurements of electric current, voltage, and chemical composition to determine the characteristic properties of cells, electrodes, ionic junctions and individual ionic species.

7.2 Electrochemical Systems

7.2.1 The features of a practical system

Our concern is with the information which can be gained by study of the transfer of *electric charge in association* with *electrochemical changes.* To carry out measurements it is necessary to have a complete electrical circuit.

Figure 7.1 shows the arrangement of such a circuit. It consists of an electrochemical cell with terminals R and L connected by wiring through an operational device which may be used to harness or drive the flow of current. An ammeter may be included to measure the rate of flow of charge, $I_{R \to L}$ or $I_{L \to R}$. Changes in amounts of chemical substances take place in the cell and may be measured by the methods of chemical analysis. A voltmeter may be connected between the terminals to measure the *terminal voltage*, $\Delta\mathcal{V} = \mathcal{V}_R - \mathcal{V}_L$, of the cell.

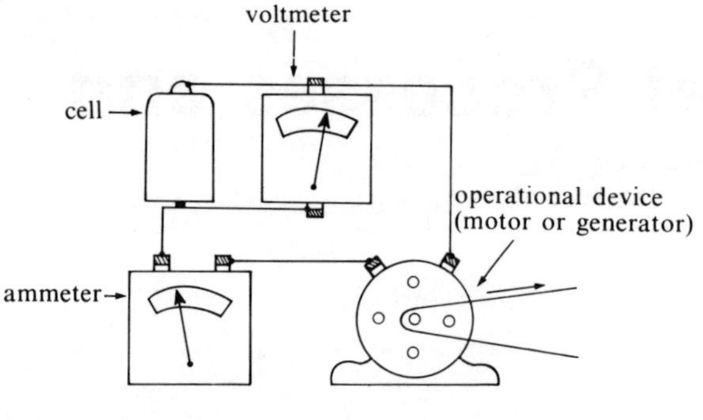

Fig. 7.1 Arrangement for electrochemical measurements.

Our objective in the immediately following sections will be to describe electrochemical cells, to find equations for the chemical changes which occur when charge passes through the wiring, and to find the potentials for these changes.

7.2.2 Electronic conductors

Metals, and a small number of other substances such as graphite, transport electric charge from one place to another within themselves almost entirely by movement of electrons. Movement of other charged species contributes negligibly. These "electronic conductors" need only very slight change in the distribution of the electrons to develop local electric charges of the amounts which are involved in ordinary chemical situations. These slight changes occur readily. As a result, the electrons in metals reach equilibrium quickly when current ceases to flow in them.

We assume that the electrons in a metal can be assigned a chemical potential, μ_{e^-}. To say that the electrons reach equilibrium, then, is the same thing as saying that the chemical potentials of the electrons are equal.

Query So two pieces of copper have equal values of the chemical potential of their electrons if current does not flow when they are touched together?

Reply Yes.

Query And also if one piece is of copper and the other of iron?

Reply Yes.

7.2.3 Terminal voltage

Various sorts of voltmeter are used to measure the terminal voltage of cells. In Fig. 7.1, this means measuring the difference in voltage between the "terminals", L and R. A voltmeter is suitable for measurements on a particular cell if the current it draws does not upset the equilibria in that cell enough to change the reading. The terminal voltage, $\Delta\mathcal{V}$, is defined as the potential for transfer of electric charge between the terminals. That is, for example,

$$\Delta\mathcal{V} = \mathcal{V}_R - \mathcal{V}_L \equiv \frac{w_{♀}(\textit{minimum})\text{, for transfer of charge } dQ}{\text{amount } dQ \text{ of charge transferred}} \quad (1)$$

The process of electron transfer can be described by the equation

$$e^-(L) = e^-(R)$$

Consequently the potential for transfer can also be described in terms of the chemical potentials of the electrons, by (2)

$$\Delta\bar{G} \equiv \mu_{e^-(R)} - \mu_{e^-(L)} \equiv \frac{w\ (\textit{minimum})\text{, for transfer of electrons}}{\text{amount } dn_{e^-} \text{ of electrons transferred}} \quad (2)$$

These two potentials describe the same phenomenon and so must be related. The charge on n_{e^-} mol of electrons is

$$Q_{e^-} = -F\,n_{e^-}$$

so

$$dQ_{e^-} = -F dn_{e^-} \quad (3)$$

and the relationship between the chemical potentials of the electrons in any phases, α, β, and the voltage, $\mathcal{V}(\beta) - \mathcal{V}(\alpha)$, is

$$\mu_{e^-(\beta)} - \mu_{e^-(\alpha)} = -F(\mathcal{V}_{(\beta)} - \mathcal{V}_{(\alpha)}) \quad (4)$$

7.2.4 Electric potentials within phases

The electric potential, Φ, within a phase was used in Section 6.6.2 as a measure of the variation of the chemical potential of a substance due to electric charges.

A convenient starting point is the definition of the standard chemical potential of the electrons in an electronic conductor or of one kind of ion in an ionic conductor. We do this in a way somewhat similar to that used for the standard state of a solvent (Section 4.4.3). That is, we choose as the standard state that condition of the phase in which the electrons in metals, or ions in ionic conductors, are exactly equivalent to the species of opposite sign in the substance, at the experimental temperature. The whole phase then bears zero charge. We call this the *condition of electroneutrality*. For an electron or ion, B, the chemical potential under any other conditions can then be regarded as having three components (equation (5)),

$$\mu_B = \mu_B^\circ + RT \ln a_B + z_B F\Phi \qquad (5)$$

arising from the standard chemical potential, μ_B°, the relative activity, a_B, and a term involving the electric potential and the charge number, z_B of the electron or ion. For electrons in metals, the charges developed are so small that there is never significant variation in electron concentration, and hence never significant change in the relative activity of electrons. Thus, for electrons in metals it is sufficient to write

$$\mu_{e^-} = \mu_{e^-}^\circ - F\Phi \qquad (6)$$

Query Will electrons in different metals have the same standard chemical potential?

Reply The difference in the metals provides a difference in environment. Theories of metallic structure strongly suggest that this alters the standard chemical potential. So metals in electronic contact tend to have the same chemical potential for their electrons, but are not in their standard states.

Two pieces of the same metal at the same temperature have the same value of $\mu_{e^-}^\circ$. In all other cases, the values of $\mu_{e^-}^\circ$ are likely to be different. We can measure terminal voltages for such situations, and thus obtain the differences in the chemical potentials of electrons. But we cannot get values for the differences in the electric potentials, Φ, unless the electronic conductor is the same and the temperature is the same. On combining equations (4) and (6) for two phases, α and β, we can deduce that

$$\Delta\mathscr{V} = \mathscr{V}_{(\beta)} - \mathscr{V}_{(\alpha)} = \Phi_{(\beta)} - \Phi_{(\alpha)} - [\mu_e^\circ{}_{(\beta)} - \mu_e^\circ{}_{(\alpha)}]/F \qquad (7)$$

The specification of Φ depends on the phase in a way that the specification of $\mathscr{V}$ does not.

Statements analogous to equation (7) can be made for the ions in ionic conductors.

7.2.5 Electromotive force

We can often change the operational device of Fig. 7.1, so that we can proceed smoothly from conditions of the device for which current flows spontaneously and its flow is harnessed, to other conditions in which the flow of current is driven in the opposite direction. Somewhere between these conditions there is a state of the equipment in which the electrochemical cell is in electrochemical equilibrium, with no flow of current. This is normally a state in which the terminal voltage, Δ , is not zero. It is called the electromotive force (e.m.f.) *E*, of the cell. Thus,

$$E_{cell} \equiv \Delta\mathscr{V}_{eq}$$

The e.m.f. of a cell is determined by the sequence and composition of its phases and not by their shapes and sizes. We will see that the e.m.f. is directly related to the potential for the overall electrochemical process occurring within the cell.

7.3 Descriptions of Cells

7.3.1 Common features of electrochemical cells

Electrochemical cells commonly have the features shown in Fig. 7.2. The interfaces within the half cells at which conduction by electrons gives way to conduction by ions are the places where chemical transformations associated with passage of charge occur. They are known as the *electrodes* of the cell. The *ionic junction phase* provides a pathway for passage of charge between the electrodes without direct contact between the electrode materials. It need not undergo change of composition when current flows. As we shall see, some cells can be arranged to have just one phase of uniform composition between the electrodes. These are particularly important in getting quantitative data.

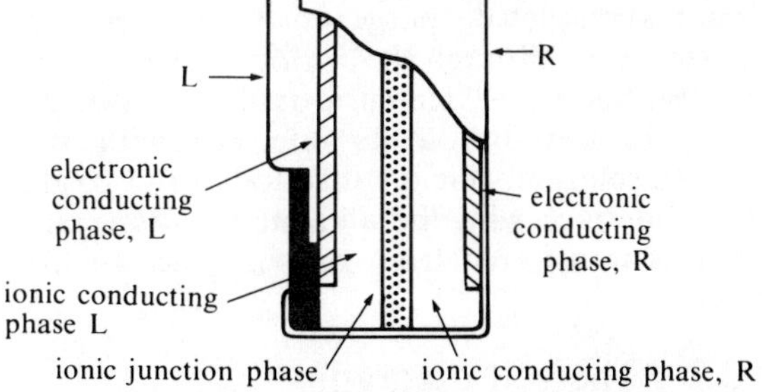

Fig. 7.2 Common features of an electrochemical cell.

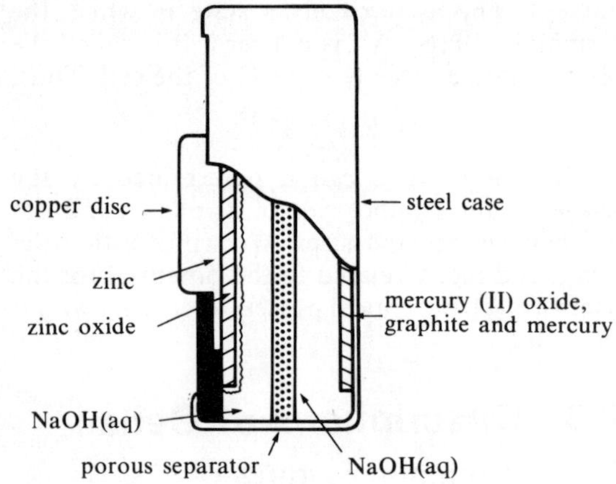

Fig. 7.3 Parts of a Mallory cell.

We shall use the Mallory cell (Fig. 7.3) to illustrate many of the principles of cells. The construction of this cell involves placing solid zinc at one end of a container and solid mercuric oxide (compressed into a pellet with graphite) at the other end. An aqueous solution of sodium hydroxide (an ionic conducting material) is inserted to make contact with both of these materials. A layer of porous fabric forms a junction which prevents contact between mercuric oxide and zinc and thus prevents the direct chemical reaction

$$HgO(s) + Zn(s) = Hg(l) + ZnO(s)$$

from occurring. The copper disc and the steel case provide convenient terminal phases of electronic conducting material, in contact with the electronic conductors zinc and graphite, to which the wiring may be connected. When the circuit is completed, charge (positive) flows from the steel case through the wiring to the copper disc. Within the cell mercuric oxide is converted to liquid mercury at its interface with the alkaline solution; zinc is converted to zinc oxide at its interface with the alkaline solution; and alkali is transferred from one interface to the other.

7.3.2 The cell diagram

The arrangement of phases in a cell may be represented concisely by means of a *symbolic cell diagram*. The appropriate diagram for the Mallory cell is written

$$M_L \mid Zn(s) \mid ZnO(s) \mid NaOH(aq,L)$$
$$\mid NaOH(aq,J) \mid$$
$$NaOH(aq,R) \mid HgO(s) \mid Hg(l) \mid M_R$$

The right hand terminal phase, M_R, and the left hand terminal phase, M_L, are not described in detail. The choice of copper, steel and graphite as electronic conducting material is purely a matter of convenience.

7.3.3 The overall electrochemical process

The changes in amounts of the chemical substances present in each of the parts of a Mallory cell can be measured using two porous separators to divide the aqueous solution into three regions. The changes are found to be *proportional to the amount of charge passed.* The relative changes are listed below the formulae in the cell diagram

$$Zn(s) \mid ZnO(s) \mid NaOH(aq,L)$$
$$\Delta n_B = -1 \qquad +1 \qquad -0.40$$
$$\mid NaOH(aq,J) \mid$$
$$0.00$$
$$NaOH(aq,R) \mid HgO(s) \mid Hg(l)$$
$$+0.40 \qquad -1 \qquad +1$$

Thus the *overall electrochemical process* can be described by the equation

$$\begin{array}{ccc} HgO(s) + Zn(s) & & Hg(l) + ZnO(s) \\ & = & \\ +0.40\,NaOH(aq,L) & & +0.40\,NaOH(aq,R) \end{array}$$

We find that the relationship between the amount of the above process and the amount of charge passed is

$$dQ_{L\to R} = -193 \times 10^3 C\, mol^{-1} \times \text{(amount of reaction)}$$
$$= -2F d\xi \text{ (electrochemical process)}$$

This corresponds to transfer of electrons outside the cell according to the equation

$$2e^-(L) = 2e^-(R)$$

The process occurring within the cell must be

$$\begin{array}{ll} 2e^-(R) & 0.40\ NaOH(aq,R) \\ +\ HgO(s) + Zn(s) & = +Hg(l) + ZnO(s) \\ +\ 0.40\ NaOH(aq,L) & +\ 2e^-(L) \end{array}$$

For the general case, we refer to the number of electrons in each of these equations as the *Faraday number* or *electron* number, z, of the process, such that

$$dQ_{L\to R} = -zFd\xi$$

Query Why are fractional coefficients used for sodium hydroxide?

Reply If there were no transfer of ions, solution L would quickly become depleted in OH^- and so become positively charged and the solution R would become negatively charged due to the accumulation of excess OH^-. Electrical neutrality of the solutions is maintained by
(a) transfer of OH^- from R to L, and
(b) transfer of Na^+ from L to R resulting in a net transfer of NaOH from L to R. In practice both these transfer processes occur. In section 7.6, we infer that the coupled changes in amounts of NaOH in the half cells indicate that the hydroxyl ions conduct electricity four times as rapidly as do sodium ions.

Observations such as described above are typical of electrochemical cells. The electrochemical process is a combination of chemical transformation and chemical transfer processes.

Query Surely the accumulation and depletion of chemical substances leads to transfer by spontaneous diffusion.

Reply Yes. The rate of spontaneous diffusion can be measured when the wiring is disconnected. We may assume in practice that the spontaneous diffusion and the electrochemical process occur independently (at least when both are occurring slowly).

7.4 Cell e.m.f.

7.4.1 Potential of cell process

The potential of a cell process is related to the electromotive force (E) of the cell by the relation

$$\Delta\bar{G}(\text{cell process}) = -zFE_{\text{cell}}$$

The origin of the relationship can be seen if the cell process is considered as being made up of two parts. The first is the *internal* cell process which for the Mallory cell (Section 7.3.3) is represented by the equation

$$2e^-(R) + HgO(s) + Zn(s) + 0.4NaOH(aq,L)$$
$$= ZnO(s) + Hg(l) + 0.4NaOH(aq,R) + 2e^-(L)$$

The second part is the transfer of electrons

$$2e^-(L) = 2e^-(R)$$

The potential for this part is related to the cell voltage by the equation

$$\Delta\bar{G}(\text{electron transfer}) = -zF\Delta V$$

The overall potential for the cell process is just the sum of the potentials of the two parts. If the internal cell process is in equilibrium, its potential is equal to zero and the cell voltage is called the cell e.m.f., or

$$\begin{aligned}\Delta\bar{G}(\text{overall cell process}) &= 0 - zFE \\ &= -zFE\end{aligned}$$

7.4.2 The Nernst equation

The potential for the overall cell process can be written in terms of the electrochemical potentials of the species involved. In the Mallory cell

$$\begin{aligned}-2FE(\text{cell}) &= \Delta\bar{G}(\text{overall cell process}) \\ &= (\mu_{ZnO(s)} + \mu_{Hg(l)} - \mu_{Zn(s)} - \mu_{HgO(s)}) \\ &\quad + 0.40(\mu_{Na^+(aq,R)} + \mu_{OH^-(aq,R)} \\ &\quad - \mu_{Na^+(aq,L)} - \mu_{OH^-(aq,L)})\end{aligned}$$

This expression may be expanded in terms of standard potentials and relative activities, as

$$\begin{aligned}-2FE = &(\mu^\circ + RT\ln a)_{ZnO(s)} \\ &+ (\mu^\circ + RT\ln a)_{Hg(l)} \\ &- (\mu^\circ + RT\ln a)_{Zn(s)} \\ &- (\mu^\circ + RT\ln a)_{HgO(s)} \\ &+ 0.40(\mu^\circ + RT\ln a)_{Na^+(aq,R)} \\ &+ 0.40(\mu^\circ + RT\ln a)_{OH^-(aq,R)} \\ &- 0.40(\mu^\circ + RT\ln a)_{Na^+(aq,L)} \\ &- 0.40(\mu^\circ + RT\ln a)_{OH^-(aq,L)}\end{aligned}$$

or,

$$-2FE = \Delta\bar{G}^\circ\ (\text{overall cell process})$$

$$+\ RT\ln \frac{a_{ZnO(s)}a_{Hg(l)}}{a_{Zn(s)}a_{HgO(s)}} \left| \frac{a_{Na^+(aq,R)}a_{OH^-(aq,R)}}{a_{Na^+(aq,L)}a_{OH^-(aq,L)}} \right|^{0.40}$$

For convenience we divide by $-zF$ and *define* the *standard cell e.m.f.*, E° by the relation

$$-zFE^\circ \equiv \Delta\bar{G}^\circ \text{ (overall cell process)}$$

We can then write

$$E(\text{cell}) = E^\circ(\text{cell}) + \frac{RT}{-2F}\ln\frac{a_{ZnO(s)}a_{Hg(l)}}{a_{Zn(s)}a_{HgO(s)}}\left\{\frac{a_{Na^+(aq,R)}a_{OH^-(aq,R)}}{a_{Na^+(aq,L)}a_{OH^-(aq,L)}}\right\}^{0.40}$$

This equation is known as the *Nernst equation* for the cell e.m.f.

Query What is the purpose of the Nernst equation?

Reply The Nernst equation describes the relationship between

(a) the e.m.f. of a cell which is related to $\Delta\bar{G}$ for the overall cell process,

(b) the standard e.m.f. of the cell which is related to $\Delta\bar{G}^\circ$ and hence K_{eq} for the overall cell process

(c) the relative activities of the species involved in the cell reaction.

The practical application of the Nernst equation will be illustrated during the remainder of this chapter.

Query Can the Nernst equation for a cell be written down by "inspection"?

Reply Yes, if you know the equation describing the overall electrochemical process of the cell and the electron number (z) for the process. The equation is

$$E_{cell} = E^\circ + (RT/-zF)\ln(a_{products}/a_{reactants})$$

or, more concisely

$$E_{cell} = E^\circ + \frac{RT}{-zF}\ln\prod_B a_B^{\nu_B}$$

for the electrochemical cell process, where $\prod_B a_B^{\nu_B}$ is, as usual, the product of the relative activities of the species raised to the power of their coefficients in the cell equation, with ν_B being positive for products (top line) and negative for reactants (bottom line). The value of z, the *left-to-right electron number* of the electrochemical cell process, is positive when the direction of the coupled flow of electric charge is from right to left through the wiring.

Example 7.1

The "hydrogen, bromine" cell is represented by the diagram

$$Pt(s)|H_2(g,p_{H_2})|HBr(aq,c_L)$$
$$|\text{Junction}|$$
$$Br_2(saturated),HBr(aq,c_R)|Br_2(l,pure)|Pt(s)$$

The changes associated with the passage of charge through the cell are found to be described by the equation,

$$Br_2(l) + H_2(g) = 0.365\ HBr(aq,L) + 1.635\ HBr(aq,R)$$

The electron number of the process is

$$z = 2$$

(a) Write down the Nernst equation for the cell e.m.f..

Answer

In ionic form the overall equation is

$$Br_2(l) + H_2(g) = 0.365\{H^+(aq,L) + Br^-(aq,L)\} + 1.635\{H^+(aq,R) + Br^-(aq,R)\}$$

The corresponding Nernst equation is

$$E = E^\circ + \frac{RT}{-2F}\ln\frac{\{a_{H^+(aq,R)}a_{Br^-(aq,R)}\}^{1.635}\{a_{H^+(aq,L)}a_{Br^-(aq,L)}\}^{0.365}}{a_{Br_2(l)}a_{H_2(g)}}$$

(b) What is the e.m.f. of such a cell at 300 K if $E^\circ(300K) = 1.066$ V,

$c_{HBr(L)} = c_{HBr(R)} = 0.5$ mol dm^{-3}

($y_{H^+} = y_{Br^-} = 0.8$) and $p_{H_2} = 50$ kPa?

Answer

$$a_{H^+} = (yc/c^\circ)_{H^+} \text{ and } a_{Br^-} = (yc/c^\circ)_{Br^-}$$

$\therefore$

$$a_{H^+(L)} = 0.4,\ a_{Br^-(L)} = 0.4,$$
$$a_{H^+(R)} = 0.4,\ a_{Br^-(R)} = 0.4$$
$$a_{H_2(g)} = p_{H_2}/P^\circ = 0.5$$
$$a_{Br_2(l)} = 1 \text{ (pure liquid)}$$

$$\therefore \quad E = 1.066\ \text{V} + \frac{8.314 \times 300}{-2 \times 96480} \ln \frac{(0.4 \times 0.4)^{1.635}\,(0.4 \times 0.4)^{0.365}}{0.5 \times 1}\ \text{V}$$

$$= 1.066 - 0.0129 \ln \frac{(0.4)^4}{0.5}\ \text{V}$$

$$= 1.10\ \text{V}$$

(c) What is the e.m.f. of the cell at 300 K if

$E°(300\text{K}) = 1.066$ V, $c_{\text{HBr(L)}} = 0.1$ mol dm^{-3},

$c_{\text{HBr(R)}} = 0.5$ mol dm^{-3}

$(y_{\text{H}^+(\text{L})} = y_{\text{Br}^-(\text{L})} = 0.8,\ y_{\text{H}^+(\text{R})} = y_{\text{Br}^-(\text{R})} = 0.8)$

$p_{\text{H}_2} = 50$ kPa?

Answer
Using the same method as for (b) we have

$$a_{\text{H}^+(\text{L})} = 0.1 \times .8 = 0.08$$

$$a_{\text{Br}^-(\text{L})} = 0.08$$

$$a_{\text{H}^+(\text{R})} = 0.5 \times .8 = 0.4$$

$$a_{\text{Br}^-(\text{R})} = 0.4$$

$$a_{\text{H}_2(\text{g})} = 0.5$$

$$a_{\text{Br}_2(\text{l})} = 1$$

$$\therefore \quad E = 1.066 + \frac{8.314 \times 300}{-2 \times 96480} \ln \frac{(0.4 \times 0.4)^{1.635}\,(0.08 \times 0.08)^{0.365}}{0.5 \times 1}$$

$$= 1.066 - 0.0129 \ln .016\ \text{V}$$

$$= 1.12\ \text{V}$$

Query Why didn't you split the equation into a chemical reaction part and an ionic transfer part?

Reply Such a division is completely arbitrary. The equation as given is the *only* experimental result available. Three equivalent ways of interpreting it are

(a) $\text{H}_2(g) + \text{Br}_2(l) = 2(\text{H}^+\text{Br}^-)$

$1.635\,(\text{H}^+\text{Br}^-)_{\text{aq,L}} = 1.635\,(\text{H}^+\text{Br}^-)_{\text{aq,R}}$

(b) $\text{H}_2(g) + \text{Br}_2(l) = (\text{H}^+\text{Br}^-)_{\text{aq,L}} + (\text{H}^+\text{Br}^-)_{\text{aq,R}}$

$0.635\,(\text{H}^+\text{Br}^-)_{\text{aq,L}} = 0.635\,(\text{H}^+\text{Br}^-)_{\text{aq,R}}$

(c) $\text{H}_2(g) + \text{Br}_2(l) = 2(\text{H}^+\text{Br}^-)_{\text{aq,R}}$

$0.365\,(\text{H}^+\text{Br}^-)_{\text{aq,R}} = 0.365\,(\text{H}^+\text{Br}^-)_{\text{aq,L}}$

Check to see that they all give the same form for the Nernst equation.

Query Why write HBr(*aq*) as H$^+$(*aq*) + Br$^-$(*aq*)?

Reply This is an important question which is relevant to Chapters 5 and 6. It will be taken up in Section 7.5. Basically ionic equations are used because the form of the Nernst equation—an equation designed to describe experimental behaviour—is simpler.

PROBLEMS

7.1. The design of a cell which may be used for estimation of partial pressures of hydrogen is represented by the diagram

$$\text{Pt}(s)\,|\,\text{H}_2(g,\text{pure},P°)\,|\,\text{HCl}(\text{aq},c_\text{L})$$
$$\nu_\text{B} = \quad -1 \qquad +0.35$$

$$\|\,\text{HCl}(aq,c_\text{R})\,|\,\text{H}_2(g,p_{\text{H}_2})\,|\,\text{Pt}(s)$$
$$-0.35 \qquad +1$$

with electron number, $z = 2$.

(a) Write the equation for the overall electrochemical process.
(b) Write the Nernst equation for the cell.
(c) What is the expression for the standard e.m.f. for this cell?
(d) Write the Nernst equation for the cell in the special case that $c_\text{L} = c_\text{R}$.
(e) Estimate the partial pressure of hydrogen at the right hand side corresponding to a measured e.m.f. ($c_\text{L} = c_\text{R}$) of +88.6 mV.

7.2. The Weston Cadmium cell is represented by the diagram

$$\text{Cd}(\text{Hg,sat})\,|\,\text{CdSO}_4.\tfrac{8}{3}\text{H}_2\text{O}(s,pure)\,|$$
$$\nu_\text{B} = \quad -1 \qquad +1$$
$$\text{CdSO}_4(aq,\text{sat})\,\|\,\text{CdSO}_4(aq,\text{sat})\,|$$
$$0.00 \qquad 0.00$$
$$\text{Hg}_2\text{SO}_4(s,\text{pure})\,|\,\text{Hg}(l,\text{pure})$$
$$-1 \qquad +2$$

(a) Write the equation for the overall electrochemical process.
(b) Write the Nernst equation for the cell e.m.f.
(c) Suggest reasons for the common use of this cell as the practical reference source of voltage ($E_\text{cell} = 1.01832$ V at 298.15 K).

7.4.3 Diffusion and Polarisation

In cells where the electrolyte composition is not uniform throughout the cell, special precautions need to be taken to reduce diffusion of ions (and hence mixing) between the different parts of the cell. When the cell is operated, the chemical compositions of all the solutions change in a manner described by the cell equation. This results in a change in the values of the relative activities of the various species and a change in the e.m.f. of the cell.

In cells where the electrolyte composition is uniform, the situation is rather more complex. As an example we take a fresh Mallory cell. The equation describing the overall electrochemical process is

$$\mathrm{HgO}(s) + \mathrm{Zn}(s) + 0.40\{\mathrm{Na^+}(aq,\mathrm{L}) + \mathrm{OH^-}(aq,\mathrm{L})\}$$
$$= \mathrm{Hg}(l) + \mathrm{ZnO}(s) + 0.40\{\mathrm{Na^+}(aq\mathrm{R}) + \mathrm{OH^-}(aq,R)\}$$

electron number, $z=2$

and

$$E = E^\circ - \frac{RT}{2F}\ln\frac{a_{\mathrm{ZnO(s)}}a_{\mathrm{Hg(l)}}}{a_{\mathrm{Zn(s)}}a_{\mathrm{HgO(s)}}}\left\{\frac{a_{\mathrm{Na^+(aq,R)}}a_{\mathrm{OH^-(aq,R)}}}{a_{\mathrm{Na^+(aq,L)}}a_{\mathrm{OH^-(aq,L)}}}\right\}^{0.40}$$

In a fresh commercial cell the electrolyte has the same concentration of sodium hydroxide throughout the cell. Hence $a_{\mathrm{OH^-(aq,L)}} = a_{\mathrm{OH^-(aq,R)}}$ and $a_{\mathrm{Na^+(aq,L)}} = a_{\mathrm{Na^+(aq,R)}}$ and the Nernst equation becomes

$$E = E^\circ - \frac{RT}{2F}\ln\frac{a_{\mathrm{ZnO(s)}}a_{\mathrm{Hg(l)}}}{a_{\mathrm{Zn(s)}}a_{\mathrm{HgO(s)}}} \times 1$$

or, as

$a_{\mathrm{ZnO(s)}} = 1$, $a_{\mathrm{Hg(l)}} = 1$, $a_{\mathrm{Zn(s)}} = 1$ and $a_{\mathrm{HgO(s)}} = 1$

$$E_{\mathrm{cell}} = E^\circ{}_{\mathrm{cell}} = 1.347\ \mathrm{V}$$

Note: this result implies that the e.m.f. of a fresh Mallory cell is not dependent on the composition of the aqueous sodium hydroxide. In use it supplies a terminal voltage which remains close to 1.347 V until the zinc or mercuric oxide is all used. Hence its suitability for powering hearing aids and heart pacemakers, and as a reference voltage. When the cell is operated under conditions of high current, there is an increase of concentration of sodium hydroxide at the right and a decrease at the left. The relative activities of the ions change so that

$$a_{\mathrm{OH^-(aq,R)}} > a_{\mathrm{OH^-(aq,L)}}$$

and

$$a_{\mathrm{Na^+(aq,R)}} > a_{\mathrm{Na^+(aq,L)}}$$

From the Nernst equation we can see that this will result in a *reduction* in terminal voltage. The cell has become *polarised.*

Opposing this change in composition is the spontaneous diffusion of sodium and hydroxyl ions from a region of high relative activity to that of low relative activity. The terminal voltage of the polarised cell will thus return to its initial value when the cell is rested. Another result is that if the diffusion rate is high enough, the imbalance in electrolyte composition due to the electrochemical process is completely annulled by the diffusion, and polarisation does not occur. Thus power cells such as the lead–acid battery are designed so that the plates are as close together as possible and have large surface area—features which increase the value of current at which polarisation becomes important.

7.4.4 Cell e.m.f. and $\Delta\bar{G}$ for chemical reactions

Another property of cells in which the electrolyte concentration is uniform throughout the cell is that the e.m.f. of the cell is directly related to the potential for an ordinary chemical transformation. Take, for example, the hydrogen, bromine cell of Example 7.1.

$$\mathrm{Pt}(s)\ \mathrm{H_2}(g) | \mathrm{HBr}(aq,\mathrm{L})$$
$$\| \mathrm{HBr}(aq,\mathrm{R}), \mathrm{Br_2}(saturated) | \mathrm{Br_2}(l) | \mathrm{Pt}(s)$$

Here, the symbol $\|$ in the cell diagram represents a junction phase or region of aqueous HBr in which steady diffusion is established. The overall equation for the electrochemical process was

$$\mathrm{Br_2}(l) + \mathrm{H_2}(g) = 0.365\ \{\mathrm{H^+} + \mathrm{Br^-}(aq,\mathrm{L})\}$$
$$+ 1.635\{\mathrm{H^+} + \mathrm{Br^-}(aq,\mathrm{R})\}$$

with $z=2$.

The potential for the cell process is given by

$$-zFE = \Delta\bar{G}$$
$$= 0.365\{\mu_{\mathrm{H^+(aq,L)}} + \mu_{\mathrm{Br^-(aq,L)}}\}$$
$$+ 1.635\{\mu_{\mathrm{H^+(aq,R)}} + \mu_{\mathrm{Br^-(aq,R)}}\}$$
$$- \mu_{\mathrm{Br_2(l)}} - \mu_{\mathrm{H_2(g)}}$$

For a cell with a uniform HBr composition c_{HBr}

$$\mu_{H^+(aq,R)} \simeq \mu_{H^+(aq,L)}$$

and

$$\mu_{Br^-(aq,R)} \simeq \mu_{Br^-(aq,L)}$$

that is, we can write the potential for the cell process

$$-zFE = \Delta\bar{G} = 2\{\mu_{H^+(aq)} + \mu_{Br^-(aq)}\} - \mu_{Br_2(l)} - \mu_{H_2(g)}$$

It is the right hand side of this equation which is of particular interest. It is just $\Delta\bar{G}$ for the reaction

$$H_2(g,p_{H_2}) + Br_2(l) = 2H^+(aq,c_{H^+}) + 2Br^-(aq,c_{Br^-})$$

hence

$$\Delta\bar{G}(\text{chemical reaction}) = -zFE_{cell}$$

Query What use is this relation?
Reply It is used in one of two different ways. The first use is to determine values of $\Delta\bar{G}$ for reactions. The method is to construct a cell (either real or imaginary) for which the equation for the cell process is the same as that for the reaction of interest. One then measures or estimates the e.m.f. of the cell and hence determines $\Delta\bar{G}$ for the reaction of interest.

The second is the reverse of that just described. Measurements or estimates of $\Delta\bar{G}$ for chemical reactions may be used to predict values for the e.m.f. of cells.

Query Can we use the e.m.f. of a cell to infer directions of spontaneous chemical change?
Reply This is their most common use in chemistry!

$$\Delta\bar{G}_{reaction} \lesseqgtr 0$$

corresponds to

$$E_{cell} \gtreqless 0$$

hence spontaneous change implies

$$\Delta\bar{G} < 0 \quad E_{cell} > 0$$

equilibrium corresponds to

$$\Delta\bar{G} = 0 \quad E_{cell} = 0$$

and for a chemical reaction which requires driving

$$\Delta\bar{G} > 0 \quad E_{cell} < 0$$

PROBLEM

7.3. The cell for refining of raw copper is represented by

$$Cu(s,\text{raw}) | CuSO_4(aq) || CuSO_4(aq) | Cu(s,\text{pure})$$

$$\nu_B = -1 \qquad +0.5 \qquad -0.5 \qquad +1$$

with electron number, $z = 2$.

(a) Write the Nernst equation for the overall electrochemical process.
(b) What value has the standard molar change of Gibbs energy for the process?
(c) The compositions of the solutions in the two half cells can be kept equal by stirring. Estimate the e.m.f. of the cell for a sample of raw copper which contains only inclusions of inert solid particles.
(d) How much operator work is applied, per mole of pure copper recovered, when a terminal voltage of 1 volt is applied to obtain a reasonable *rate* of production? How much work is "wasted"?

7.5 Standard Chemical Potentials and Relative Activities of Ionic Species

7.5.1 Standard chemical potentials and relative activities

In order to go on with our analysis of cells, we need to extend our treatment of the chemical potentials and activities of ionic species. Consider the ions $H_3O^+(aq)$ and $Br^-(aq)$. These species cannot be formed *separately* from the elements. We shall examine the potential for their formation together, in aqueous hydrobromic acid, from the elements H_2 and Br_2 in their standard states and water, that is,

$$\tfrac{1}{2}H_2(g) + \tfrac{1}{2}Br_2(l) + H_2O(l) = H_3O^+(aq) + Br^-(aq) = HBr(aq)$$

Since $\mu_{H_2} = \mu_{Br_2} = 0$ under these conditions,

$$\Delta\bar{G} \equiv \mu_{H_3O^+(aq)} + \mu_{Br^-(aq)} - \mu_{H_2O(l)}$$
$$= \mu^{\circ}_{H_3O^+(aq)} - \mu^{\circ}_{H_2O(l)} + \mu^{\circ}_{Br^-(aq)} + RT\ln\{a_{H_3O^+}a_{Br^-}/a_{H_2O}\}$$

A convention is usually used for H_3O^+ in such examples, because they are common and because there has been much debate about the chemical structure of these species. It involves defining chemical potentials and relative activities for the (solvated) proton as

$$\mu_{H^+} \equiv \mu_{H_3O^+} - \mu_{H_2O}$$

$$a_{H^+} \equiv a_{H_3O^+}/a_{H_2O}$$

Corresponding definitions are used for any other solvent. Thus, in liquid ammonia, $NH_3(l)$,

$$\mu_{H^+} \equiv \mu_{NH_4^+} - \mu_{NH_3}$$

$$a_{H^+} \equiv a_{NH_4^+}/a_{NH_3}$$

With this convention

$$\Delta\bar{G}\{\tfrac{1}{2}H_2(g) + \tfrac{1}{2}Br_2(l) = H^+(aq) + Br^-(aq)\}$$
$$= \mu^\circ_{H^+(aq)} + \mu^\circ_{Br^-(aq)} + RT\ln a_{H^+}a_{Br^-}$$

Exercise 7.1
Write corresponding expressions for $K_2SO_4(aq)$.

Answer
The potential of $K_2SO_4(aq)$ is the potential for formation of $2K^+$ and SO_4^{2-} ions in water, $\Delta\bar{G}\{2K(s) + S\ (s,\ \text{monoclinic}) + 2O_2(g, P^\circ)$

$$= 2K^+(aq) + SO_4^{2-}(aq)\}$$

$$= 2\mu^{\circ+}_{K(aq)} + \mu^{\circ\,2-}_{SO_4} + RT\ln a^{2\,+}_{K(aq)}$$
$$a^{2-}_{SO_4(aq)}$$

$$= \mu_{K_2SO_4(aq)}$$

with other terms disappearing because the elements in their standard states are the zero of chemical potential.

For any ionic substance in solution, the standard chemical potential can be written in terms of similar expressions for the sum of the standard chemical potentials for the ions while the activity of the ionic substance can be written as the product of the activities of the ions.

Example 7.2
For a 10^{-3} mol dm^{-3} solution of potassium sulfate.

$$\mu^\circ_{K_2SO_4} = 2\mu^\circ_{K^+} + \mu^\circ_{SO_4^{2-}}$$

and $\quad a_{K_2SO_4} = a^2_{K^+}a_{SO_4^{2-}}$

Since the solution is very dilute, we may use relative concentrations as estimates of relative activities of the ionic species

$$a_{K^+} \simeq 2 \times 10^{-3}$$

$$a_{SO_4^{2-}} \simeq 1 \times 10^{-3}$$

Then $\quad a_{K_2SO_4} \simeq 4 \times 10^{-9}$

Note the very low value of the activity of such a substance, by comparison with the concentration. This *is* the number which is needed in calculations of equilibrium but it often seems uncomfortably small.

7.5.2 Standard chemical potentials of individual ions

So that we can tabulate values of the standard chemical potentials of individual ions, we give the standard chemical potential of the hydrated proton the value zero. That is, $\mu^\circ_{H^+(aq)} \equiv 0$. Hence

$$\mu^\circ_{Br^-(aq)} = \{\mu^\circ_{H^+(aq)} + \mu^\circ_{Br^-(aq)}\}$$

The same convention is adopted for any other solvent. Consequently all ionic species can be assigned values of μ°.

7.5.3 Activities of individual ionic species

The concentrations of ionic species in a solution are not independent. The effect of the requirement of "electroneutrality" is that in solutions like sea water or blood serum, which contain Na^+, K^+, Cl^-, Mg^{2+}, SO_4^{2-} and PO_4^{3-}, we can only design experiments to determine products of activities for ions of opposite charge and ratios of activities for ions of like charge. One set of such combinations for the six ions mentioned above is

$$a_{K^+}a_{Cl^-}$$

$$a_{Na^+}/a_{K^+}, \quad a_{SO_4^{2-}}/a^2_{Cl^-}$$

$$a_{Mg^{2+}}/a^2_{K^+}, \quad a_{PO_4^{3-}}/a^3_{Cl^-}$$

There are other ways of combining the activities which agree with the above requirements. Any of these other combinations can be written as combinations of the members of the above set.

Exercise 7.2
Show that the members of the above set can be used to calculate the values of the following.

$$a_{K^+}a_{Cl^-}$$

$$a^2_{K^+}a_{SO_4^{2-}}, \quad a^3_{K^+}a_{PO_4^{3-}}$$

$$a_{Na^+}a_{Cl^-}, \quad a_{Mg^{2+}}a^2_{Cl^-}$$

Answer The set includes $a_{K^+} a_{Cl^-}$, while

$$a^2_{K^+} a_{SO_4^{2-}} = (a_{K^+} a_{Cl^-})^2 \times a_{SO_4^{2-}}/a^2_{Cl^-}$$

$$a^3_{K^+} a_{PO_4^{3-}} = (a_{K^+} a_{Cl^-})^3 \times a_{PO_4^{3-}}/a^3_{Cl^-}$$

$$a_{Na^+} a_{Cl^-} = a_{K^+} a_{Cl^-} \times a_{Na^+}/a_{K^+}$$

$$a_{Mg^{2+}} a^2_{Cl} = (a_{Mg^{2+}}/a^2_{K^+}) \times (a_{K^+} a_{Cl^-})^2$$

In fact, if we look closely at such sets, we will find that the number of independent results is always one less than the number of activities for which we wish to know the values. We need one more equation if we are to get values of activities for individual ions.

An acceptable procedure, which has been recommended, involves giving the same value to y_{K^+} and y_{Cl^-}. Then

$$y_{K^+} = y_{Cl^-} = (y_{K^+} y_{Cl^-})^{\frac{1}{2}}$$

and activity coefficients for all other ionic species which are present can be determined relative to these.

This procedure is consistent with the theoretical model developed by Debye and Hückel which fits the experimental patterns up to moderate concentrations of ionic substances (see Section 4.4.3). According to this model, the logarithm of the activity coefficient of an ionic species of charge number, z, is approximately proportional to the square of z. That is, in a given solution

$$\log y_{B^{z+}} \simeq z^2 \log y_1$$

or

$$y_{B^{z+}} \simeq y_1^{z^2}$$

Exercise 7.3

A sample of blood plasma is found to have $y_{K^+} y_{Cl^-} = 0.49$.

(a) What values are assigned to the activity coefficients of K^+ and Cl^-?

(b) What are the approximate values of the activity coefficients of Na^+, Mg^{2+}, SO_4^{2-} and PO_4^{3-} in this material?

Answer

(a) 0.70, 0.70.

(b) 0.70, 0.24, 0.24, 0.041.

Tables of *experimental* activity coefficients for ionic substances contain values for combinations of ions, because that is the only way they can be measured. The values are most often listed as *mean activity coefficients* $y_\pm$. Thus for aqueous solutions of sodium chloride, $y_\pm$ is defined so that

$$y^2_\pm(\text{NaCl}, aq) \equiv y_{Na^+(aq)}\, y_{Cl^-(aq)}$$

In the case of sodium sulfate,

$$y^3_\pm(\text{Na}_2\text{SO}_4, aq) \equiv y^2_{Na^+(aq)}\, y_{SO_4^{2-}(aq)}$$

The general case for an ionic substance, M_aX_b is

$$y_\pm^{(a+b)} \equiv y_M^a\, y_X^b$$

Example 7.3

A value of 0.15 for $y_\pm$ for 0.1 mol dm^{-3} $CuSO_4$ means that

$$y_{Cu^{2+}}\, y_{SO_4^{2-}} \equiv y^2_\pm = 0.15^2 = 0.023$$

Similarly, a value of 0.268 for the mean activity coefficient of $K_3Fe(CN)_6$ in 0.5 mol dm^{-3} solution means that

$$y^3_{K^+}\, y_{Fe(CN)_6^{3-}} \equiv y^4_\pm = 0.005$$

Exercise 7.4

Interpret the following mean activity coefficients, reported for 0.05 mol dm^{-3} solution:

NaCl	0.82	$BaCl_2$	0.55
H_2SO_4	0.34	$ZnSO_4$	0.20

In practice it is not difficult to use mean activity coefficients. It only involves taking suitable notice of the ions which are present.

Example 7.4

The equation for the formation of the copper(II) tetramine ion is usually written

$$Cu^{2+} + 4NH_3 = Cu(NH_3)_4^{2+}$$

and thus ignores the negative ions. In the laboratory we have to choose to work with a particular copper(II) compound and this choice may be important in a quantitative treatment. Suppose we choose copper(II) chloride:

$$Cu^{2+} + 2Cl^- + 4NH_3 = Cu(NH_3)_4^{2+} + 2Cl^-$$

Equilibrium is established when

$$\frac{a_{Cu(NH_3)_4^{2+}}\, a^2_{Cl^-}}{a_{Cu^{2+}}\, a^2_{Cl^-}\, a^4_{NH_3}} = K_{eq}$$

This equation can be rewritten in terms of concentrations and activity coefficients as

$$\frac{c_{Cu(NH_3)_4^{2+}}\, c^2_{Cl^-}\, y^3_{\pm Cu(NH_3)_4 Cl_2}}{c_{Cu^{2+}}\, c^2_{Cl^-}\, c^4_{NH_3}\, y^3_{\pm\, CuCl_2}\, y^4_{NH_3}} = K_{eq}$$

and simplified by removing the terms in c_{Cl^-}

$$\frac{c_{Cu(NH_3)_4^{2+}}\, y^3_{\pm Cu(NH_3)_4Cl_2}}{c_{Cu^{2+}}\, c^4_{NH_3}\, y^3_{\pm\, CuCl_2}\, y^4_{NH_3}} = K_{eq}$$

Activity coefficients for ionic solutes are obtained either from measurements on electrochemical cells (Chapter 7) or by analysis of measurements on properties of the solvent (for example, its vapor pressure) in the solutions.

We have based this discussion on the *concentration* ($mol\,dm^{-3}$) scale. Either the *molality* or the *mole fraction* scales could have been used equally well. In fact, most accurate measurements for ionic solutes are expressed on the molality scale and use mean activity coefficients, $\gamma_{\pm}$, defined just like $y_{\pm}$.

Exercise 7.5
The mean activity coefficients, $\gamma_{\pm}$, at 20°C of some 17 $mol(kg\,H_2O)^{-1}$ solutions are:

	$\gamma_{\pm}$
NaOH	15.82
H_2SO_4	1.604
LiCl	43.8

What are the activities of these solutes?

PROBLEM

7.4. The symbolic diagram for the cell in the lead acid battery is represented by

$$\text{L.}\,Pb(s)|PbSO_4(s)|H_2SO_4(aq)$$
$$\nu_B = -1 \quad +1 \quad -1.2$$
$$||H_2SO_4(aq)|PbSO_4(s)|PbO_2(s)|Pb(s).\text{R}$$
$$-0.8 \quad +1 \quad -1 \quad 0$$

with electron number, $z = 2$.

(a) Write the equation for the electrochemical process.

(b) Write the Nernst equation for the e.m.f. of this cell.

(c) The standard e.m.f., based on the equation for the chemical reaction,

$$Pb(s) + PbO_2(s) + 4H^+(aq) + 2SO_4^{2-}(aq) = 2PbSO_4(s) + 2H_2O$$

is 2.041 V. What is the standard Gibbs energy, $\Delta\bar{G}°$, for this reaction?

(d) When an automobile battery is fully charged and rested, the aqueous solutions in the cells have a density equal to $1.26\,g\,cm^{-3}$, and contain $4.5\,mol\,dm^{-3}$ sulfuric acid. The e.m.f. of each cell is +2.100 V. What is the value of

$$a^2_{H^+(aq)}\,a_{SO_4^{2-}(aq)}/a_{H_2O}$$

in the solution?

(e) Immediately after rapid recharging, until the density of the solution in the junction region is $1.26\,g\,cm^{-3}$, the e.m.f. of a lead–acid cell may be as high as 2.6 V. Suggest an explanation.

7.6 Processes of Charge Transfer within Cells

7.6.1 Interpretation of electrochemical processes

In preceding sections we have used the basic approach to analysing the changes within cells. That is, we used only the measured amount of charge passed through the wiring and the associated changes in amounts of substances in the phases within the cell to obtain equations for electrochemical processes.

In this section we interpret these overall electrochemical processes in terms of the transfer of charge at electrodes (the boundaries within the half cells at which conduction by electrons gives way to conduction by ions) and the transfer of charge across junctions between half cells. This approach enables us to design half cells and junctions which may be used to construct complete cells and to forecast the electrochemical changes.

As an example, we consider the transfer of charge within a cell which produces gaseous hydrogen and aqueous chlorine from aqueous hydrochloric acid. The relative amounts of change in the phases, found by measuring volumes, doing titrations and monitoring charge transfer, are listed below the formulae in the cell diagram,

$$Pt(s)|Cl_2, HCl(aq, c_L)||HCl(aq, c_R)|H_2(g)|Pt(s)$$
$$\nu_B = +1.00 - 1.65 \quad -0.35 \quad +1.00$$

with electron number, $z = 2$.

Transfer of charge through the wiring is represented by the equation

$$2 \times \{e^-(L) = e^-(R)\}$$

At the interface (electrode) at the right, $Pt(s)|HCl(aq)$, the electrochemical process involves consumption of 2 electrons and release of 1 hydrogen molecule. Hydrogen ions can supply the necessary balance of elements and electric charge. So an equation for reaction at the right hand electrode can be written,

$$1 \times \{2e^-(R) + 2H^+(aq, R) = H_2(g)\}$$

In the aqueous solution associated with the

right hand electrode we observe net loss of only 0.35 hydrogen ions. This can be explained by transfer of $2 - 0.35 = 1.65$ hydrogen ions from the half cell at the left. Thus, the equation for transfer of hydrogen ions is multiplied by this factor to give its relative amount in the electrochemical process,

$$1.65 \times \{ H^+(aq,L) = H^+(aq,R) \}$$

Chloride ions are not involved in the above electrode reaction. The loss of 0.35 chloride ions from the aqueous solution adjacent to the right hand electrode corresponds to a *negative* amount of transfer of chloride ions from left to right. The contribution to the electrochemical process is represented by the equation

$$-0.35 \times \{ Cl^-(aq,L) = Cl^-(aq,R) \}$$

Together, these equations for transfer of ionic species in the electrochemical process correspond to transport of 2 faraday of electric charge from left to right within the cell, matching the transport of −2 faraday from left to right in the wiring.

At the interface (electrode) $\{ Pt(s) | HCl(aq) \}$ in the left hand half cell, we observe production of 1 chlorine molecule and release of 2 electrons. Chloride ions can supply the necessary balance of elements and electric charge in the electrode reaction. Thus we can write the equation for the contribution to the electrochemical process in the form

$$-1 \times \{ 2e^-(L) + Cl_2(aq) = 2Cl^-(aq,L) \}$$

The sum of the equations for the charge transfer processes, as written above, is the equation for the overall electrochemical process. Thus by rearranging these equations, for simplicity, and adding them up,

$$\begin{aligned} 2e^-(R) + 2H^+(aq,R) &= H_2(g) \\ + 1.65\, H^+(aq,L) &= 1.65\, H^+(aq,R) \\ + 0.35\, Cl^-(aq,R) &= 0.35\, Cl^-(aq,L) \\ + 2\, Cl^-(aq,L) &= Cl_2(aq) + 2e^-(L) \\ 2e^-(L) &= 2e^-(R) \end{aligned}$$

$$\left. \begin{array}{l} 0.35\, H^+(aq,R) + 0.35\, Cl^-(aq,R) \\ + 1.65\, H^+(aq,L) + 1.65\, Cl^-(aq,L) \end{array} \right| = \left\{ \begin{array}{l} H_2(g) \\ + Cl_2(aq) \end{array} \right.$$

we obtain the result which comes directly from the experiment.

By the reasoning above, we have developed a mechanism which accounts for the chemical changes and for the transfer of electric charge. This mechanism involves four stages:

(a) reaction at the electrode in the right hand half cell;

(b) transfer of ionic species through the material of the junction;

(c) reaction at the electrode in the left hand half cell; and

(d) transfer of electrons through the wiring.

Of course, each stage may itself be the result of a sequence of smaller steps.

7.6.2 Electrodes and electrode reactions

The usefulness of interpreting electrochemical cells in terms of electrodes and junctions comes from the finding that a given half cell arrangement has the same electrode reaction when it is incorporated in different cells.

The equation for an electrode reaction represents the conservation of elements and the conservation of electric charge. Such equations can be worked out on the basis of observed or proposed changes. Thus, in the right hand half cell of the Mallory system, we observe reduction of mercuric oxide to mercury. An equation for the process can be built up by considering the directly observed change, additions to conserve chemical elements and additions to conserve electric charge.

$HgO(s)$		$Hg(l)$	required change
$+ H_2O$	=	$+ 2OH^-(aq)$	conservation of elements
$+ 2e^-(M)$			conservation of charge

The equation for the electrode reaction is thus

$$2e^-(M) + HgO(s) + H_2O = Hg(l) + 2OH^-(aq)$$

corresponding to transfer of charge, $-2F = -2 \times 96.5 \times 10^3\, C\, mol^{-1}$, from the terminal phase (M) to the ionic conducting phase (*aq*). This electrode may be represented by a *symbolic electrode diagram*,

$$Hg(l) | HgO(s) | OH^-(aq)$$

showing only the *specific ions* which are involved in the electrode reaction. The above equation corresponds to the direction of chemical change coupled with supply of electrons from the terminal and wiring. When this process advances we refer to the electrode as a *cathode*. Chemical reduction is taking place at any electrode which is acting as a cathode. When the electrode process occurs in the direction to release electrons to the terminal and wiring, we refer to the electrode as an *anode*. Chemical oxidation occurs at any electrode which is acting as an anode.

A similar equation can be written for an electrode reaction to reduce zinc oxide to zinc,

$$2e^-(M) + ZnO(s) + H_2O = Zn(s) + 2OH^-(aq)$$

The corresponding electrode diagram is

$$Zn(s)|ZnO(s)|OH^-(aq)$$

Query In the Mallory cell the reaction at this electrode goes backward according to the above equation. Do we reverse the equation?

Answer It is common practice to write *individual* electrode reactions in the direction of cathodic reduction, using electrons as reactants. Equations for processes in cells are written to correspond to the right hand electrode reaction going forward (cathodic reduction) and the left hand electrode reaction going backward (anodic oxidation, reversal of cathodic reduction). The conventional equation for an electrochemical cell process then includes the equation for the right hand cathodic reduction *minus* the equation for the left hand cathodic reduction.

Electrodes are often described according to the following classifications.

(a) *Metal electrodes*, $M|M^{z+}(aq)$. An example is the "copper electrode", $Cu(s)|Cu^{2+}(aq)$, formed by dipping a piece of copper into a solution of, say, copper(II) sulfate. The electrode reaction is

$$2e^-(M) + Cu^{2+}(aq) = Cu(s)$$

(b) *Metal/compound electrodes*, $M|MX(s)|X^-(aq)$. This class is typified by the "silver–silver chloride electrode", $Ag(s)|AgCl(s)|Cl^-(aq)$, which may be formed by depositing a layer of silver chloride on a silver wire and dipping it in an aqueous chloride solution. The electrode reaction is

$$e^-(M) + AgCl(s) = Ag(s) + Cl^-(aq)$$

An alternative way of viewing this electrode process is to consider it as the sum of the two processes

$$e^-(M) + Ag^+(aq) = Ag(s)$$
$$AgCl(s) = Ag^+(aq) + Cl^-(aq)$$

Another commonly used metal/compound electrode is the "calomel electrode", $Hg(l)/Hg_2Cl_2(s)/Cl^-(aq)$, formed by adding mercury (I) chloride (calomel) at the interface between liquid mercury and aqueous potassium chloride.

The electrodes of the Mallory cell belong to the class of metal/compound electrodes.

(c) *Gas electrodes*, $M|X_2(g)|X^-(aq)$. The "oxygen electrode", $M(s)|O_2(g)|OH^-(aq)$, involves the electrode reaction

$$4e^-(M) + O_2(g) + 2H_2O = 4OH^-(aq)$$

When the aqueous solution is acidic rather than alkaline, we may prefer to write the equation in terms of the predominant species, $H^+(aq)$, thus

$$4e^-(M) + O_2(g) + 4H^+(aq) = 2H_2O$$

Under these circumstances the appropriate electrode diagram is $M|O_2(g)|H^+(aq)$. The "hydrogen electrode", $M|H_2(g)|H^+(aq)$ is used as the reference system for values of electrode potentials (Section 7.8.1).

(d) *Redox electrodes*, $M|A,A'(aq)$. This term refers to electrodes in half cells which have products and reactants of the electrode reaction contained in the ionic conducting phase. Thus iron(III) ions and iron(II) ions in aqueous solution may be interconverted at an electrode, $M|Fe^{3+}, Fe^{2+}(aq)$, without change in the electronic conducting material, M.

$$e^-(M) + Fe^{3+}(aq) = Fe^{2+}(aq)$$

Similarly we may have the reduction of dichromate ions to chromium(III) ions,

$$\begin{matrix} Cr_2O_7^{2-}(aq) \\ +\,14\,H^+(aq) \\ +\,6e^-(M) \end{matrix} = \begin{matrix} 2Cr^{3+}(aq) \\ +\,7H_2O \end{matrix}$$

occurring at the electrode represented by $M|Cr_2O_7^{2-},H^+,Cr^{3+}(aq)$.

Exercise 7.6

Write equations for electrode reactions which are allowed at the following electrodes

(a) $Al(s)|Al^{3+}$(molten cryolite)

(b) $Pt(s)|Fe(CN)_6^{3-}, Fe(CN)_6^{4-}(aq)$

(c) $Pt(s)|MnO_4^-,Mn^{2+},H^+(aq)$

(d) $Pt(s)|$ $\begin{array}{c}CH_2COOH\\|\\CH_2COOH,\end{array}$ $\begin{array}{c}CHCOOH\\||\\CHCOOH,\end{array}$ $H^+(aq)$

succinic acid maleicacid

(e) $M|CO_2(g)|CH_3OH,OH^-(aq)$

7.6.3 Junctions between half cells

When a cell is designed by choosing two half cells which will have particular electrode reactions, it is necessary to give thought to the arrangement of the junction between the ionic conducting phases. The junction in a cell must serve two functions:

(a) providing the pathway for transfer of electric charge between the ionic conducting materials of the half cells; and
(b) acting as a mechanical barrier to direct spontaneous reaction between substances in the half cells.

A common type of situation involves ionic conducting materials of different compositions in the two half cells. A junction may be formed by inserting a layer of porous or permeable material which will prevent bulk mixing while allowing transfer of ionic species. A given junction arrangement (including the adjacent ionic conducting phases) has the same ionic transfer processes when used with different electrodes.

The arrangement of a junction in a cell is just like the experimental arrangement for study of diffusion. But with cells we are more interested in the relative rates of migration of individual ionic species in the transport of charge. These are represented by the coefficients which appear in the equation for transport of 1 faraday of charge.

An aqueous hydrochloric acid junction is represented by the *symbolic junction diagram*

$$\underset{c_L}{HCL(aq,L)}|\underset{\text{porous disc}}{HCL(aq,J)}|\underset{c_R}{HCl(aq,R)}$$

The equation for transport of 1F is found to be

$$+0.825\{H^+(aq,L)=H^+(aq,R)\}$$
$$-0.175\{Cl^-(aq,L)=Cl^-(aq,R)\}$$

Thus aqueous hydrochloric acid is partially selective (82.5%) toward transport of charge by transfer of hydrogen ions.

It is useful to classify junctions according to the selectivity of the ionic conducting material. We may group them as follows.

(a) *Steady diffusion junctions.* Many liquid junction materials, such as aqueous solutions, allow migration of both positive and negative ions. Junctions formed with these materials also permit continuous steady diffusion of ionic species (without net transfer of charge) whenever there are differences of activity between the two sides. We consider the spontaneous diffusion and transport of charge by transfer of ionic species as separate processes.

The example of the aqueous hydrochloric acid junction was described above. The approximately five-fold predominance of transport of charge by hydrogen ions is typical of aqueous acid solutions.

Aqueous alkali solutions show similar partial selectivity (approximately four-fold) toward transport of charge by transfer of hydroxide ions.

Aqueous salt solutions are more nearly non-selective. Thus, for aqueous barium chloride, the equation for transport of 1F is

$$+0.22\{Ba^{2+}(aq,L)=Ba^{2+}(aq,R)\}$$
$$-0.56\{Cl^-(aq,L)=Cl^-(aq,R)\}$$

Exercise 7.7
Show that this equation does correspond to transport of +1F, from left to right.

In the specially prepared cells for making accurate measurements of potentials for reactions, both half cells contain ionic conducting material of precisely the same composition. The junction in such a cell may be no more than a region of the same material which is of sufficient width to separate the electrodes.

Cells without solid barriers are more efficient to operate. Thus power cells and cells for production of materials are constructed in this way whenever possible. Ionic polarisation (decrease in the cell potential (e.m.f.) as a result of transfer of ionic species) may be minimised, or eliminated while the cell is rested, by diffusion and bulk mixing.

(b) *Salt bridge junctions.* When two half cells are selected which contain different ionic species, it is often necessary to avoid diffusion of these species from one half cell to the other. An arrangement which is commonly used in laboratories has an intermediate phase, or *salt bridge*, of concentrated aqueous salt solution which is made into a jelly with agar to prevent bulk flow. The salt bridge greatly increases the time for significant amounts of species from one half cell to reach the other half cell. The concentrated salt makes the dominant contribution to transfer of charge when current is passed.

When we wish to measure and interpret the e.m.f. of such a cell, the preferred choice of salt is potassium chloride, for which the equation for transport of 1F is

$$+0.490\{K^+(aq,L) = K^+(aq,R)\}$$

$$-0.510\{Cl^-(aq,L) = Cl^-(aq,R)\}$$

The nearly complete lack of selectivity toward transport of charge by cations (K^+) and anions (Cl^-) allows us to neglect the contribution of the junction to the potential of the cell (the e.m.f.) to the level of ± 1mV precision.

(c) *Specific ion selective junctions.* A few solid (or plastic) materials, which have been found to be highly selective toward transport of charge by one ionic species, have important applications for forming *specific-ion selective junctions* between half cells.

In these materials one ion is free to move through a fixed lattice or network which holds the second ion. Fixed lattices of anions may exist in glass or sulfonated polystyrene, through which only cations (Na^+, H^+, etc.) can migrate. In solid silver sulfide, the silver cations form a fixed lattice through which sulfur anions may move. Solid lanthanum fluoride is selectively permeable to fluoride ions and is the key component of cells for monitoring fluoridated water.

The construction of a hydrogen–oxygen fuel cell with a hydrogen ion exchange resin as the ionic conducting material has special advantages. The cell diagram is

$$Pt(s)|H_2(g)|H^+(resin)|O_2(g)|H_2O(l)|Pt(s)$$

The electrode reactions are represented by the equations

$$4e^-(R) + O_2(g) + 4H^+(resin,R) = 2H_2O(l)$$

and $2H_2(g) = 4H^+(resin,L) + 4e^-(L)$

The equation for ionic transport of charge is

$$4H^+(resin,L) = 4H^+(resin,R)$$

and the equation for electronic transport of charge is

$$4e^-(L) = 4e^-(R)$$

So the equation for the overall electrochemical process is

$$2H_2(g) + O_2(g) = 2H_2O(l)$$

The system may be operated continuously without loss of efficiency due to accumulation and depletion of ionic substances in the half cells.

7.7 Forecasting of Cell Reactions

We have shown that the overall electrochemical process in a cell system can be interpreted as the sum of separate electrode reactions and ionic transfer processes. It is commonly possible to write down equations for the most likely electrode reactions by "inspection" of the given half cells. The equation for a *cell reaction* may then be written by combining the equations for the electrode reactions and transfer of electrons.

Example 7.5
A "Daniell cell",

$$Zn(s)|ZnSO_4(aq)||CuSO_4(aq)|Cu(s)$$

may be designed. The likely reaction at the electrode in the right hand half cell is

$$2e^-(M) + Cu^{2+}(aq) = Cu(s)$$

The likely reaction at the electrode in the left hand half cell is

$$2e^-(M) + Zn^{2+}(aq) = Zn(s)$$

These electrode reactions may be coupled by transfer of electrons through the wiring,

$$2e^-(L) = 2e^-(R)$$

with the reaction at the left hand electrode going backward. The "sum" of the electrode reactions and transfer of electrons, as they should occur in the cell, is the equation for the *cell reaction.*

$$\begin{aligned} 2e^-(R) + Cu^{2+}(aq,R) &= Cu(s) \\ Zn(s) &= Zn^{2+}(aq,L) + 2e^-(L) \\ 2e^-(L) &= 2e^-(R) \end{aligned}$$

$$\left.\begin{matrix} Cu^{2+}(aq,R) \\ +Zn(s) \end{matrix}\right\} = \left\{\begin{matrix} Cu(s) \\ +Zn^{2+}(aq,L) \end{matrix}\right.$$

or

$$Zn(s) + Cu^{2+}(aq,R) = Zn^{2+}(aq,L) + Cu(s)$$

The equation for the cell reaction is *part* of the equation for the overall electrochemical process. It includes only the chemical substances and ionic species which participate in the electrode reactions. The equation for transport of charge by ionic species must be added to the equation for the cell reaction to obtain the complete equation for the electrochemical process. However, there are many applications of cells in which we are interested only in the chemical transformations. The details of ionic transfer are then unimportant.

7.8 Subdivision of the Cell e.m.f.

We have shown that the e.m.f. of a cell is directly related to the potential for the overall electrochemical process,

$$-zFE_{cell} = \Delta \bar{G}(\text{overall electrochemical process})$$

which may be expressed as the "sum" of $\mu^\circ + RT\ln a$ for the chemical substances which appear in the equation corresponding to the electron number, z.

We have also shown that the equation for the overall process may be represented as the sum of the equation for cell reaction and the equation for transport of zF by ionic species. The terms in the potential for the overall process may be collected in the same way to express the potential for the process as

$$\Delta\bar{G} = \text{"sum" of } (\mu^\circ + RT\ln a) \text{ for the cell reaction} + \text{"sum" of } (\mu^\circ + RT\ln a) \text{ for the transport process}$$

Thus we may write

$$-zFE_{cell} = -zFE_{\text{cell reaction}} - zFE_{\text{transport process}}$$

or $$E_{cell} = E_{\text{cell reaction}} + E_{\text{transport process}}$$

The terms in the "sum" of $(\mu^\circ + RT\ln a)$ for the cell reaction may be further divided up into the "sums" of $(\mu^\circ + RT\ln a)$ for chemical substances and ionic species which are involved in the separate electrode reactions to obtain *electrode potentials*, $E_{\text{electrode}}$, such that

$$E_{\text{cell reaction}} = E_{\text{right hand electrode}} - E_{\text{left hand electrode}}$$

These definitions are devised in such a way that we may calculate the electrode potentials for given half cells and the *junction potential*, E_{junction} (referred to above as $E_{\text{transport process}}$), for a given arrangement of the junction. The e.m.f., E_{cell}, is thus equal to the "sum" of the two individual electrode potentials plus the junction potential

$$E_{cell} = E_{\text{right hand electrode}} - E_{\text{left hand electrode}} + E_{\text{junction}}$$

This subdivision should be regarded as a convenient way of expressing the potential for the overall electrochemical process. When we consider only the cell reaction it is possible to interpret the junction potential as an electric potential difference, $\Delta\Phi$(junction), which affects the potentials of the ionic species.

7.8.1 Electrode potentials

The manner in which electrode potentials are defined may be illustrated by reference to the "silver/silver chloride–hydrogen cell"

$$Pt(s)|H_2(g)|HCl(aq)|HCl(aq)|HCl(aq)|AgCl(s)|Ag(s)$$

The equation for the silver/silver chloride electrode reaction is

$$e^-(M) + AgCl(s) = Ag(s) + Cl^-(aq)$$

The potential of the silver/silver chloride electrode is defined by

$$-1FE_{\text{electrode}} = (\mu^\circ + RT\ln a)_{Cl^-(aq)} + (\mu^\circ + RT\ln a)_{Ag(s)} - (\mu^\circ + RT\ln a)_{AgCl(s)}$$

The corresponding expression for defining the standard electrode potential, E°, is

$$-1FE^\circ_{Ag(s)|AgCl(s)|Cl^-(aq)} = \mu^\circ_{Cl^-(aq)} + \mu^\circ_{Ag(s)} - \mu^\circ_{AgCl(s)}$$

So the electrode potential may be expressed in terms of the standard electrode potential and the relative activities of the chemical substances and ionic species,

$$E_{\text{electrode}} = E^\circ_{Ag(s)|AgCl(s)|Cl^-(aq)} + \frac{RT}{-1F}\ln\frac{a_{Cl^-(aq)}a_{Ag(s)}}{a_{AgCl(s)}}$$

It will be observed that the properties of the

electrons involved in the electrode reaction do not appear in these definitions. Electrons do not appear in the equation for the overall electrochemical process, so it is convenient to ignore them in defining electrode potentials.

Now we transfer our attention to the hydrogen electrode. The equation for reaction at this electrode is

$$2e^-(M) + 2H^+(aq) = H_2(g)$$

The potential of the hydrogen electrode is defined by

$$-2FE_{\text{electrode}} = (\mu^\circ + RT\ln a)_{H_2(g)} - 2(\mu^\circ + RT\ln a)_{H^+(aq)}$$

The corresponding standard electrode potential, E°, is then given by

$$-2FE^\circ_{Pt(s)|H_2(g)|H^+(aq)} = \mu^\circ_{H_2(g)} - 2\mu^\circ_{H^+(aq)}$$

We have already set both the standard potential of gaseous hydrogen and the standard potential of $H^+(aq)$ equal to zero. Thus, the standard potential of the hydrogen electrode becomes the reference value for electrode potentials

$$E^\circ_{Pt(s)|H_2(g)|H^+(aq)} \equiv \text{zero}$$

Tables of standard electrode potentials (also known as standard reduction potentials) provide the basic data for calculating electrode potentials.

As general representation of an electrode reaction we may write

$$\text{electro-reactants} + ze^- = \text{electro-products}$$

The corresponding statement of the Nernst equation for electrode potential is then

$$E_{\text{electrode}} = E^\circ + (RT/-zF)\ln(a_{\text{electro-products}}/a_{\text{electro-reactants}})$$

PROBLEMS

7.5. Use the table of standard electrode potentials to calculate the potentials of the following electrodes at 298 K:

(a) $Ag(s)|Ag^+(aq, a = 0.100)$

(b) $Mg(s)|Mg^{2+}(aq, a = 0.100)$

(c) $Pt(s)|H_2(g)|H^+(aq, a = 10^{-7})$

(d) $Pt(s)|O_2(g)|OH^-(aq, a = 10^{-7})$

(e) $Pt(s)|O_2(g)|H^+(aq, a = 10^{-7})$

(f) $Pt(s)|Cr_2O_7^{2-}, Cr^{3+}, H^+(aq)$

(i) $a_{Cr_2O_7^{2-}} = 10^{-2}$, $a_{Cr^{3+}} = 10^{-2}$, $a_{H^+} = 1$

(ii) $a_{Cr_2O_7^{2-}} = 10^{-2}$, $a_{Cr^{3+}} = 10^{-2}$, $a_{H^+} = 10^{-4}$

Answer (a) +0.8582 V; (b) −2.334 V; (c) +0.414 V; (d) +0.82 V; (e) +0.82 V; (f) (i) 1.35 V, (ii) 0.80 V

7.6. Use the table of standard electrode potentials to calculate the standard potentials (μ°) of the ionic species

Table 7.1 Standard Electrode Potentials with Aqueous Solutions at 298 K.

Electrode reaction	E°/V	*Electrode reaction*	E°/V
$e^- + Li^+(aq) = Li(s)$	−3.01	$2e^- + Cu^{2+}(aq) = Cu(s)$	+0.34
$e^- + Na^+(aq) = Na(s)$	−2.714	$4e^- + O_2(g) + 2H_2O = 4OH^-(aq)$	+0.401
$2e^- + Mg^{2+}(aq) = Mg(s)$	−2.363	$2e^- + I_2(s) = 2I^-(aq)$	+0.536
$3e^- + Al^{3+}(aq) = Al(s)$	−1.663	$2e^- + O_2(g) + 2H^+(aq) = H_2O_2(aq)$	+0.6824
$2e^- + Zn^{2}(aq) = Zn(s)$	−0.763	$e^- + Fe^{3+}(aq) = Fe^{2+}(aq)$	+0.771
$2e^- + S(s) = S^{2-}(aq)$	−0.51	$e^- + Ag^+(aq) = Ag(s)$	+0.7991
$2e^- + Fe^{2+}(aq) = Fe(s)$	−0.441	$2e^- + Br_2(aq) = 2Br^-(aq)$	+1.087
$2e^- + 2H^+(aq) = H_2(g)$	0.000	$4e^- + O_2(g) + 4H^+(aq) = 2H_2O$	+1.23
$2e^- + CH_2O + 2H^+(aq) = CH_3OH(aq)$	+0.19	$6e^- + Cr_2O_7^{2-}(aq) + 14H^+(aq) = 2Cr^{3+}(aq) + 7H_2O$	+1.33
$e^- + AgCl(s) = Ag(s) + Cl^-(aq)$	+0.222	$2e^- + Cl_2(g) = 2Cl^-(aq)$	+1.358
$2e^- + Hg_2Cl_2(s) = 2Hg(l) + 2Cl^-(aq)$	+0.2676	$2e^- + F_2(g) = 2F^-(aq)$	+2.87

(a) $Li^{+}(aq)$

(b) $S^{2-}(aq)$

(c) $I^{-}(aq)$

(d) $F^{-}(aq)$

Answer (a) -299 kJ mol^{-1};

(b) $+98$ kJ mol^{-1};

(c) -33 kJ mol^{-1};

(d) -277 kJ mol^{-1}

7.7. Use the table to calculate the standard potentials, $\Delta\bar{G}°$, and equilibrium constants, K_{eq}, for the reactions below. ($\Delta\bar{G}° = -RT\ln K_{eq}$).

(a) $Br_2(aq) + 2I^{-}(aq) = 2Br^{-}(aq) + I_2(s)$

(b) $Ag^{+}(aq) + Fe^{2+}(aq) = Ag(s) + Fe^{3+}(aq)$

(c) $F_2(g) + 2Li(s) = 2Li^{+}(aq) + 2F^{-}(aq)$

Answer (a) -106.3 kJ mol^{-1}, $10^{+18.6}$
(b) -2.8 kJ mol^{-1}, $10^{+0.49} = 3.1$
(c) -1135 kJ mol^{-1}, 10^{+199}

7.8.2 Junction potentials

The junction potential for a given junction arrangement is defined by

$$-FE_{\text{junction}} = \text{"sum" of } (\mu° + RT\ln a)$$

for the equation representing transport of charge, zF, from the left hand half cell through the junction to the right hand half cell.

The processes which transfer charge across the junction in the silver/silver chloride–hydrogen cell are the same as those in the cell for production of hydrogen and chlorine (Section 7.6.1). For $+1$F of charge transferred from L to R through the junction, the appropriate equation is

$$0.825\,\{H^{+}(aq,L) = H^{+}(aq,R)\}$$
$$-0.175\,\{Cl^{-}(aq,L) = Cl^{-}(aq,R)\}$$

corresponding to transfer of -1F of charge from L to R through the wiring. This equation is used to define the *junction potential*, by

$$-1FE_{\text{junction}} = 0.825 \times \left\{\begin{matrix}(\mu° + RT\ln a)_{H^{+}(aq,R)} \\ -(\mu° + RT\ln a)_{H^{+}(aq,L)}\end{matrix}\right\} -0.175\left\{\begin{matrix}(\mu° + RT\ln a)_{Cl^{-}(aq,R)} \\ -(\mu° + RT\ln a)_{Cl^{-}(aq,L)}\end{matrix}\right\}$$

Thus the junction potential has contributions from each of the ions, weighted according to the relative amounts transferred. We may define two standard junction potentials

$$-1FE°_{H^{+}(aq)|H^{+}(aq)} = \mu°_{H^{+}(aq,R)} - \mu°_{H^{+}(aq,L)}$$

and

$$-1FE°_{Cl^{-}(aq)|Cl^{-}(aq)} = \mu°_{Cl^{-}(aq,R)} - \mu°_{Cl^{-}(aq,L)}$$

Then the potential of the junction,

$$HCl(aq)|HCl(aq)|HCl(aq)$$

is given by

$$E_{\text{junction}}$$
$$= +0.825\{E°_{H^{+}(aq)|H^{+}(aq)} + (RT/-1F)\ln a_{H^{+}(aq,R)}/a_{H^{+}(aq,L)}\}$$
$$-0.175\{E°_{Cl^{-}(aq)|Cl^{-}(aq)} + (RT/-1F)\ln a_{Cl^{-}(aq,R)} \times /a_{Cl^{-}(aq,L)}\}$$

In the present case (and in most cells) we have the same phase type (solvent) at each side of the junction. So the standard junction potentials for the individual ionic species are zero. The simplified expression for the potential of the aqueous hydrochloric acid junction is

$$E_J = \frac{RT}{-1F}\ln\left\{\frac{a_{H^{+}(aq,R)}}{a_{H^{+}(aq,L)}}\right\}^{+0.825}\left\{\frac{a_{Cl^{-}(aq,R)}}{a_{Cl^{-}(aq,L)}}\right\}^{-0.175}$$

The concentrations of $H^{+}(aq)$ and $Cl^{-}(aq)$ in each solution are equal. If we *assume* that their activities are also equal, that is,

$$a_{H^{+}(aq)} \simeq a_{Cl^{-}(aq)} \simeq (y_1 c/c°)_{HCl(aq)}$$

an estimate of the junction potential is

$$E_J \simeq 0.650\,\frac{RT}{-1F}\ln\frac{(y_1 c/c°)_{HCl(aq,R)}}{(y_1 c/c°)_{HCl(aq,L)}}$$

where $0.650 = +0.825 - 0.175$

The significance of aqueous potassium chloride junctions,

$$\underset{c_L}{KCl(aq)}|KCl(aq)|\underset{c_R}{KCl(aq)},$$

can now be explained. The equation for transfer of ionic species is

$$+0.490\,\{K^{+}(aq,L) = K^{+}(aq,R)\}$$
$$-0.510\,\{Cl^{-}(aq,L) = Cl^{-}(aq,R)\}$$

The junction potential is given by

$$E_J = \frac{RT}{-1F} \ln \left\{\frac{a_{K^+(aq,R)}}{a_{K^+(aq,L)}}\right\}^{+0.490} \left\{\frac{a_{Cl^-(aq,R)}}{a_{Cl^-(aq,L)}}\right\}^{-0.510}$$

When we can assume that

$$a_{K^+(aq)} \simeq a_{Cl^-(aq)} \simeq (y_1 c/c^\circ)_{KCl(aq)}$$

the estimate of the junction potential is given by

$$E_J = -0.020 \frac{RT}{-1F} \ln \left\{\frac{(y_1 c/c^\circ)_{KCl(aq,R)}}{(y_1 c/c^\circ)_{KCl(aq,L)}}\right\}$$

where $\quad -0.020 = +0.490 - 0.510$

Because of the virtual non-selectivity, the junction potential is small for quite large ratios of concentrations.

Example 7.6
Use the above equation to estimate the junction potential for a 100-fold ratio of concentrations of potassium chloride

$$KCl(aq) \| KCl(aq) \| KCl(aq)$$
$$0.010 \text{ mol dm}^{-3} \qquad 1.00 \text{ mol dm}^{-3}$$

Answer

$$E_J \simeq -0.020 \times 2.303 \frac{RT}{-1F} \log \frac{1.00}{0.010}$$

$$= \frac{-0.020 \times 2.303 \times 8.314 \text{ JK}^{-1}\text{mol}^{-1} \times 298 \text{ K} \times (+2)}{-1 \times 96.5 \times 10^3 \text{ C mol}^{-1}}$$

$$= 1.18 \times 10^{-3} \text{ JC}^{-1}$$

Thus the junction potential will be approximately + 1.2 mV.

PROBLEM

7.8. What should be the approximate value of the junction potential for the following arrangement?

$$\begin{array}{l|l|l} \text{blood} & H^+\text{-glass} & HCl(aq) \\ pH = 7 & & 10^{-1} \text{ mol dm}^{-3} \end{array}$$

Answer -0.355 V

The accuracy of prediction of junction potentials is limited by three factors:

(a) the accuracy of measurement of the coefficients for transfer of charge by individual ions;
(b) the reliability of the procedure for estimating the relative activities of individual ionic species; and
(c) the reliability of the assumption that the coefficients for transfer of charge by ions are not affected by the rate of spontaneous diffusion of ionic substances.

However, the value for a particular junction arrangement may be established by measuring the e.m.f. of a cell when we know the values of the electrode potentials.

A result of particular importance is the finding that junctions formed by using concentrated KCl salt bridges between aqueous solutions of markedly different compositions have potentials in the range ±2 millivolt. When results are required only to this degree of precision, the junction potentials in salt bridged cells may be ignored.

$$E_{\text{salt bridged cell}} = E_R - E_L \pm 2\text{mV}$$

Electrochemical cells may be formed by inserting one platinum probe into the interior of a biological cell and another in the extracellular fluid. They often show e.m.f. values of 50–100 mV. If it is assumed that both electrode reactions involve the same species at the same relative activities (whatever they may be), then the electrode potentials may be ignored. The e.m.f. corresponds to the potential of the junction formed by the cell wall. It may be interpreted in terms of the selectivity of the cell wall toward transfer of such ions as sodium and potassium. Nerve action appears to involve abrupt changes in the permeability of cell walls to particular ions.

PROBLEMS CONCERNING ELECTROCHEMICAL CELLS

7.9. A cell is constructed with zinc dipping into a solution of zinc sulphate at the bottom of which is a pool of mercury covered with a precipitate of Hg_2SO_4. Construct the cell diagram appropriate to this cell.

7.10. Describe the essential features in the design of a hydrogen electrode.
(a) Calculate the potential of an hydrogen electrode in which the activity of $H^+(aq)$ is 10^{-8} and the activity of $H_2(g)$ is unity.
(b) If two hydrogen electrodes are connected so that charge is trans-

ferred for a long time what will eventually happen to the hydrogen ion activity in the two electrode systems?

7.11. Calculate the electrode potentials for the system

$$Cu|Cu^{2+}(aq) \text{ and } Zn|Zn^{2+}(aq)$$

at the following values of the ionic activities:

0.01, 0.1, 1.0 and 2.0

Given that

$$E^{\circ}_{Cu^{2}|Cu} = 0.34\text{V and } E^{\circ}_{Zn^{2+}|Zn} = -0.763\text{V}$$

If a salt bridged cell is constructed from these two electrode systems such that the initial activity of $(c/c^{\circ})_{Zn^{2+}} = 1.00$ and of $(c/c^{\circ})_{Cu^{2+}} = 0.01$ in equal volumes, show graphically how the cell e.m.f. varies as it runs down.

7.12 Using the half cells represented by

$Zn^{2+}|Zn$ $E^{\circ} = -0.763$ volts

$H^{+}|H_2$ $E^{\circ} = 0.000$ volts

$Ag^{+}|Ag$ $E^{\circ} = 0.799$ volts

set up three salt bridged electrochemical cells. For each cell write down

(a) the electrode reactions,
(b) the cell reaction,
(c) the standard e.m.f. of the cell,
(d) $\Delta\bar{G}^{\circ}$ for the cell reaction, and
(e) the direction of spontaneous electron flow in an external circuit when all relative activities are approximately unit.

7.13. Calculate the potential of an electrode formed by a platinum wire dipping into a solution containing $Fe(NO_3)_3$ ($2.0\ \text{mol dm}^{-3}$) and $Fe(NO_3)_2$ ($1.0 \times 10^{-4}\ \text{mol dm}^{-3}$) relative to the standard hydrogen electrode.
(It appears, by accident, that the activity coefficients of the salts in these solutions are close to 0.9).

7.14. It is said that it is not possible to measure the potential of a single electrode. Only the terminal voltage between a pair of electrodes in an electrochemical cell can be determined. On the other hand standard electrode potentials are listed in textbooks. We make use of them to estimate the potentials of individual electrodes and thus to predict the e.m.f.'s of cells. How does this become possible?

7.15. Consider the cell

$$Na\ (\text{in Hg}, x_{Na})|\,NaCl\ (\text{in } H_2O, c_{NaCl})$$
$$|\,AgCl\ (s)|\,Ag(s)$$

(a) Write symbolic representations of the *electrodes* in this cell system.
(b) Write equations for the reactions at these electrodes.
(c) Write the equation for the cell reaction, assuming that the aqueous solution is continually stirred.
(d) Write the equation which connects the equilibrium voltage of this cell with the compositions of the phases.
(e) Write the equilibrium condition (in terms of potentials of chemical substances, ionic species and electrons) for each electrode.

8. Energies of Materials

The energy changes which accompany chemical reaction are important to chemistry because they are related to the ways in which change of temperature alters the nature and extent of reactions. Control of the temperature is a significant feature of practical chemistry. It is used to control reactions when we wish to form some products and not others. A simple example is the formation of oxides of nitrogen in internal combustion engines. The proportion of nitric oxide, NO, in the cool exhaust gas is a function of the maximum temperature in the combustion flame and therefore also a function of the amount of energy which is given up by the hot gas to the piston and to the cooling system. Sometimes we have to control the rate of the reaction in order to keep the temperature of a reacting system at an acceptable value. It is highly desirable to know in advance how closely control has to be maintained. In all such cases, a knowledge of the changes in the energy is involved. We shall deal with the changes which occur in three quantities—the internal energy, the enthalpy and the entropy. Our treatment will be applied to substances, to materials and to systems.

Until this chapter we have been almost entirely concerned with the direction of chemical change and this led us to examine the ideas of absolute activity, rate of reaction and chemical potential. With such a starting point, energy is a topic of secondary importance. In our experience, it is better to separate the discussion of the ideas of chemical potential and energy rather than to deal with them together, as is usually done. We find that the separation helps to avoid confusing the ideas with one another. Many good chemists and biologists are very accurate in their use of the general concepts which we have presented so far. Accurate, that is, except that some of them believe they are talking about energies when, in fact, they are talking about chemical potentials. Because such confusion is common, we have given a more complete treatment in this chapter than in others of this book.

8.1 Energy

Let us begin with a general introduction to the concept of energy.

8.1.1 The basic concept of energy

The word "energy" is used widely and conveys a range of ideas. "Energy" is what a person "has" and "uses" in working toward some desired result—the construction of a building, the presentation of a report, the completion of a work of art... The values which we attach to such achievements may be gauged in different ways—how much personal satisfaction is gained, how much the product is admired, how much money is offered in payment... These values are much concerned with what we want as a result of the work. In one sense or another, the energy of the worker is transferred to the product of his work. We speak of great effort which leads only to a result of little value as "a waste of energy".

During the last three centuries the term "energy" has been taken over into physical science and given restricted meanings. The first adoption of the concept of energy in physical science occurred in simple mechanics. This deals only with systems in which there are no thermal effects (changes of temperature and related phenomena associated with friction, resistance, etc.) or chemical effects (changes of composition and amounts of material). For such simple mechanical systems it is possible to make the following two statements about energy:

I. In any change which may occur when two or more systems interact in isolation from the rest of the universe, the sum of their energies is unaltered.

II. In any change in the state of a system, the increase in its energy is equal to the work applied to the system.

Statement I implies that energy may be transferred from one system to another, but is conserved. A decrease of energy in one part of a system must be matched by an equal increase of energy in another part of the system, or in its surroundings. This statement is retained and is true for all systems. Statement II does not apply to the majority of systems which are chemically or biologically interesting.

8.1.2 The transfer of energy

In order for energy to transfer from one system to another, there must be some interaction between the systems. The amount of energy transferred is simply related to the amount of interaction which occurs.

Comment I don't think I understand what you mean by that!
Reply The examples illustrated in Fig. 8.1 might help.

In Fig. 8.1a, the sun and the flower interact through the emission and absorption of radiation (the sunlight). The amount of the interaction (the amount of energy transferred) depends on the number of photons transferred and on their energy.

The man in Fig. 8.1b interacts with the bag of flour by applying a force to it. The amount of energy transferred from the man to the flour is equal to the amount of work done on the flour by the man.

The man in Fig. 8.1c interacts with his environment by eating food. The amount of energy transferred to the man depends on the amount and the nature of the food.

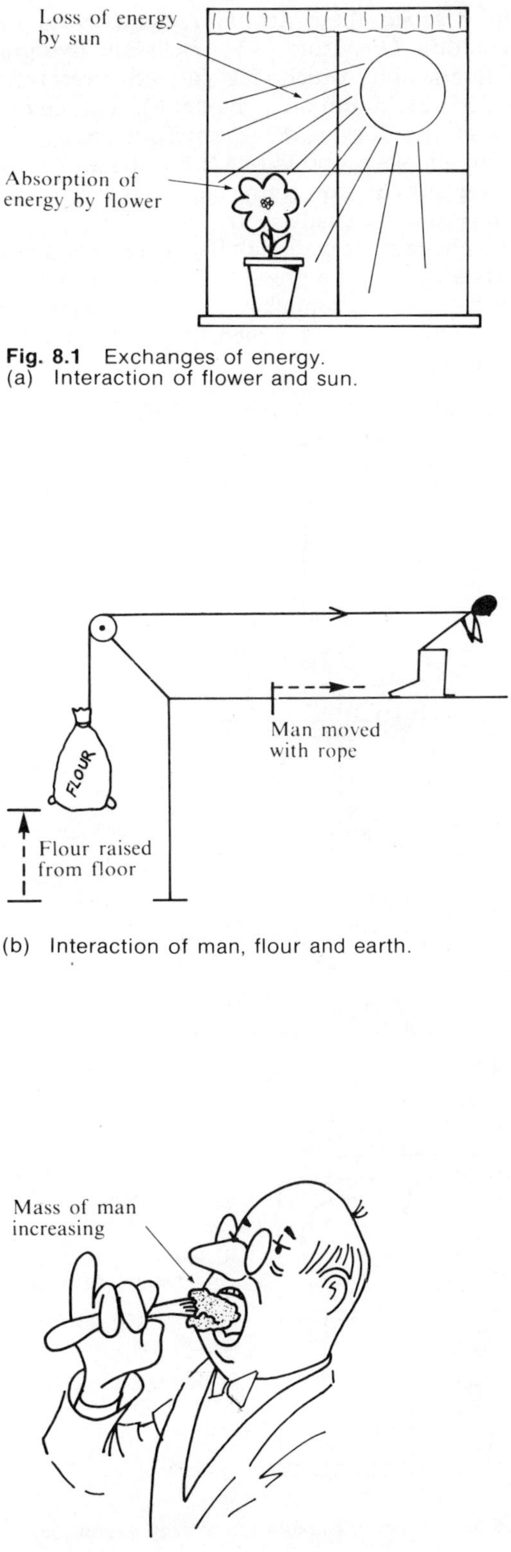

Fig. 8.1 Exchanges of energy.
(a) Interaction of flower and sun.
(b) Interaction of man, flour and earth.
(c) Interaction of man and environment.

In practice there are *four* main types of interaction. Operator work, $w_{\text{♀}}$, is one method of interaction which has already received considerable attention (Chapter 6). The environment also does work, w (*environment*), on a system simply by providing a constraint on its motion or a constraint in expansion or compression. Usually w (*environment*) arises from changes of volume. If the environment exerts a hydrostatic pressure (P_e) on the system (Fig. 8.2) and the volume of the system *increases* by an amount $V_2 - V_1$, then the work done by the environment

$$w(environment) = -P(environment) \times (V_2 - V_1)$$

or

$$= -P(environment) \times \Delta V$$

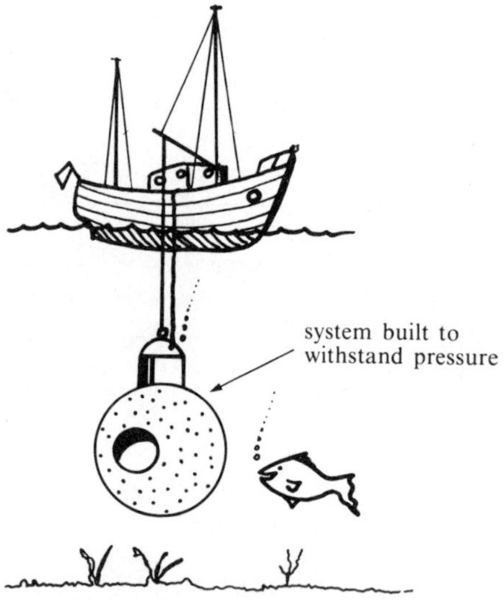

Fig. 8.2 Hydrostatic pressure of environment.

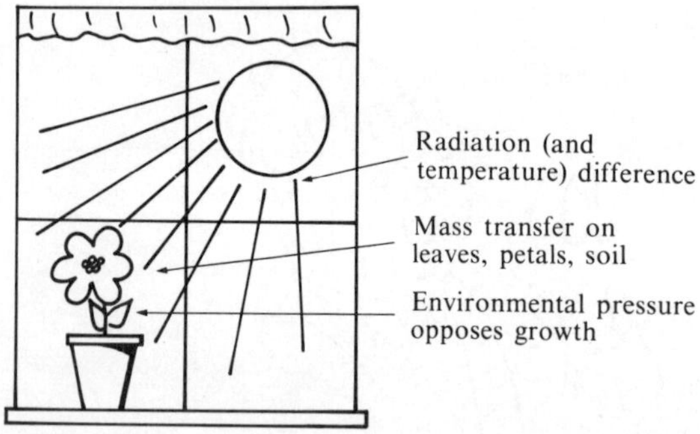

Fig. 8.3 Interactions leading to energy exchange.

The third type of interaction occurs wherever bodies differ in temperature. In this case the interaction may involve radiation of energy, conduction of energy or convective transfer. The transfer of energy by any, or all, of these processes is called *heat* (symbol q).

Sometimes we have to work with "open" systems which can exchange matter with others, as opposed to "closed" (constant mass) systems. Material which comes into, or goes out of, an open system does so bearing energy which depends on the state of the material, that is, on composition, phase type, temperature and pressure. This is a fourth method of interaction.

The interactions which give rise to energy transfers in one particular case are sketched in Fig. 8.3.

Exercise 8.1
List the energy exchanges which are present in the following systems, and classify them as above:
(a) a cat, sleeping in the sunshine;
(b) an electric hotplate, switched on;
(c) a tree in a gusty wind;
(d) a plate of zinc dissolving in an acidic solution;
(e) a transistor radio;
(f) a firing rocket motor;
(g) a hot air balloon.
In some of these cases there may well be more than one way of describing the "system", and so different ways of classifying the energy exchanges may be adopted.

The energy exchanges may be expressed as

$$\Delta E = w_{\text{♀}} + w(environment) + q + E_m$$

In this equation

ΔE = amount of energy transferred to the system—the increase in the energy of the system

$w_{\text{♀}}$ = the total amount of operator work done on the system

$w(environment)$ = the total amount of work done on the system by the environment

q = the amount of heating of the system

and

E_m = the amount of energy associated with the transfer of additional matter to the system.

8.1.3 Examples of energy transfer

Example 8.1
Consider a kettle of cold water whose temperature is to be raised to about 370K. We can put the kettle on a hot surface (a "hot plate") or in contact with a gas flame (Fig. 8.4). Heating will be done on the kettle and the water. There is a very small (negative) amount of environmental work as the kettle and its contents expand. There is no operator work, and, if the lid is a good fit, no significant change in the amount of the contents so

$$\Delta E(\text{kettle} + \text{contents}) = q$$

Example 8.2
An electric radiator (Fig. 8.5) provides a further example which is worth thought. When the radiator is at the same temperature as the room and we turn on the switch, electrical work begins to be done. In time, d*t*, the work done is

$$w_{♀} = \Delta \mathscr{V} I \,\mathrm{d}t$$

where ΔV is the applied voltage and I is the current. At first, this only increases the energy of the radiator

$$w_{♀} = \Delta E(\text{radiator})$$

But immediately the temperature of the radiator begins to be higher than that of the surroundings, heating begins to be done on the surroundings. As the temperature rises, so the heat loss increases in importance, and eventually the temperature of the radiator and its energy reach a steady value. The constancy of the energy implies that, in this state,

$$w_{♀} = -q$$

When the switch is turned off, work ceases to be done but transfer of energy by heat continues as long as the radiator is at a higher temperature than the surroundings.

Kettles with "heating elements" built into them are related to radiators. When connected to the power supply, electrical work is done but there is no heating. To be a little more precise, the only heat transfer in this case is a term which is small for a well-designed kettle—the heat loss to the surroundings when the kettle is at a higher temperature than they are. So, in operating this type of kettle, the heat may be neglected, and

$$\Delta E(\text{kettle} + \text{contents}) = w_{♀}$$

Query But the kettle gets hot. Surely work is turned into heat?

Reply "Work" and "heat" are only *ways* of transferring energy. It is the *energy* and the *temperature* of the kettle which increase.

Example 8.3
There are many ways in which we could react hydrogen with oxygen, starting from gases at 298K and 101 kPa and ending with liquid water at 298K and at 101 kPa.

$$2H_2(g,298\text{K},P^\circ) + O_2(g,298\text{K},P^\circ) = 2H_2O(l,298\text{K},P^\circ)$$

Let us look at the energy transfers which accompany 1 mol of this reaction.

(a) We might cause the gases to react on a solid catalyst surface under conditions in which the temperature stays constant because

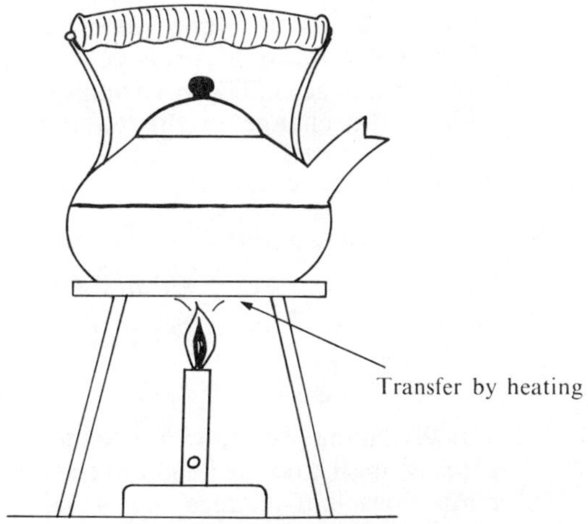

Fig. 8.4 Energy transfers in a kettle.

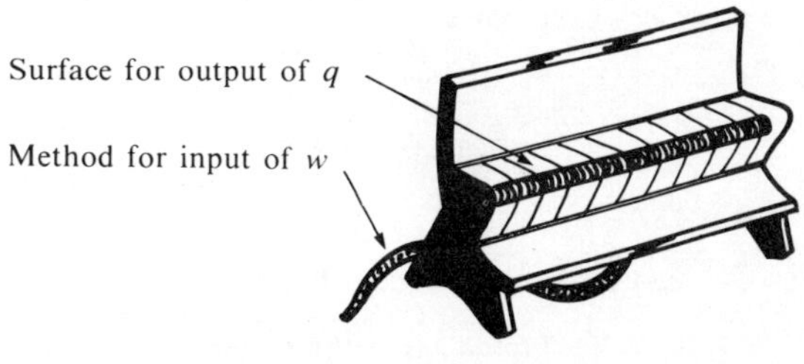

Fig. 8.5 Energy transfers in an electric radiator.

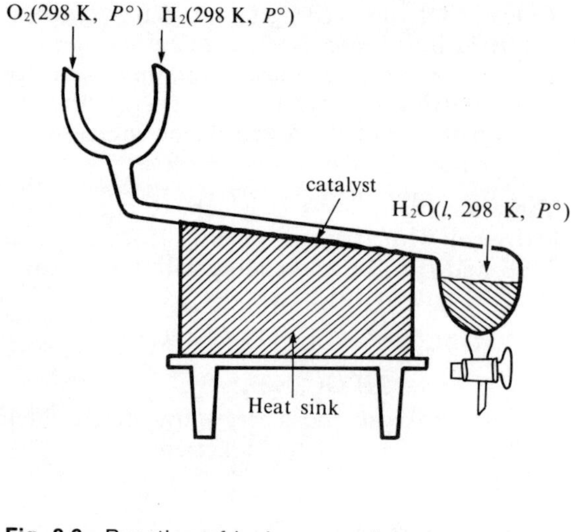

Fig. 8.6 Reaction of hydrogen and oxygen on a catalyst surface.

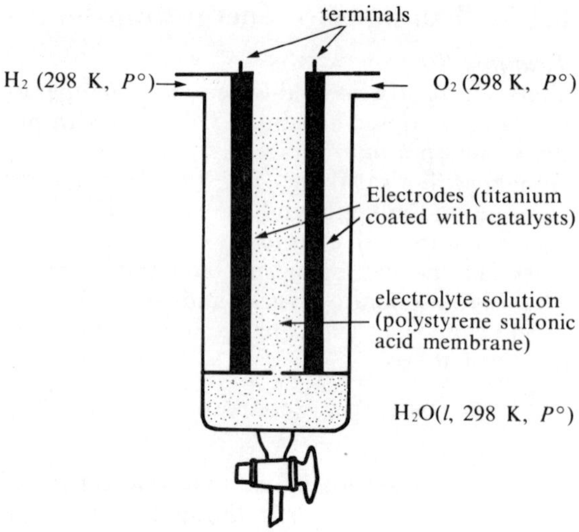

Fig. 8.7 Reaction of hydrogen and oxygen in an electrochemical cell.

there is efficient heat transfer to a large "heat sink" at a temperature just below 298K (Fig. 8.6). In this process, the operator work, $w_{♀}$, is zero. The environmental work reflects the change in the volume of the system.
Thus,

$$w(\text{environment}) = -P_e\Delta V$$
$$= 3\ \text{mol} \times RT^{‡}$$
$$= 7.4\ \text{kJ}$$

$$V(\text{products}) = 36\ \text{cm}^3$$

If the change in the kinetic energy of the gases is small, and the height of the centre of mass does not change by a significant amount, then

$$\Delta E = -563\ \text{kJ}$$

‡For a gas at low pressures

$$V \simeq nRT/P$$

$$n(\text{reactants}) = 3\ \text{mol}$$

hence

$$V(\text{reactants}) \simeq 3\ \text{mol} \times RT/P$$

$$\Delta V \simeq V(\text{products}) - V(\text{reactants})$$

$$= 36\ \text{cm}^3 - 73.6\ \text{dm}^3$$

and $-P_e\Delta V \simeq 7430\ \text{J}$

Hence, in this case, using

$$\Delta E = w_{♀} + w(\textit{environment}) + q$$
$$q = (-563 - 7.4)\text{kJ}$$
$$= -570\ \text{kJ}$$

Thus, the largest part of the energy change is the energy transfer to the "heat sink". Some of the troubles most of us have with the concept "heat" can be traced to our failure to appreciate what is being done in a "heat sink". In practice we carry out many operations where there is a great deal of energy transferred by heat with only slight temperature change. However, the transfer cannot occur without *some* temperature difference. Suitable choice of materials and devices may make a very small temperature difference big enough to be useful.

(b) Consider now the same reaction carried out in an electrochemical cell (Fig. 8.7). The minimum operator work to carry out the process is equal to the change in Gibbs energy, ΔG. For 1 mol of reaction, using data from Table 6.1,

$$w_{♀}(\textit{minimum}) = -474\ \text{kJ}$$

As for the previous example, the environmental work will be almost entirely due to the change in volume which accompanies the removal of the gases.

Hence $w(\textit{environment}) = 7.4\ \text{kJ}$

Also as before, providing there is little change in the kinetic energy of the gases, or of the height of their centre of mass

$$\Delta E = -563\ \text{kJ}$$

Hence, in a completely efficient electrochemical cell,

$$\Delta E = w_{♀} + w(environment) + q$$

$$-563\ \text{kJ} = -474\ \text{kJ} + 7.4\ \text{kJ} + q$$

That is, $q = -96\ \text{kJ mol}^{-1}$

In practice, fuel cells like this are not used at 100% efficiency. A realistic figure is 60%. That means that

$$w_{♀} = 0.6\ w_{♀}\ (minimum)$$
$$= -284\ \text{kJ mol}^{-1}$$

and thus

$$q = (-563 + 284 - 7.4)\ \text{kJ mol}^{-1}$$
$$= -286\ \text{kJ mol}^{-1}$$

Note, in particular, that the amount of energy exchange as heat depends very greatly on the amount of operator work. The more efficiently we use the cell to produce operator work, the less energy is transferred as heat.

(c) As a third case, we might consider the use of hydrogen as a fuel in an internal combustion engine. The sequence of events as illustrated in Fig. 8.8 is a gross oversimplification, but contains the general features of the operation of such an engine.

For the overall process,

$$w_{♀}\ (minimum) = -138.5\ \text{kJ}$$

$$w\ (environment) = 7.4\ \text{kJ mol}^{-1}$$

$$\Delta E = -563\ \text{kJ}$$

and hence $q = -432\ \text{kJ}$

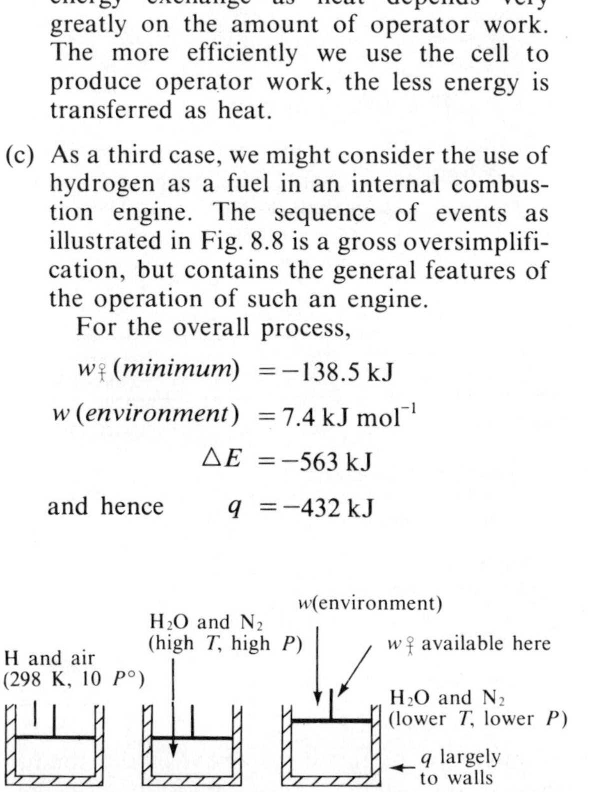

Fig. 8.8 Hydrogen and oxygen reacting in an internal combustion engine.

Exercise 8.2

(a) A sleeping man loses energy to his environment at about 70 J s^{-1}. Examine the modes in which this loss occurs.

(b) A man working hard (breaking stones) lifts a 5 kg hammer to a height of 2 m six times per minute. His total energy loss to the environment is about 500 J s^{-1}. Examine the modes in which this loss occurs.

(c) A man travels along a level road. By considering the interactions between his legs and the environment, show that normal walking at 5 km hr^{-1} uses more energy than cycling at 5 km hr^{-1} which uses more energy than a shuffling gait where the feet barely clear the ground.

PROBLEM

8.1. Consider the operation of the Mallory cell at 298 K, in which the cell reaction is

$$Zn(s) + HgO(s) = ZnO(s) + Hg(l)$$
$$(\Delta \bar{E} = -258\ \text{kJ mol}^{-1}$$
$$\Delta \bar{G} = -259\ \text{kJ mol}^{-1})$$

Deduce the amount of energy lost to the environment as heat

(a) when it operates as a power source at 80% efficiency (the temperature remains approximately constant);

(b) when the same reaction occurs at constant temperature without power production.

8.1.4 Energy and internal energy

The total energy, E, of a body can conveniently be represented as the sum of three parts; namely kinetic, potential and internal energies.

Any motion of the whole body causes it to have *kinetic energy*, E_k. This depends only on the mass, m, of the body and on its velocity, v

$$E_k = \tfrac{1}{2} m v^2$$

External gravitational, electric or magnetic fields cause a body to have *potential energy*, E_p. The magnitude of E_p depends on the position of the masses, electric charges and magnetic dipoles of the body in the field. For example, the gravitational potential energy depends on the mass of the body, its distance from the earth, and the local value of the gravitational acceleration, g.

$$E_p = mgh$$

The internal energy, U, is associated with the material within the body. It is the only part of the total energy which depends on the composition properties of the materials which make up the body. Thus, it is the part of the total energy, E, which is of chemical interest.

8.2 The Internal Energy of a System and Its Materials

We have seen how the kinetic energy of a body is described by its motion and the potential energy of a body is described by its position. In the next two sections we will examine the way in which the internal energy of a body is related to the properties of its constituent materials. First we shall consider the four ways in which this is done—namely, as the internal energy of bodies; as the specific internal energy of materials; as the partial molar internal energies of chemical substances within materials; and as the internal energy of interaction of materials with one another.

8.2.1 A particular system

It may be easier to see what is being done if we take a particular example and follow it through where we can. Consider a reaction vessel (Fig. 8.9) in which toluene and nitric acid are reacting to form nitrotoluenes.

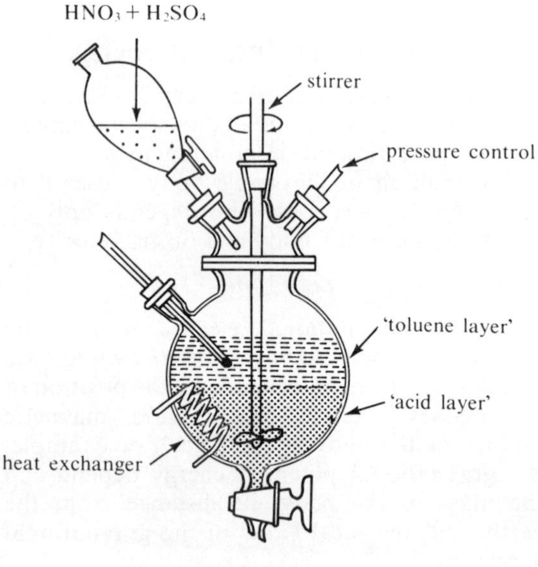

Fig. 8.9 A reaction vessel for nitrating toluene.

The main reactions are:

$$\text{toluene }(l)\text{ (in }H_2SO_4, l) \quad o\text{-nitrotoluene (in }H_2SO_4, l) \tag{1}$$

$$\langle\bigcirc\rangle\text{—}CH_3 + HNO_3 = \langle\bigcirc\rangle\text{—}CH_3 + H_2O$$

$$\text{toluene }(l)\text{ (in }H_2SO_4, l) \quad (p\text{-nitrotoluene (in }H_2SO_4, l) \tag{2}$$

$$\langle\bigcirc\rangle\text{—}CH_3 + HNO_3 = NO_2\text{—}\langle\bigcirc\rangle\text{—}CH_3 + H_2O$$

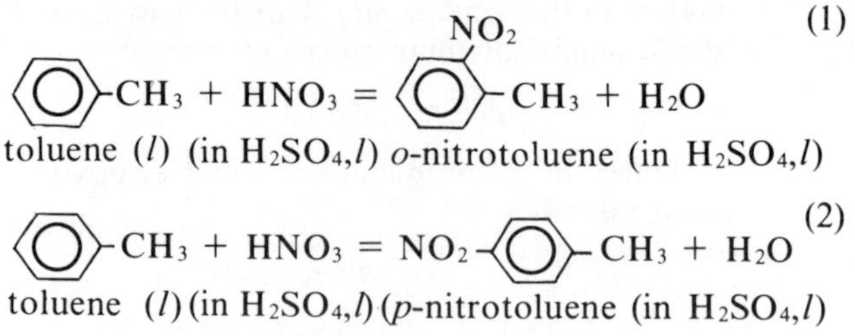

During the reaction, a mixture of nitric acid and sulfuric acids is added slowly to the vessel. If the mononitrotoluenes are to be produced, it is necessary to control the temperature (for example, to 25–30°C) and to stir the mixture. Typically this process then gives:

56% o-nitrotoluene

30% p-nitrotoluene

4% m-nitrotoluene

10% other products.

In what follows we will consider reactions conducted under conditions similar to these, and the changes of internal energy which they involve.

8.2.2 The internal energy of a system

Consider as a system, the reaction vessel (Fig. 8.9) with its heat exchanger, stirrer and thermometer, together with the reacting fluids. The system interacts with its surroundings in a number of well defined ways (Table 8.1).

Table 8.1
Interactions between the System and Its Surroundings.

Interaction	Relative Size
work done on stirrer	fairly small
work done by atmosphere as the volume of the system changes	very small
cooling through the heat exchanger	large
heat lost through the walls of the reaction vessel	very small

We use the interactions to control the conditions of the reaction. Their sum is equal to the change in the internal energy of the system.

The energy of this system has three parts

which are treated separately. The internal energy of the system is just the sum of the internal energies of the reaction vessel, including its accessory equipment, and of the reacting fluids. There is also kinetic energy of motion of the reacting fluids, the stirrer and other moving parts and potential energy due to the relative positions (in particular, of the fluids in their reservoirs before mixing and in the main reactor).

8.2.3 The internal energy of a body

The reaction vessel, with its heat exchanger and thermometer, are likely to be used repeatedly in related experiments. For such apparatus it is usually convenient to deal with the internal energy, U(vessel), and changes, ΔU(vessel), in the internal energy of the whole vessel.

8.2.4 Specific internal energy

If we need to compare the internal energy of two such reaction vessels, or to predict the values for a new one, we are likely to use the values of the *specific internal energy*, u, and masses, m, of the materials. For a vessel made of glass, steel, etc., we would write

$$U(\text{vessel}) = u(\text{steel}) \times m(\text{steel}) + u(\text{glass}) \times m(\text{glass}) + \dots$$

Changes in U(vessel) can similarly be written in terms of the changes in the specific internal energies and the changes in mass.

Example 8.4
A well-insulated copper tank (mass 5 kg) contains 70 kg of paraffin oil. Calculate the change in the internal energy of the system (tank plus oil) when its temperature is raised to 170°C if

$$\Delta u(\text{oil}) = 1.3 \text{ kJ g}^{-1}$$

$$\Delta u(\text{copper}) = 220 \text{ J g}^{-1}$$

Answer

$$\Delta U = \Delta u(\text{oil}) \times m(\text{oil}) + \Delta u(\text{copper}) \times m(\text{copper})$$

$$= 1.3 \times 10^3 \text{ J g}^{-1} \times 70 \times 10^3 \text{ g} + 220 \text{Jg}^{-1} \times 5 \times 10^3 \text{g}$$

$$= 9.2 \times 10^7 \text{ J}$$

8.2.5 Molar internal energies

In the reaction vessel, there are two liquid phases and one gas phase. One liquid phase contains mainly toluene and the nitrotoluenes, and the other contains mainly nitric acid and sulfuric acid. The chemical compositions of each of the liquid phases (measured by the mole fractions or concentrations of its constituents) change greatly during the reaction, and so, presumably, does the internal energy of these phases. Initially, one of them contains almost entirely toluene. By the end of the reaction it is largely a mixture of the three mononitrotoluenes. The method used to describe the dependence of the internal energy of such a phase on the chemical composition is the same as that used for volume. We define partial molar internal energies, $\bar{U}_B$, for each substance, B, so that the internal energy of the phase can be expressed in terms of the $\bar{U}_B$ values and the amounts, n_B of the substances. That is, for the "toluene" phase,

$$U = \bar{U}_1 \times n_1 + \bar{U}_2 \times n_2 + \bar{U}_3 \times n_3 + \dots$$

where substance 1 is toluene; substance 2 is *o*-nitrotoluene; substance 3 is *p*-nitrotoluene, etc.

Example 8.5
Estimate the internal energy of a mixture of 20 mol of ethanol and 15 mol of water if

$$\bar{U}(\text{ethanol}) = -235 \text{ kJ mol}^{-1}$$

$$\bar{U}(\text{water}) = -286 \text{ kJ mol}^{-1}$$

Answer

$$U_{(\text{solution})} = \bar{U}(\text{ethanol}) \times n(\text{ethanol}) + \bar{U}(\text{water}) \times n(\text{water})$$

$$= (-235 \times 20) + (-286 \times 15) \text{ kJ}$$

$$= -9 \times 10^6 \text{ J}$$

The values of the partial molar internal energies will not be constant, but vary with the composition of the phase. The gain which results from their use is that they are properties of the "toluene" phase and do not depend on the presence of the other phases, and hence on the presence of these particular chemical reactions.

Exercise 8.3
Write equations for the internal energy of the other liquid phase and for the gas phase, assuming that the gas has the composition of ordinary air.

8.2.6 Other issues

The remaining terms in the internal energy of the system depicted in Fig. 8.9 involve the interaction of the phases with one another and with the reaction vessel. These terms are likely to be small in the case now under discussion. In other situations, as for example in studies of adhesives, they will be much more important.

8.3 Changes in Internal Energy

We shall now consider the ways in which the internal energy changes with changing temperature, pressure, chemical composition (chemical reaction) and phase. In any practical situation, most of these variables will be altering. However it turns out that we need only consider the effect of changing one variable at a time.

8.3.1 Change of temperature

We will begin by considering the change in internal energy with temperature. For small changes in temperature, changes in internal energy are expressed in terms of the ratio of the change in internal energy to the change in temperature. That is, we use $\delta U/\delta T$. If the pressure and composition are kept constant, then the effect of very small changes in temperature on the value of the internal energy is described by

$$(\partial U/\partial T)_{P,\text{ composition}}$$

This is the partial derivative of internal energy with respect to temperature, at constant temperature and amount of material. Values of $(\partial U/\partial T)_{P,x}$ for a number of pure substances are listed in Table 8.2.

For a given change in temperature (T_1 to T_2), the change in internal energy is given by the integral of $(\partial U/\partial T)_P$,

$$\Delta U = \int_{T_1}^{T_2} (\partial U/\partial T)_{P,x} \mathrm{d}T \qquad (1)$$

or

$$U(T_2) = U(T_1) + \int_{T_1}^{T_2} (\partial U/\partial T)_{P,x} \mathrm{d}T$$

For small changes in temperature it is often sufficiently accurate to treat $(\partial U/\partial T)_{P,x}$ as if it were independent of temperature. Then the evaluation of (1) is straightforward and is given by

$$\Delta U = (\partial U/\partial T)_{P,x} \times (T_2 - T_1)$$

Example 8.6
Determine the change in energy of 1 tonne of

Table 8.2
Variation of Internal Energy with Temperature at 298 K *and* 101 kPa

Substance	$(\partial U/\partial T)_P$ / $\mathrm{J\,mol^{-1}K^{-1}}$	C_P / $\mathrm{J\,mol^{-1}K^{-1}}$	C_V / $\mathrm{J\,mol^{-1}K^{-1}}$
Gases			
Helium	12.47	20.79	12.48
Hydrogen	20.52	28.84	20.52
Oxygen	21.02	29.36	21.05
Nitrogen	20.79	29.12	20.81
Nitrous oxide, N_2O	30.28	38.71	30.40
Carbon dioxide, CO_2	28.72	37.65	28.81
liquids			
Carbon disulfide, CS_2	75.7	75.7	47.5
Water	75.34	75.34	74.86
Methanol	80.8	80.8	64.8
Ethanol	66.5	66.5	40.7
n-pentane	120.2	120.2	93.8
Benzene	136	136	93
Mercury	27.82	27.82	24.16
Solids			
Aluminium	24.34	24.34	
Copper	24.47	24.47	
Lead	26.82	26.82	26.6
Aluminium oxide, Al_2O_3	78.99	78.99	79.0
Calcium fluoride, CaF_2	67.02	67.02	67.0
Silicon oxide, SiO_2	44.35	44.35	44.3

tallow when it is heated from 20°C to 170°C. For 1 tonne of tallow

$$(\partial U/\partial T)_P = 8 \times 10^5 \text{ J K}^{-1}$$

$$\Delta U \approx (\partial U/\partial T)_P (T_2 - T_1)$$

$$= 8 \times 10^5 \text{ J K}^{-1} \times 150 \text{ K}$$

$$= 12 \times 10^7 \text{ J}$$

For large changes of temperature, or when answers of higher precision are required, it is necessary to take into account changes in the value of $(\partial \bar{U}/\partial T)_P$ with temperature. If $(\partial \bar{U}/\partial T)_P$ is known as an explicit function of temperature then equation (1) can be evaluated directly.

Example 8.7
At atmospheric pressure and at temperatures between 300 K and 800 K, the change in the specific internal energy of aluminium is given by the relation

$$(\partial u/\partial T)_{P,x} = 0.766 + 0.00046\, T/K \text{ J g}^{-1} \text{ K}^{-1}$$

Estimate the change in the specific internal energy of aluminium when heated from 300 to 500 K at $P = 101$ kPa.

$$\Delta u = \int_{T_1}^{T_2} (\partial u/\partial T)_{P,x}\, \mathrm{d}T$$

$$\Delta u = \int_{300}^{500} (0.766 + 0.00046\ T)\, \mathrm{d}T$$

$$= \Big[(0.766\ T + \tfrac{1}{2} 0.00046\ T^2)\Big]_{300}^{500}$$

$$= 190 \text{ J g}^{-1}$$

Often values of $(\partial \bar{U}/\partial T)_P$ are tabulated at a number of different temperatures and the algebraic equation for $(\partial \bar{U}/\partial T)_P$ is not known. In this case the integration of equation (1) is carried out graphically.

Example 8.8
Table 8.3 lists values of $(\partial \bar{U}/\partial T)_P$ of triuranium octaoxide (U_3O_8) at atmospheric pressure and a number of temperatures. The change in internal energy of U_3O_8 on cooling from 550 K to 300 K can be determined by plotting the graph of $(\partial \bar{U}/\partial T)_P$ against T (Fig. 8.10) and evaluating the area under the graph from $T = 550$ to 300 K.

Table 8.3
Values of $(\partial \bar{U}/\partial T)_P$ of $U_3O_8(s)$

T/K	$(\partial \bar{U}/\partial T)_P$/(J mol^{-1}K^{-1})
300	239
350	255
400	265
450	271
479	281
482.7	289
489	279
500	276
550	281

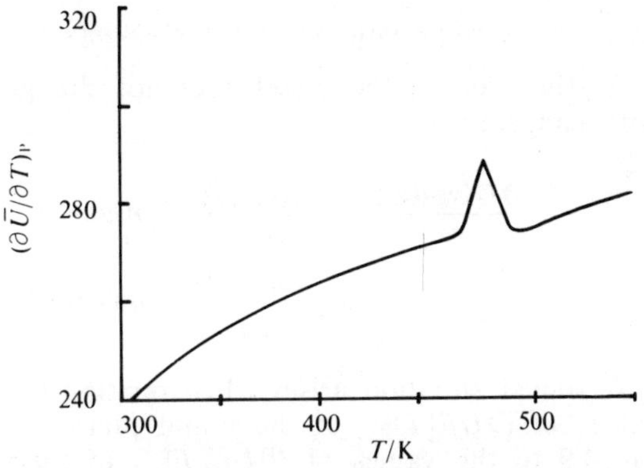

Fig. 8.10 Values of $(\partial \bar{U}/\partial T)_P$ for triuranium octaoxide (U_3O_8) at atmospheric pressure.

Exercise 8.4
The internal energy of gaseous HCl is −92 kJ mol^{-1} at 10^5 Pa and 300 K. Estimate the internal energy of HCl at

(a) 500 K
(b) 1300 K.

Assume that $(\mathrm{d}\bar{U}/\mathrm{d}T)_P$ has the values:

(i) zero
(ii) 29 J K^{-1} mol^{-1} (the value at 300 K)
(iii) 29 J K^{-1} mol^{-1} for (a) (the value at 400 K) and 31 J K^{-1} mol^{-1} for (b) (the value at 800 K)
(iv) $28.17 + 1.82 \times 10^{-3}\ T/K + 15.5 \times 10^{-7}\ (T/K)^2$ J K^{-1} mol^{-1}

Answer

	(a)	(b)
(i)	−92 kJ mol^{-1}	−92 kJ mol^{-1}
(ii)	−86 kJ mol^{-1}	−63 kJ mol^{-1}
(iii)	−86 kJ mol^{-1}	−61 kJ mol^{-1}
(iv)	−86 kJ mol^{-1}	−61 kJ mol^{-1}.

The answers to the above exercise provide an example of the very common situation in which the changes with temperature $(dU/dT)_P$ need only very approximate treatment.

In some of the above examples we have talked about the change in the internal energy of a system with temperature while in others we have talked about the change in the specific internal energy of a material with temperature, or the change in the partial molar internal energy with temperature. It is important to see how these quantities are related.

For example, the internal energy of a vessel made of glass and steel is related to the specific internal energy of its materials by

$$U_{(\text{vessel})} = u(\text{glass}) \times m(\text{glass}) + u(\text{steel}) \times m(\text{steel})$$

As the mass of the vessel does not change with temperature,

$$\frac{\partial U(\text{vessel})}{\partial T} = \frac{\partial u(\text{glass})}{\partial T} \times m(\text{glass}) + \frac{\partial u(\text{steel})}{\partial T} \times m(\text{steel})$$

A similar situation arises when relating the values of $(\partial U/\partial T)_{P,x}$ of the liquid phases in Fig. 8.9 to the values of $(\partial \bar{U}/\partial T)_{P,x}$ of their constituents. For the toluene phase

$$\left(\frac{\partial U(\text{"toluene" phase})}{\partial T}\right)_{P,x}$$

$$= \left(\frac{\partial \bar{U}_1}{\partial T}\right)_{P,x} \times n_1 + \left(\frac{\partial \bar{U}_2}{\partial T}\right)_{P,x} \times n_2 + \left(\frac{\partial \bar{U}_3}{\partial T}\right)_{P,x} \times n_3 \ldots$$

where substance 1 is toluene; substance 2 is *o*-nitrotoluene; and substance 3 is *p*-nitrotoluene.

In practice, tables of values of $(\partial \bar{U}/T)_{P,x}$ are rarely compiled. Instead we tend to find tables of the more frequently needed and closely related, "heat capacity at constant pressure" (C_P), developed in Section 8.6.1. These are related (Section 8.6) by

$$(\partial \bar{U}/\partial T)_{P,x} = \bar{C}_P - P\bar{V}\alpha.$$

in which V is the volume, and α is the coefficient of thermal expansion (Section 2.8).

More rarely, tables are compiled for the "heat capacity at constant volume" (C_V)‡ for the various substances which make up the system.

‡ $\bar{C}_V = (\partial \bar{U}/\partial T)_{V,x}$

That is, it describes the way in which the internal energy changes with temperature when the volume is kept constant. It is of considerable theoretical importance as it is more easily related to the molecular structure of a substance or material than either C_P or $(\partial U/\partial T)_{P,x}$

Inspection of Table 8.2 shows that for gases $(\partial \bar{U}/\partial T)_{P,x} \approx \bar{C}_V$, while for condensed phases $(\partial \bar{U}/\partial T)_{P,x} \approx \bar{C}_P$.

8.3.2 Change of pressure

We will now consider the way in which the internal energy of a phase changes with change of pressure when the temperature is kept constant. For small changes of pressure this is conveniently expressed in terms of the ratio of the change in internal energy to the change in pressure. That is, we use $(\partial U/\partial P)_T$. Some examples are given in Table 8.4.

The actual change in internal energy for any particular change of pressure is given by

$$\Delta U = \int_{P_1}^{P_2} (\partial U/\partial P)_T \, dP \qquad (2)$$

For relatively small changes in pressure, $(\partial U/\partial P)_T$ can be treated as being constant. Under these conditions equation (2) is readily evaluated, giving

$$\Delta U = (\partial U/\partial P)_T (P_2 - P_1)$$

Table 8.4
Variation of Internal Energy with Pressure at 298 K and 101 kPa

Substance	$10^6 \dfrac{(\partial \bar{U}/\partial P)_T}{\text{J mol}^{-1}\text{Pa}^{-1}}$
Gases	
Helium	+1.0
Hydrogen	−7
Oxygen	−59
Nitrogen	−51
Nitrous oxide, N_2O	−310
Carbon dioxide, CO_2	−300
Liquids	
Carbon disulfide, CS_2	−20.8
Water	−1.13
Methanol	−14.4
Ethanol	−18.9
n-pentane	−53.6
Benzene	−32.5
Mercury	−0.80
Solids	
Aluminium	−0.258
Copper	−0.126
Lead	−0.537
Aluminium oxide, Al_2O_3	−0.182
Calcium fluoride, CaF_2	−0.41
Silicon oxide, SiO_2	*ca.* −0.008

Table 8.4 lists values for the change in molar internal energy of a number of substances when the pressure is changed.

The values are small enough to make the differences in internal energy smaller than those which arise from chemical reaction, at least for pressure changes up to a few times atmospheric pressure.

8.3.3 Internal energy of different phases

The transfer of substances from one phase ot another is often used, biologically and technologically, to maintain an almost constant temperature despite changes of chemical composition or transfers of energy as heat or work. Most of these processes occur at constant temperature and receive more detailed treatment later (Section 8.7.3). Some examples of the differences of internal energy of substances when in different phases are given in Table 8.5.

The transfer or transformation of a substance from one phase into another is best considered as a change in the chemical content of the two phases.

Consider the solution of 4.0 g (~0.10 mol) of sodium hydroxide in 1.0 kg (55.56 mol) of water, at 25°C and at atmospheric pressure. Under these conditions,

$$\bar{U}(\text{NaOH}, \textit{solid}) = -426.73 \text{ kJ mol}^{-1}$$

$$\bar{U}(\text{Na}^+, \textit{aqueous}) = -239.655 \text{ kJ mol}^{-1}$$

$$\bar{U}(\text{OH}^-, \textit{aqueous}) = -229.94 \text{ kJ mol}^{-1}$$

$$\bar{U}(\text{H}_2\text{O}, \textit{liquid}) = -285.839 \text{ kJ mol}^{-1}$$

The initial condition is two phases: one being 0.1 mol of solid sodium hydroxide with an internal energy:

$$U(\textit{solid}) = \bar{U}(\text{solid NaOH}) \times n(\text{NaOH}, \textit{solid})$$

the other being 1 kg of water with an internal energy of

$$U(\text{water}) = \bar{U}(\text{water}) \times n(\text{water}).$$

Hence the internal energy initially is

$$U(\text{initial}) = U(\text{solid NaOH}) + U(\text{water})$$

$$= -426.73 \times 0.1 - 285.839 \times 55.56 \text{ kJ}$$

$$= -15924 \text{ kJ}$$

Table 8.5
Variation of Internal Energy in Some Changes of Phase at 101 kPa.

Substance	*Phase Change*	T/K	ΔU/kJ mol^{-1}
oxygen	$s = l$	54.8	0.44
water	$s = l$	273	6.01
camphor	$s = l$	449	6.8
magnesium	$s = l$	924	9.2
oxygen	$l = g$	90	6.07
acetone	$l = g$	329	27.6
chloroform	$l = g$	334	20.3
water	$l = g$	373	37.6
naphthalene	$l = g$	484	36.5
sodium hydroxide	s = dilute aqueous solution	298	−43
sodium nitrate	s = dilute aqueous solution	298	+20
ammonia	g = dilute aqueous solution	298	−32

After dissolution, one phase only remains: a solution consisting of 0.10 mol of sodium and hydroxide ions and 55.56 mol of water. The internal energy of this solution is therefore

$$U(\text{final solution}) = \bar{U}(\text{Na}^+, \textit{aqueous}) \times n_{\text{Na}} + \bar{U}(\text{OH}^-, \textit{aqueous}) \times n_{\text{OH}^-} + \bar{U}(\text{H}_2\text{O}, \textit{liquid}) \times n(\text{H}_2\text{O})$$

$$= -239.655 \text{ kJ mol}^{-1} \times 0.1 \text{ mol} - 229.94 \text{ kJ mol}^{-1} \times 0.1 \text{ mol} - 285.839 \text{ kJ mol}^{-1} \times 55.56 \text{ mol}$$

$$= -15928 \text{ kJ}$$

The change in internal energy is simply

$$\Delta U = U(\text{final}) - U(\text{initial})$$

$$= -15928 - (-15924)$$

$$= -4 \text{ kJ}$$

Query Does the value of $\bar{U}$ of the substances in solution change with concentration?

Reply Yes. However, as the solutions become more and more dilute, values of $\bar{U}$*(solutes)* tend to become constant and values of $\bar{U}$*(solvents)* tend towards the value for the pure solvent. Thus for dilute solutions the calculation is straightforward. However for concentrated solutions, such as the mixture of sulfuric and nitric acids used in the nitration of toluene, it is necessary to know the values of $\bar{U}$ of each of the components at each concentration.

Exercise 8.5
Estimate the change in internal energy when 2 kg of ice is transformed into water at 0°C.

$$u(\text{ice}) = -16.068\ \text{kJ g}^{-1}$$
$$u(\text{water}) = -15.985\ \text{kJ g}^{-1}$$

Answer

$$\Delta U = 166\ \text{kJ}$$

8.3.4 Chemical reaction

To be clear about the internal energy changes in a chemical reaction, let us go back to the particular system of Section 8.2.1 (Fig. 8.11).

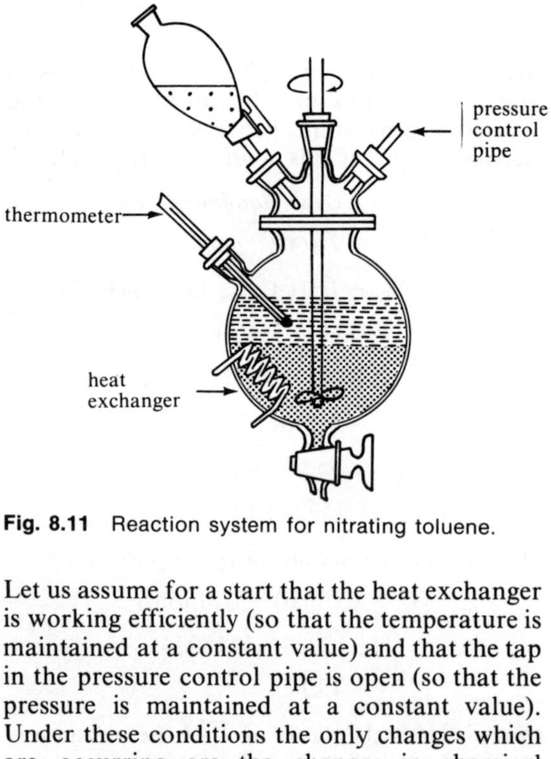

Fig. 8.11 Reaction system for nitrating toluene.

Let us assume for a start that the heat exchanger is working efficiently (so that the temperature is maintained at a constant value) and that the tap in the pressure control pipe is open (so that the pressure is maintained at a constant value). Under these conditions the only changes which are occurring are the changes in chemical composition in the two liquid phases which result from the reactions taking place.

We will consider first changes due to reaction (1), that is

$$\text{toluene} + HNO_3 = o\text{-nitrotoluene} + H_2O \quad (1)$$

In the "toluene" phase, reaction (1) leads to changes in the amounts of toluene and of *o*-nitrotoluene. For a small amount of reaction the change in the internal energy of this phase is

$$\begin{aligned} dU(\text{"toluene" phase, due to (1)}) &= \bar{U}_{\text{toluene}} \times dn_{\text{toluene}}(\text{due to (1)}) \\ &+ \bar{U}_{o\text{-nitrotoluene}} \times dn_{o\text{-nitrotoluene}} \end{aligned}$$

Similarly for the "acid" phase, there is a change

$$\begin{aligned} dU(\text{"toluene" phase, due to (1)}) &= \bar{U}_{HNO_3} \times dn_{HNO_3}(\text{due to (1)}) \\ &+ \bar{U}_{H_2O} \times dn_{H_2O}(\text{due to (1)}) \end{aligned}$$

The internal energy change due to reaction (1) is the sum of the changes in the two liquid phases. This sum can be simplified since the changes in amounts are all connected through the equations

$$d\xi_1 = \frac{dn_{o\text{-nitrotoluene}}}{1} = \frac{dn_{\text{toluene}}(\text{due to (1)})}{-1}$$
$$= \frac{dn_{H_2O}(\text{from (1)})}{1} = \frac{dn_{HNO_3}(\text{due to (1)})}{-1}$$

In these we have recognised that three species appear also in equation (2). (Page 150)

$$\begin{aligned} dU(\text{due to (1)}) = dU(&\text{"toluene" phase, due to (1)}) \\ + dU(&\text{"acid" phase, due to (1)}) \end{aligned}$$

That is,

$$dU(\text{due to (1)}) = (\bar{U}_{o\text{-nitrotoluene}} + \bar{U}_{H_2O} - \bar{U}_{\text{toluene}} - \bar{U}_{HNO_3})d\xi_1$$

The term in brackets on the right hand side of this equation is of considerable importance. It is the difference between the partial molar energies of the products and the reactants and is given the special symbol $\Delta\bar{U}$ (delta-U-bar)

$$\boxed{\Delta\bar{U}(\text{reaction 1}) = \bar{U}_{o\text{-nitrotoluene}} + \bar{U}_{H_2O} - \bar{U}_{\text{toluene}} - \bar{U}_{HNO_3}}$$

or, if we write the equation for a reaction in the form (Section 1.4.2)

$$0 = \Sigma\nu_B B$$

$$\boxed{\Delta\bar{U} = \Sigma\nu_B\bar{U}_B}$$

The change in internal energy corresponding to a small amount of reaction ($d\xi$) is given by

$$\Delta U = (\Delta\bar{U})d\xi$$

Exercise 8.6

(a) Define $\Delta \bar{U}$ for the formation of *p*-nitrotoluene (equation 2). (Page 150)

(b) Assume that *o*-nitrotoluene and *p*-nitrotoluene are being formed in the ratio (65:35) and write

(i) the relationship between changes in the extent of reaction by equations (1) and (2);

(ii) the change in internal energy per mole of toluene reacting by the reactions represented in equations (1) and (2).

Comment So far you have only talked about the changes in internal energy for infinitesimal amounts of reaction. Surely in the example the objective is to carry out large amounts of reaction?

Reply Yes, of course. There are two cases. The first is the very simple one where the value of $\Delta \bar{U}$ does not change as the reaction proceeds. In this case we write

$$dU = \Delta \bar{U} d\xi$$

or $$\Delta U = \int \Delta \bar{U} d\xi$$

or ($\bar{U}$ constant) $$\Delta U = \Delta \bar{U} \Delta \xi$$

That is, the change in the internal energy due to a reaction is simply the product of $\Delta \bar{U}$ for the reaction and the amount of reaction which took place.

Example 8.9

The calcining of lime,

$$CaCO_3(s) \rightarrow CaO(s) + CO_2(g)$$

The partial molar internal energies of each of the three substances involved are independent of the amounts present, hence, at a given temperature

$$\Delta \bar{U} \equiv \bar{U}(CO_2) + \bar{U}(CaO) - \bar{U}(CaCO_3)$$

is constant, and hence for the production of n_{CaO} of quicklime ($dn_{CaO} = d\xi$)

$$\Delta U = \Delta \bar{U} \times n_{CaO}$$

Example 8.10

Water from brine by reverse osmosis,

$$H_2O(\text{brine}) \rightarrow H_2O(\text{pure liquid})$$

In this example only a small fraction of the water is extracted from the brine. The concentration of the brine leaving the plant is therefore only slightly higher than when it enters. Hence, inside the plant, $\bar{U}(H_2O, \text{brine})$ is unchanged and

$$\Delta \bar{U} \equiv \bar{U}(H_2O, \textit{pure}) - \bar{U}(H_2O, \text{brine})$$

is constant. The change in internal energy on production of pure water is therefore

$$\Delta U = \Delta \bar{U} \times n$$

If, as in the example of Section 8.2.1, $\Delta \bar{U}$ is not constant during the reaction, the change in internal energy must be found by evaluating the integral

$$\Delta U = \int \Delta \bar{U} d\xi$$

This is not nearly as difficult as it looks. It is done simply by evaluating the internal energy after reaction, using the values of partial molar energies and composition at the completion of reaction and subtracting the value of the internal energy prior to reaction. This was the method used in Section 8.3.3 to evaluate the change in internal energy associated with changes in phase.

8.3.5 Correlation of changes in internal energy

In the preceding sections we have considered separately the effects on the internal energy of changes in temperature, pressure, chemical composition and phase. In this section we will correlate these changes with transfers of energy to the surroundings by heat and work. We also will examine strategies for determining changes in internal energy, when, as is usually the case, several of the factors considered in the preceding sections vary simultaneously.

Let us examine the system of Section 8.2.1 (Fig. 8.12). When reaction occurs, we measure the change in internal energy of the system as the total of the energy transferred to the system. The heat exchanger is intended to be the main method for heat transfers. Work is done by the stirrer. It imparts kinetic energy to the contents of the vessel, and later, as viscosity slows this motion, it increases the temperature of both vessel and contents. Work is also done by the

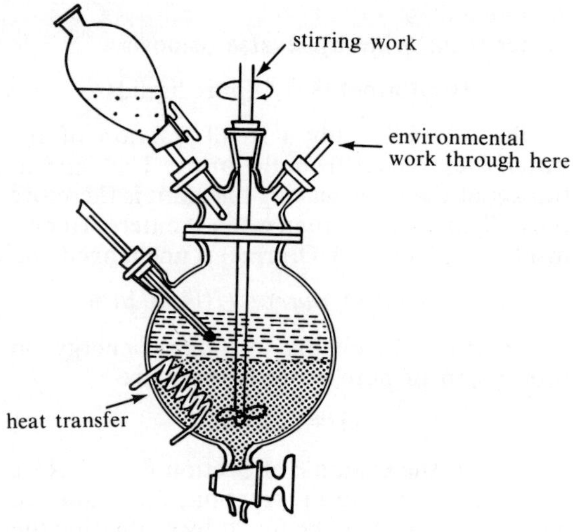

Fig. 8.12 Energy exchanges in nitrating toluene.

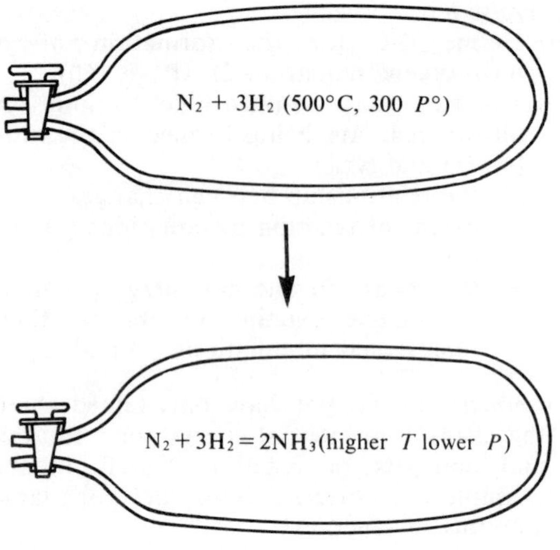

Fig. 8.13 Nitrogen and hydrogen reacting to form ammonia.

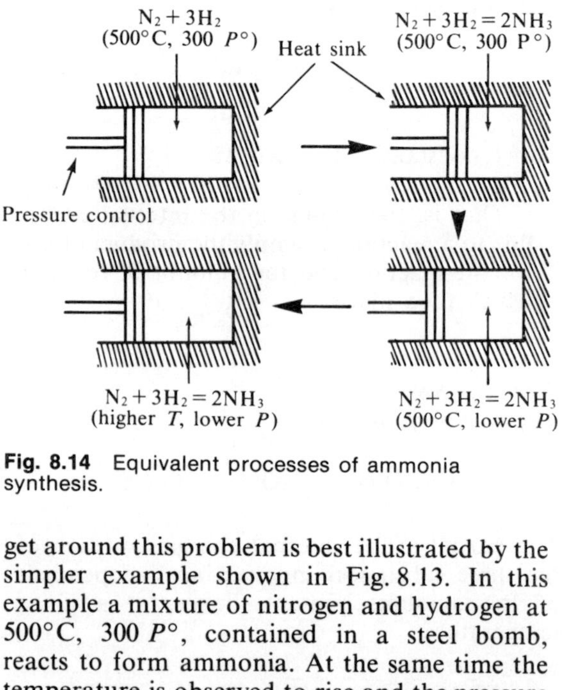

Fig. 8.14 Equivalent processes of ammonia synthesis.

environment due to the change in total volume which occurs during reaction.

Query Suppose the heat exchanger was not working and we turned off the tap in the pressure control pipe and the stirrer. What happens to the internal energy?

Reply It seems that that stops heat and work. So the internal energy of the system stays constant.

$$\Delta U(\text{system}) = 0$$

Query But the reaction keeps going on. Won't the system get hot?

Reply Yes, the temperature will rise. The pressure will probably change, too. The changes in internal energy corresponding to these changes of temperature and pressure will just balance the change in internal energy corresponding to the reactions and phase transfers.

There is one catch. The effect on the internal energy of changes in temperature, pressure and chemical composition are not independent of one another. That is, the values of ΔU (corresponding to the change in pressure) and ΔU (corresponding to the change in temperature) both depend on the chemical composition of the various phases. Similarly values of both ΔU (corresponding to reaction) and ΔU (corresponding to change in phase) depend on the temperature and the pressure. The way we get around this problem is best illustrated by the simpler example shown in Fig. 8.13. In this example a mixture of nitrogen and hydrogen at 500°C, 300 $P°$, contained in a steel bomb, reacts to form ammonia. At the same time the temperature is observed to rise and the pressure to fall. We would have got the *same result* if we had carried out the set of processes shown in Fig. 8.14. So the total internal energy change under the conditions of Fig. 8.13 is the same as the total under those of Fig. 8.14.

Query Does the order in which the effects of the different changes are computed matter?

Reply No. We pick the most convenient order. (That is, the one for which we have the most information). The important thing is that we treat the process as though it occurred in a number of separate steps.

We have looked at an extreme case for the change of internal energy which may accompany a chemical reaction. We supposed that conditions were chosen so there were no exchanges of energy (by heat or by work, including environmental work) between the system and its surroundings, and there was no exchange of material with the surroundings either.

Query What does that "no exchange of material" mean? Figure 8.12 shows nitric and sulfuric acids entering the vessel from a reservoir.

Reply We mark out the boundaries of the system so that the reservoir is inside, not outside, the system. We include the properties of those acids in accounting for the system.

If we relax these conditions so that there are exchanges by heat and work between the system and its surroundings, then we can calculate the change in the internal energy for the system from the sum of such exchanges. Alternatively, we can compute the change in internal energy from the observed changes in the temperature, pressure and chemical composition within the system. Comparison between the two answers constitutes an *energy balance* for the system. Usually we can measure the operator work without great difficulty. If the pressures and volumes are known, the environmental work can be calculated for most cases. If we can calculate the change in internal energy, for example, from values of the partial molar internal energies and from the heat capacities and temperatures of the bodies and phases, then we can calculate the heating. Or, if we know the heat term, we can calculate some other term in the total sum.

The restriction to systems where there is no exchange of material with its surroundings (closed systems) is irksome and unnecessary. It can be removed by including with heat and work the internal energy associated with the substances exchanged between the system and its surroundings. Take for example the reaction between zinc and sulfuric acid depicted in Fig. 8.15. The boundaries of the system may be drawn as shown by the "dashed" lines. In this case energy is exchanged between the system and its surroundings by

(a) thermal transfer of energy;
(b) work done by the environment as the system changes volume; and
(c) the internal energy which accompanies the hydrogen. This is equal to the partial molar energy of the hydrogen multiplied by the amount of hydrogen supplied.

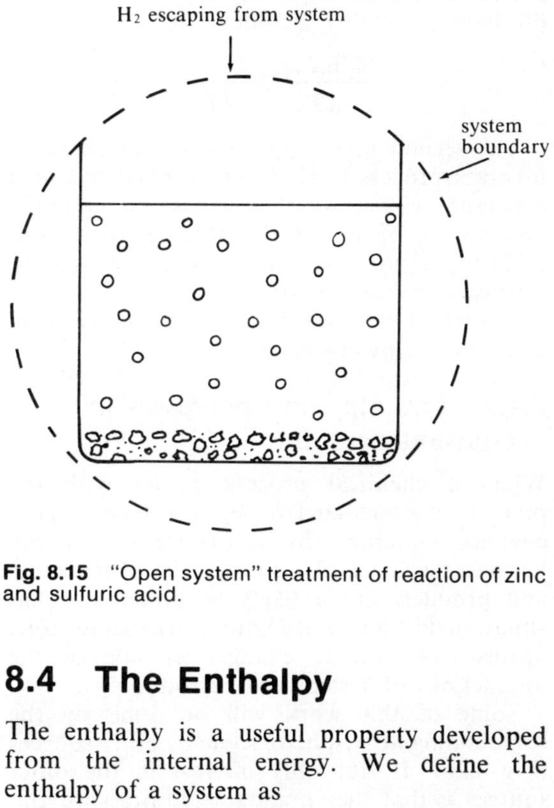

Fig. 8.15 "Open system" treatment of reaction of zinc and sulfuric acid.

8.4 The Enthalpy

The enthalpy is a useful property developed from the internal energy. We define the enthalpy of a system as

$$H(\text{system}) \equiv U(\text{system}) + P(\text{system}) \times V(\text{system})$$

8.4.1 Reasons for generating the function enthalpy

Very often in the analysis of chemical systems the internal energy, pressure and volume appear in the combination $(U + PV)$. The situations in which this happens are so numerous and of such practical importance that we will use values of H more often than we will use values of U. Thus it is values of enthalpies of substances which are tabulated rather than the values of their internal energies. There are three types of situation where the use of enthalpy rather than internal energy is particularly convenient.

The most important use of enthalpy in chemistry is in treatment of the variation of equilibrium constants with temperature. Empirically it is found (Section 4.8) that the variation of the equilibrium constant with temperature is related to the difference between the standard partial molar enthalpies of the reactants and products ($\Delta \bar{H}°$) by the equation.

$$\frac{d \ln K_{eq}}{dT} = \frac{\Delta \bar{H}°}{RT^2}$$

The second most important use of enthalpy involves processes which are carried out at a constant pressure, whether in vessels open to the atmosphere or in vessels at some other controlled pressure, like the autoclaves used to sterilise hospital laundry.

The third important use of enthalpy is in analyses of flowing systems.

8.4.2 Enthalpy and processes at constant pressure

When a chemical process occurs with the pressure at some fixed value, the system usually changes volume (by ΔV(system)) simply because the space-filling properties of reactants and products are unlikely to be equal. That situation inevitably leads to work being done against the volume changes as one of the interactions of system and surroundings.

Some of this work will be done by the surrounding atmosphere, some by other sources (Fig. 8.16). If our only interest in the other sources is that they maintain the pressure and not how much work they do, then they are included as part of the environment.

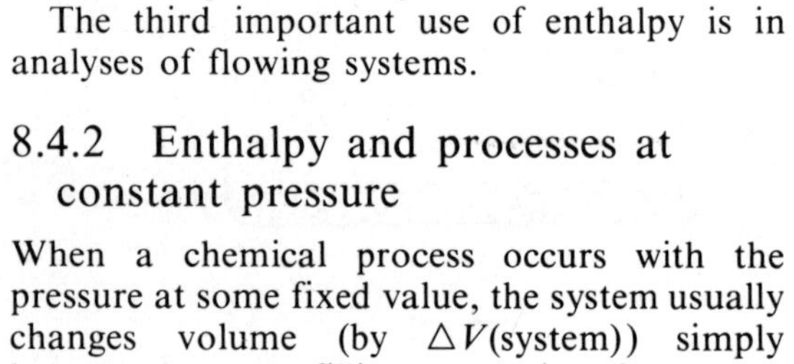

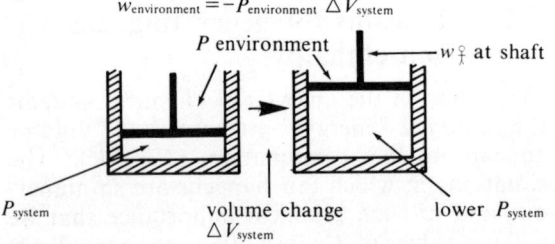

Fig. 8.16 Expansion usually involves some environmental work and some operator work.

When there is mechanical equilibrium between system and environment then, at the zones where they are in contact, the total pressure applied by the environment, P(environment), is equal to the pressure of the system, P(system). So we have an environmental work contribution

$$w(\text{environment}) = -P(\text{environment})\,\Delta V(\text{system}) = -P(\text{system})\,\Delta V(\text{system})$$

Thus changes in any system in which the pressure is kept steady involve finite exchange by environmental work. Changes in enthalpy automatically include this contribution as part of the energy balance for the system. Thus we have

$$H \equiv U + PV$$

and $$\Delta H = \Delta U + \Delta(PV)$$

For any change in which the pressure does not alter, the change in enthalpy will be

$$\Delta H = \Delta U + P(\text{system})\,\Delta V(\text{system})$$

When, in addition, the pressures of system and environment are equal, this requires that

$$\Delta H = \Delta U + P(\text{environment})\,\Delta V(\text{system})$$

or $\Delta H = \Delta U - w(\text{environment})$

We have seen (Section 8.1.4) that the internal energy change can be written as

$$\Delta U = q + w(\text{environment}) + w_♀$$

Thus the value of ΔH for such a change (constant pressure process) will be equal to the heating, q, plus the operator work $w_♀$

$$\Delta H = q + w_♀ \text{ (in a constant pressure process)}$$

There are four important special cases of this equation.

(a) If we choose to do no operator work, the change in enthalpy is equal to the heating. (This is the origin of the names "heat content" for H and "heat of reaction" for ΔH).
(b) If we choose to have no heating, the change in enthalpy is equal to the operator work done.
(c) If we carry out experiments in which heat and operator work are of opposed sign but numerically equal, the pressure being

constant, we have a process without change of enthalpy.

(d) If we carry out experiments at constant pressure in which we do no operator work and take care that no heating occurs then there is no change of enthalpy.

Query What happens to the enthalpy if the pressure is not constant?

Reply The change in enthalpy is still a real quantity, and one which is often useful. From the definition of enthalpy we can readily deduce the value for any change as

$$\Delta H = \Delta U + \Delta(PV)$$

which can be evaluated from a knowledge of the internal energies, volumes and pressures of the substances involved. For an infinitesimal change,

$$\mathrm{d}H = \mathrm{d}U + P\,\mathrm{d}V + V\,\mathrm{d}P$$

A familiar example of case (a) involves water boiling in an open vessel over a gas burner. The equivalent example of case (b) involves water boiling in a (thermally insulated) electric kettle.

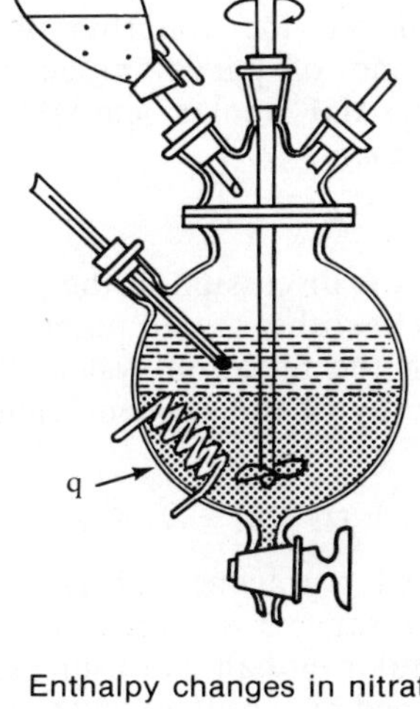

Fig. 8.17 Enthalpy changes in nitrating toluene.

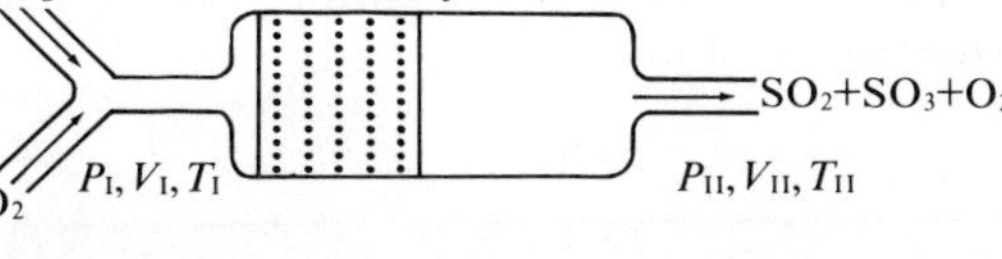

Fig. 8.18 Flow system for reaction of SO_2 and oxygen.

Another example of case (a) involves the reaction for forming nitrotoluenes from toluene (Fig. 8.17). If we choose to operate with a minimum of stirring, so that $w_{♀}$ is close to zero, and use a thermally insulated reaction vessel (with the heat exchanger operating), and also keep the pressure at any convenient constant value, the value of ΔH for the process will be equal to the heating of the system by the heat exchanger. If the same system were used with the heat exchanger *not* operating, but all other conditions unchanged, (case d) the process would occur at constant enthalpy ($\Delta H = 0$).

Many devices to measure the energy changes which occur in reactions (calorimeters) are based on constant pressure situations. Such devices are usually designed so that the way in which they function corresponds to case (b) or case (d).

8.4.3 Enthalpy and flowing systems

Many chemical processes are carried out in flowing systems similar to that sketched in Fig. 8.18. In this case, a steady flow of gas, including the reactants, into the catalyst chamber is matched by steady flow of gas, including the products, out of the reactor. These steady flows, together with energy supplied to the catalyst chamber by heat, lead to constant rates of reaction within the chamber and to constant temperatures and pressures (not usually equal) of the inlet and output gas. Similar conditions are chosen for many reactions in the liquid phase. In the case chosen for Fig. 8.18, sulfur dioxide which enters the catalyst chamber leaves it either as SO_3 or as (unreacted) SO_2. Suppose a fraction, x, of the SO_2 reacts during passage through the chamber.

Consider the change in a particular slug of input gas, which contains $n(O_2)$ as it enters the catalyst chamber. Suppose that, as it enters, the *total* pressure of entering gases is P_I and the volume of gas which contains $n(SO_2)$ is V_I. As it leaves, the gas pressure will be different, say P_{II}. The volume of exit gas, which contains nx of SO_3 and $n(1-x)$ of SO_2, is also different, say V_{II}. When it is operating properly, the catalyst and the catalyst chamber do not alter in temperature, pressure or composition and so their internal energy does not change. Thus we need only to consider changes in the internal energy of the gases as they traverse the catalyst chamber. These changes can be measured from the interactions with the surroundings. Suppose that, in the time it takes for this amount of gas to traverse the chamber, the energy entering by heat is q. The work done is the sum of the work ($+P_{\mathrm{I}}V_{\mathrm{I}}$) to introduce the gas to the catalyst chamber and the work ($-P_{\mathrm{II}}V_{\mathrm{II}}$) to extract the gas from the chamber. That is,

$$w = P_{\mathrm{I}}V_{\mathrm{I}} - P_{\mathrm{II}}V_{\mathrm{II}}$$

The change in internal energy of the gas which initially contains $n\mathrm{SO_2}$, due to passage through the catalyst chamber, is thus

$$\Delta U = U_{\mathrm{II}} - U_{\mathrm{I}} = q + w$$
$$= q + P_{\mathrm{I}}V_{\mathrm{I}} - P_{\mathrm{II}}V_{\mathrm{II}}$$

That is,

$$q = (U_{\mathrm{II}} + P_{\mathrm{II}}V_{\mathrm{II}}) - (U_{\mathrm{I}} + P_{\mathrm{I}}V_{\mathrm{I}})$$
$$= \Delta H\,(\text{gas})$$

Thus the quantity which describes the heating of a flow reactor operating in this steady way is the enthalpy, and not the internal energy.

8.5 Enthalpy of a System and Its Materials

The method of analysis of the enthalpy of a system into enthalpies of its components follows exactly along the lines of the analysis of internal energy. We shall therefore discuss exactly the same system (introduced in Section 8.2.1 and Fig. 8.19) as for the earlier function.

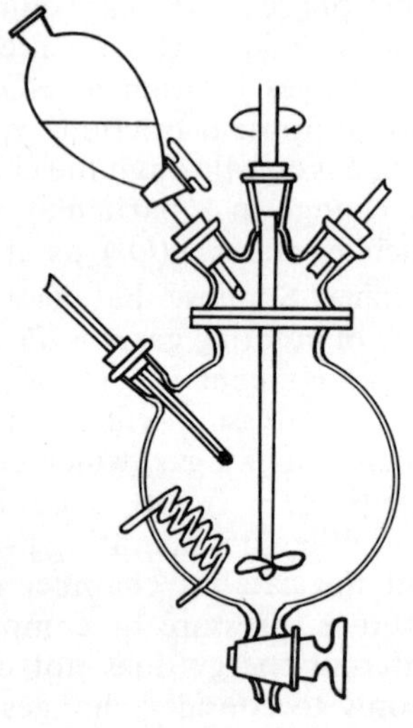

Fig. 8.19 A reaction vessel, of enthalpy, *H* vessel.

The reaction vessel depicted in Fig. 8.19 will be usefully considered as a single unit for exactly the same reasons as those previously given. It is likely to be used often. The internal energies of any component parts will always be involved together. For the enthalpy, we add the comment that the volumes of its component parts will also be involved together. We can therefore treat it as a whole and assign it an enthalpy, H_{vessel}. When conditions change, there may be changes, $\Delta H_{\mathrm{vessel}}$, which need to be examined.

8.5.1 Specific enthalpy

Materials such as those of which the vessel is constructed can be associated with a *specific enthalpy*, *h*, varying in proportion to the mass of the material. So the enthalpy of a complete body, like the vessel, can be written in terms of the specific enthalpies and masses of its materials (for example, steel, glass, etc).

$$H_{\mathrm{vessel}} = h_{\mathrm{steel}} \times m_{\mathrm{steel}} + h_{\mathrm{glass}} \times m_{\mathrm{glass}} + \ldots$$

Changes in specific enthalpy and mass of materials are associated with changes in the enthalpy of the body as the equation requires.

8.5.2 Molar enthalpy

We have introduced the concepts of partial molar volume and of partial molar internal energy. The partial molar enthalpy is a combination of them,

$$\bar{H} = \bar{U} + P\bar{V}$$

which is used in the discussion of the properties of phases just as the other molar quantities are.‡ For example, in the nitration system (Section 8.2.1), the enthalpy of the "toluene" phase can be written as

$$H = \bar{H}_1 \times n_1 + \bar{H}_2 \times n_2 + \bar{H}_3 \times n_3 + \ldots$$

where substance 1 is toluene; substance 2 is *o*-nitrotoluene; substance 3 is *p*-nitrotoluene, etc.

The partial molar enthalpy of toluene in that phase is determined by the effect on the phase of changing the amount of toluene present, keeping all other conditions such as temperature, pressures and the amounts of other substances constant.

‡ In many usages the word "partial" is deleted and $\bar{H}$ is called just the "molar enthalpy".

In algebraic terms this is written as

$$\bar{H}(\text{toluene}) \equiv \left(\frac{\partial H(\text{toluene phase})}{\partial n(\text{toluene})}\right)_{T,P,n(o\text{-nitrotoluene}),\, n(p\text{-nitrotoluene})}$$

The partial molar enthalpy of toluene in such a phase varies (though not drastically) as the composition of the phase changes.

8.5.3 Other issues

Enthalpies in the interaction of phases with one another (for example, wetting, glueing) are sometimes very important. They do not concern us at this time.

8.6 Changes in the Enthalpy of a Body

The properties which need to be considered when comparing changes in the enthalpy of a body are changes in temperature and pressure. Changes of enthalpy within a system may occur as a result of changes in temperature, changes of phase, and chemical reactions. In this section we will show how the enthalpy is affected by change in each of these properties.

8.6.1 Change of temperature

When the temperature of a body changes, there is usually a change both in its internal energy and in its volume. Both alter the value of the enthalpy. Consider an infinitesimal change, dT, in temperature and the ratio of changes in H and T.

$$\frac{dH}{dT} = \frac{dU}{dT} + \frac{d\,PV}{dT}$$

Suppose the change is to be made so that the pressure does not alter, then

$$\left(\frac{\partial H}{\partial T}\right)_P = \left(\frac{\partial U}{\partial T}\right)_P + P\left(\frac{\partial V}{\partial T}\right)_P$$

The quantity $(\partial H/\partial T)_P$ is commonly known as the "heat capacity at constant pressure", C_P

$$C_P \equiv (\partial H/\partial T)_P$$

From the argument in Section 8.4.2, and cases (a) and (b), values of C_P are appropriate in calculating the change in temperature which accompanies gain of energy due to heat or operator work in processes in which the pressure is equal to that of the environment. The quantity $(\partial U/\partial T)_P$ was introduced in Section 8.3.1. The remaining quantity,

$$P\left(\frac{\partial V}{\partial T}\right)_P$$

is related to α, the coefficient of thermal expansion.

$$P\left(\frac{\partial V}{dT}\right)_P = PV\frac{1}{V}\left(\frac{\partial V}{\partial T}\right)_P$$
$$= PV\alpha$$

Thus the equation for $(\partial \bar{H}/\partial T)_P$ may be written as

$$\bar{C}_P = \left(\frac{\partial \bar{U}}{\partial T}\right)_P + P\bar{V}\alpha$$

This relation was used in calculating the values of

$$\left(\frac{\partial \bar{U}}{\partial T}\right)_P$$

in Table 8.4. In Section 8.3.1 we met the "heat capacity at constant volume", C_V which is given by

$$C_V \equiv \left(\frac{\partial \bar{U}}{\partial T}\right)_V = \bar{C}_P - \frac{\alpha^2 T\bar{V}}{\beta}$$

For gases at low pressure

$$\bar{V} = \frac{RT}{P}$$

hence $$\alpha \equiv \frac{1}{\bar{V}}\left(\frac{\partial \bar{V}}{\partial T}\right)_P = \frac{R}{\bar{V}P}$$

$$\beta \equiv -\frac{1}{\bar{V}}\left(\frac{\partial \bar{V}}{\partial P}\right)_T = +\frac{RT}{\bar{V}P^2}$$

hence $$\bar{C}_P = \bar{C}_V + \frac{\bar{V}\alpha^2 T}{\beta}$$

or $$\bar{C}_P = \bar{C}_V + R,$$

a very simple result which, of course, applies only to gases at low pressures.

In Table 8.6 we have gathered a few values of molar heat capacity, $\bar{C}_P$, and specific heat capacity, c_P, for some substances and materials.

Exercise 8.7

(a) Calculate the sum of heating and operator work when the temperature of 1 dm^3 of a lubricating oil is varied from 25°C to 45°C. Its specific heat capacity is c_P

$= 2.4\ J\ g^{-1}\ K^{-1}$ and its density is $\rho = 0.95\ g\ cm^{-3}$

(b) The temperature of a reaction vessel (Fig. 8.12) with or without contents is to be raised from 20°C to 80°C by passing steam through the heat exchanger. How much heat is required
  (i) for the empty reaction vessel ($c_P = 3.9\ kJ\ g^{-1}$)
  (ii) for the reaction vessel when it contains 15 mol of toluene

$$(\bar{C}_P = 150\ J\ mol^{-1}\ K^{-1})$$

Answer
(a) 46 kJ
(b) 370 kJ.

8.6.2 Change of pressure

It is convenient to examine the change of enthalpy with changing pressure under conditions in which the temperature is not altered. The appropriate ratio, $(\partial H/\partial P)_T$ can be determined from the relationship‡

$$(\partial H/\partial P)_T \doteq V(1 - \alpha T)$$

Table 8.6
Variation of Enthalpy with Temperature at 298 K *and* 101 kPa

Substance	$(\partial \bar{H}/\partial T)_P = \bar{C}_P$ / $J\ K^{-1}\ mol^{-1}$
Gases	
Helium	20.79
Hydrogen	28.84
Oxygen	29.36
Nitrogen	29.12
Nitrous oxide, N_2O	38.71
Carbon dioxide, CO_2	37.65
Liquids	
Carbon disulfide, CS_2	75.7
Water	75.34
Methanol	80.8
Ethanol	66.5
n-Pentane	120.2
Benzene	136
Mercury	27.82
Solids	
Aluminium	24.34
Copper	24.47
Lead	26.82
Aluminium oxide, Al_2O_3	78.99
Calcium fluoride, CaF_2	67.02
Silicon oxide, SiO_2	44.35
Paraffin oil	2.13–2.26 ($J\ g^{-1}K^{-1}$)
Polythene	2.30 ($J\ g^{-1}K^{-1}$)
PVC	1.05 ($J\ g^{-1}K^{-1}$)

‡Guggenheim, E.A., *Thermodynamics*, fifth edition, North Holland Publishing Company, Amsterdam, 1967, p.40.

where α is the coefficient of *thermal* expansion and V is the appropriate volume (of the body, material or substance) being examined. In Table 8.7, we have gathered some values of $(\partial H/\partial P)_T$. For gases at low pressures, this quantity is very small. Almost always the effect of changing pressure leads to a much smaller change in enthalpy than the change due to any chemical reactions which are present, and hence can often be neglected.

8.6.3 Change with concentration

The variation of partial molar enthalpy with concentration of mixtures of methanol and water is illustrated in Table 8.8. The total variation is about $8\ kJ\ mol^{-1}$; not a small quantity! There is no simple way of estimating this variation of enthalpy with concentration. Fortunately in many cases important in chemistry, particularly reactions in very dilute solutions, the effect due to changing concentrations is much smaller than the above and is much smaller than that due to the other changes which are occurring.

8.6.4 Changes in phase

In practice many phase changes are carried out under conditions close to constant pressure. For this case (Section 8.4.2), the enthalpy change is

Table 8.7
Variation of Enthalpy with Pressure at 298 K *and* 101 kPa

Substance	$10^6\ (\partial \bar{H}/\partial P)_T$ / $J\ mol^{-1}Pa^{-1}$
Gases	
Helium	−50
Hydrogen	13.5
Oxygen	−16.5
Nitrogen	− 4.8
Nitrous oxide, N_2O	−133
Carbon dioxide, CO_2	−122
Liquids	
Carbon disulfide, CS_2	37.7
Water	16.9
Methanol	26.2
Ethanol	39.8
n-Pentane	62.4
Benzene	56.9
Mercury	14.0
Solids	
Aluminium	9.7
Copper	7.0
Lead	17.7
Aluminium oxide, Al_2O_3	25.3
Calcium fluoride, CaF_2	24.0
Silicon oxide, SiO_2	23.1

equal to the sum of the operator work and the heating.

$$\Delta H = w_{\text{♀}} + q \text{ (constant pressure)}$$

Typical values for a number of pure substances are given in Table 8.9.

As was the case with differences in internal energy (Section 8.3.3) the change in enthalpy on transfer of a substance from one phase into another is best considered as a change in the chemical content of the two phases.

Consider the solution of 4.0 g (~0.10 mol) of sodium hydroxide in 1.0 kg (55.56 mol) of water, at 25°C and at atmospheric pressure.

Table 8.8
Partial Molar Enthalpies of Water and Methanol in Solutions of Water and Methanol at 25°C

mole fraction of methanol	$[\bar{H}(\text{solution}) - \bar{H}(\text{pure})]$ kJ mol^{-1} *water*	*methanol*
0	0	−7.35
0.13	−.25	−3.3
0.24	−.69	−1.4
0.31	−.87	− .88
0.40	−1.0	− .60
0.57	−1.2	− .38
0.64	−1.3	− .31
0.76	−1.7	− .18
0.83	−1.9	− .10
0.91	−2.4	− .03
1.00	−3.0	0.00

Under these conditions

$\bar{H}(\text{NaOH}, \textit{solid}) = -426.73$ kJ mol^{-1}
$\bar{H}(\text{Na}^+, \textit{aqueous}) = -239.655$ kJ mol^{-1}
$\bar{H}(\text{OH}^-, \textit{aqueous}) = -229.94$ kJ mol^{-1}
$\bar{H}(\text{H}_2\text{O}, \textit{liquid}) = -285.839$ kJ mol^{-1}

The initial condition is two phases: one being 0.1 mol of solid sodium hydroxide with an enthalpy of

$$H(\textit{solid}) = \bar{H}(\text{NaOH}, \textit{solid}) \times n(\text{NaOH}, \textit{solid})$$

the other being 1 kg of water with an enthalpy given by

$$H(\text{water}) = \bar{H}(\text{water}) \times n(\text{water})$$

Hence the initial value of the enthalpy is

$$H(\text{initial}) = H(\text{solid NaOH}) + H(\text{water})$$
$$= -426.73 \times 0.1 - 285.839 \times 55.56 \text{ kJ}$$
$$= -15923.9 \text{ kJ}$$

After dissolution, one phase only remains: a solution consisting of 0.10 mol of sodium and hydroxide ions and 55.56 mol of water. The enthalpy of this solution is therefore

$$H(\text{final solution}) = \bar{H}(\text{Na}^+, \textit{aqueous}) \times n_{\text{Na}^+} + \bar{H}(\text{OH}^-, \textit{aqueous}) \times n_{\text{OH}^-} + \bar{H}(\text{H}_2\text{O}, \textit{liquid}) \times n_{\text{H}_2\text{O}}$$
$$= -239.655 \times 0.1 - 229.94 \times 0.1 - 285.839 \times 55.56 \text{ kJ}$$
$$= -15928.2 \text{ kJ}$$

Table 8.9
Variation of Enthalpy in Some Changes of Phase at 101 kPa

Substance	*Phase change*	*T*/K	ΔH/kJ mol^{-1}
oxygen	$s = l$	54.8	0.44
water	$s = l$	273	6.01
camphor	$s = l$	449	6.8
magnesium	$s = l$	924	9.2
oxygen	$l = g$	90	6.82
acetone	$l = g$	329	30.4
chloroform	$l = g$	334	23.0
water	$l = g$	373	40.6
naphthalene	$l = g$	484	40.5
sodium hydroxide	$s =$ dilute aqueous solution	298	−43.0
sodium nitrate	$s =$ dilute aqueous solution	298	+20.0
ammonia	$g =$ dilute aqueous solution	298	−34.0
bromine	$l =$ dilute aqueous solution	298	− 3.0
tin	s(grey) = s(white)	286	2.0

The change in enthalpy is simply

$$\Delta H = H(\text{final}) - H(\text{initial})$$
$$= -15928.2 + 15923.9$$
$$= -4.3 \text{ kJ}$$

PROBLEM

8.2. How much heat ($w_{\ddagger} = 0$) is required to remove the water of crystallisation from 500 g (2 mol) of copper(II) sulphate if

$$\bar{H}(\text{CuSo}_4 \cdot 5\text{H}_2\text{O}, s) = -2278 \text{ kJ mol}^{-1}$$
$$\bar{H}(\text{CuSO}_4, s) = -\ 770 \text{ kJ mol}^{-1}$$
$$\bar{H}(\text{H}_2\text{O}, g) = -\ 242 \text{ kJ mol}^{-1}$$

For small changes, the calculation is even simpler. Consider for example the transfer of a small amount (δn) of water from liquid to gas. In this case

$$\delta H(gas) = \bar{H}(\text{H}_2\text{O}, gas) \times \delta n$$

and $$\delta H(liquid) = \bar{H}(\text{H}_2\text{0}, liquid) \times (-\delta n)$$

(the change in the amount of liquid $= -\delta n$ mol), hence

$$\delta H = [\bar{H}(\text{H}_2\text{O}, gas) - \bar{H}(\text{H}_2\text{O}, liquid)] \times \delta n$$

or, for an infinitesimal change, we can write

$$\frac{\mathrm{d}H}{\mathrm{d}n} = [\bar{H}(\text{H}_2\text{O}, gas) - \bar{H}(\text{H}_2\text{O}, liquid)]$$

The difference in partial molar enthalpies in the two phases is called the *molar* enthalpy change for the process

$$\text{H}_2\text{O}(liquid) = \text{H}_2\text{O}(gas)$$

and is given the symbol $\Delta\bar{H}$. Thus

$$\Delta\bar{H}(\text{H}_2\text{O}, liquid = \text{H}_2\text{O}, gas) = \bar{H}(\text{H}_2\text{O}, gas) - \bar{H}(\text{H}_2\text{O}, liquid)$$

It is an important quantity in discussions of the shapes of phase boundaries (see Sections 3.3.5, 3.7 and 3.8.7).

Exercise 8.8
Define $\Delta\bar{H}$ for the following phase transfer processes.

$$\text{H}_2\text{O}(s) = \text{H}_2\text{O}(l)$$
$$\text{I}_2(s) = \text{I}_2(g)$$
$$\text{NaCl}(s) = \text{NaCl}(aq)$$

8.6.5 Chemical reaction

The treatment of the way in which the enthalpy changes during reaction exactly parallels that of the changes in internal energy (Section 8.3.4).
Note: in order to emphasise the parallel nature of the treatment we will use the *same* examples as used in Section 8.3.4.

Let us go back to the particular system of Section 8.2.1 (Fig. 8.20). Let us assume for a start that the heat exchanger is working efficiently (so that the temperature is maintained at a constant value) and that the tap in the pressure control pipe is open (so that the pressure is maintained at a constant value). Under these conditions the only changes which are occurring are the changes in chemical composition in the two liquid phases which result from the reactions taking place.

We will consider first, changes due to reaction (1), i.e.

$$\text{toluene} + \text{HNO}_3 = o\text{-nitrotoluene} + \text{H}_2\text{O}$$

In the "toluene" phase, reaction (1) leads to changes in the amounts of toluene and of *o*-nitrotoluene. For a small amount of reaction the change in the enthalpy of this phase is

$$\mathrm{d}H(\text{"toluene" phase, due to (1)})$$
$$= \bar{H}_{\text{toluene}} \times \mathrm{d}n_{\text{toluene}}(\text{due to (1)})$$
$$+ \bar{H}_{o\text{-nitrotoluene}} \times \mathrm{d}n_{o\text{-nitrotoluene}}$$

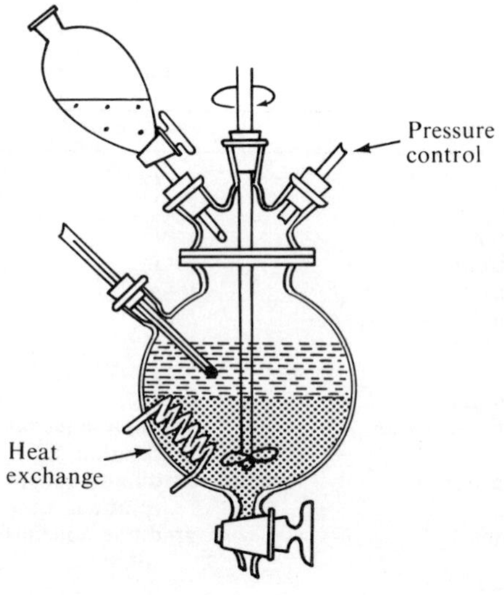

Fig. 8.20 System for nitrating toluene.

Similarly, for the "acid" phase, there is a change

$$dH(\text{"acid" phase, due to (1)}) = \bar{H}_{HNO_3} \times dn_{HNO_3}(\text{due to (1)}) + \bar{H}_{H_2O} \times dn_{H_2O}(\text{due to (1)})$$

The enthalpy change due to reaction (1) is the sum of the changes in the two liquid phases. This sum can be simplified since the changes in amounts are all connected through the equations

$$d\xi_1 = \frac{dn_{\text{o-nitrotoluene}}}{1} = \frac{dn_{\text{toluene (due to (1))}}}{-1} = \frac{dn_{H_2O \text{ (for (1))}}}{1} = \frac{dn_{HNO_3 \text{ (due to (1))}}}{-1}$$

In these we have recognised that three species appear also in equation (2), (page 150).

$$dH(\text{due to (1)}) = dH(\text{"toluene" phase, due to (1)}) + dH(\text{"acid" phase, due to (1)})$$

That is

$$dH(\text{due to (1)}) = (\bar{H}_{\text{o-nitrotoluene}} + \bar{H}_{H_2O} - \bar{H}_{\text{toluene}} - \bar{H}_{HNO_3})d\xi_1$$

The term in brackets on the right hand side of this equation is of considerable importance. It is the difference betweeen the partial molar enthalpies of the products and the reactants and is given the special symbol $\Delta\bar{H}$ (delta-H-bar)

$$\Delta\bar{H}(\text{reaction 1}) = \bar{H}_{\text{o-nitrotoluene}} + \bar{H}_{H_2O} - \bar{H}_{\text{toluene}} - \bar{H}_{HNO_3}$$

or, if we write the equation for a reaction in the form (Section 1.4.2)

$$0 = \Sigma\nu_B B$$

$$\Delta\bar{H} = \Sigma\nu_B\bar{H}_B$$

The change in enthalpy corresponding to a small amount of reaction ($d\xi$) is given by

$$\Delta H = (\Delta\bar{H})d\xi$$

Exercise 8.9

(a) Define $\Delta\bar{H}$ for the formation of *p*-nitrotoluene (equation 2).

(b) Assume that *o*-nitrotoluene and *p*-nitrotoluene are being formed in the ratio (65:35) and write

(i) the relationship between changes in the extent of reaction by equations (1) and (2);

(ii) the change in enthalpy per mole of toluene reacting by the reactions represented in equations (1) and (2).

Comment So far you have only talked about the changes in enthalpy for infinitesimal amounts of reaction. Surely in the example the objective is to carry out large amounts of reaction?

Reply Yes, of course. There are two cases. The first is the very simple one where the value of $\Delta\bar{H}$ does not change as the reaction proceeds. In this case we write

$$dH = \Delta\bar{H}d\xi$$

or

$$\Delta H = \int \Delta\bar{H}d\xi$$

or ($\bar{H}$ constant)

$$\Delta H = \Delta\bar{H}\Delta\xi$$

That is, the change in the enthalpy due to a reaction is simply the product of $\Delta\bar{H}$ for the reaction and the amount of reaction which took place.

Example 8.11

(a) The calcining of lime has equation

$$CaCO_3(s) = CaO(s) + CO_2(g)$$

The partial molar enthalpies of each of the three substances involved are independent of the amounts present, hence at a given temperature

$$\Delta\bar{H} \equiv \bar{H}(CO_2) + \bar{H}(CaO) - \bar{H}(CaCO_3)$$

is constant and hence for the production of n_{CaO} of quicklime ($dn_{CaO} = d\xi$)

$$\Delta H = \Delta\bar{H} \times n_{CaO}$$

(b) Water from brine by reverse osmosis

$$H_2O(\textit{brine}) = H_2O(\textit{pure,liquid})$$

In this example only a small fraction of the water is extracted from the brine, hence the concentration of the brine leaving the plant is only slightly higher than when it enters. Hence inside the plant $\bar{H}(H_2O, \textit{brine})$ is constant, and

$$\Delta\bar{H} \equiv \bar{H}(H_2O, \textit{pure}) - \bar{H}(H_2O, \textit{brine})$$

The change in internal energy on production of n of pure water is therefore

$$\Delta \bar{H} = \Delta \bar{H} \times n$$

If, as in the example of Section 8.2.1, $\Delta \bar{H}$ is not constant during the reaction, the change in enthalpy must be found by evaluating the integral

$$\Delta \bar{H} = \int \Delta \bar{H} \mathrm{d}\xi$$

8.6.6 Correlation of changes in enthalpy

In the preceding sections we have considered the effects on the enthalpy of changing the temperature or the pressure, or the chemical composition or the phase, or of carrying out chemical processes. In most practical situations several of these will change at the same time. Consider for example the combusion of methane (Fig. 8.21). The inlet gases (a mixture of methane and air at room temperature) differ from the exit gases (mainly carbon dioxide, water and unburnt air) both in temperature and in chemical composition. To compute the difference in the enthalpy of the inlet and exit gases, we compute separately the effect of the change in composition (due to reaction) and the effect of the change in temperature. Two ways of doing this are illustrated schematically in Fig. 8.22.

Query Does the order in which the effects of the different changes are computed matter?

Reply No. We pick the most convenient order. The important thing is that we treat the process as though it occurred in a number of separate steps. The enthalpy change for the total process is then just the total of the enthalpy changes for the individual steps.

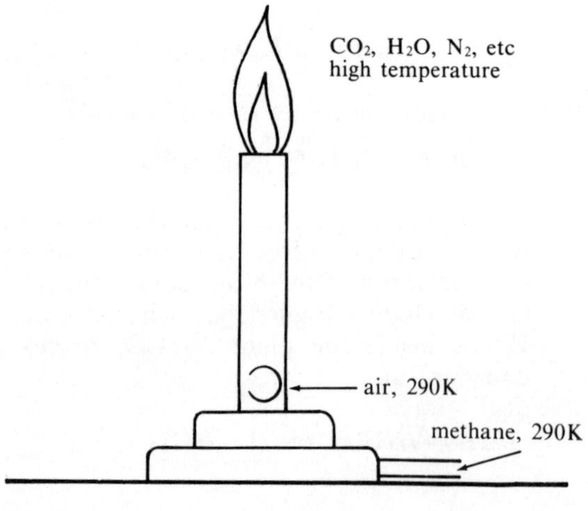

Fig. 8.21 Combustion of methane in a simple burner.

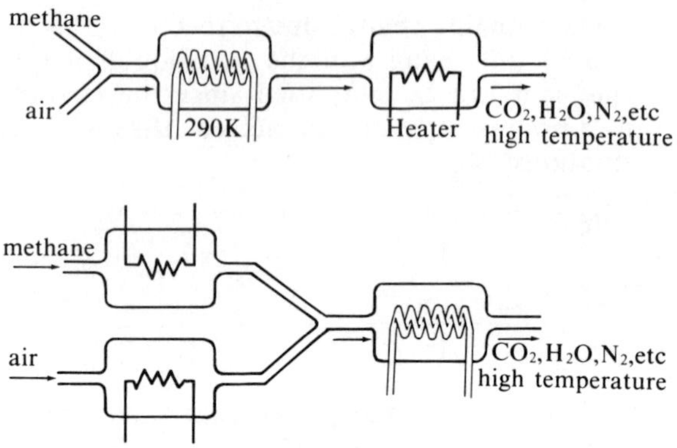

Fig. 8.22 Equivalent processes to combustion of methane.
(a) React at low temperature and heat.
(b) Heat gases and react them.

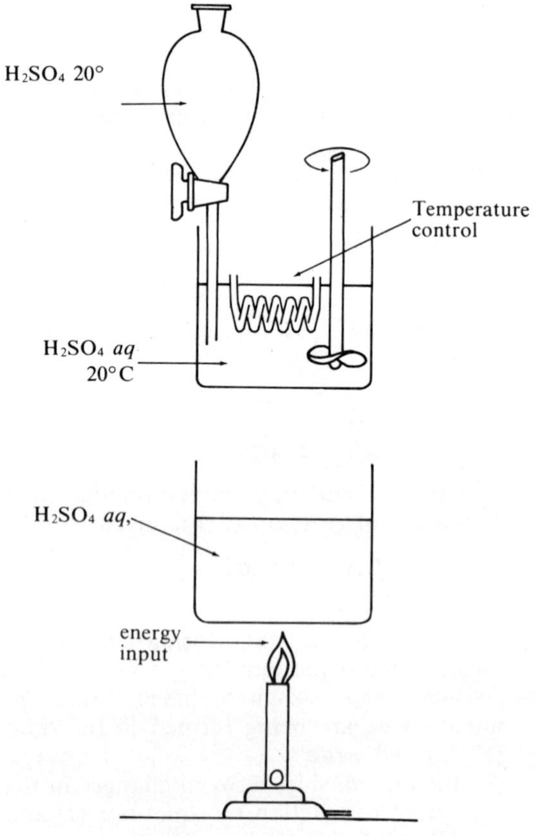

Fig. 8.23 Calculation of enthalpy change on dissolving sulfuric acid.

Example 8.12
When 184 g (100 cm³) of concentrated sulfuric acid (96 g of acid/100 g of solution) at 20°C was added to 100 g of water contained in a beaker at 20°C, the temperature rose to 80°C. The change in enthalpy is calculated as follows (Fig. 8.23).

ΔH(total) = ΔH (due to change in composition at constant temperature) + ΔH (due to the change in temperature)

$$= -113\ \text{kJ} + 35.5\ \text{kJ}$$
$$= -77.5\ \text{kJ}$$

We saw in Sections 8.4.2 and 8.4.3 that when a reaction is carried out at constant pressure or in a flow reactor we can relate the change in enthalpy to the amount of heat transferred to the system and the amount of operator work done on the system.

$$\Delta H = q + w_{♀}$$

We can use this relation to correlate the changes which occur inside the system with those interactions with the environment which are of most interest. If, in the dilution of concentrated sulfuric acid (Example 8.7) no operator work was done, we can readily compute the heat exchange between the solution and its environment.

$$\Delta H = q + w_{♀}$$

that is, $\quad -77.5\ \text{kJ} = q + 0$

This implies transfer of 77.5 kJ by heating the environment.

Example 8.13
Hydrogen and oxygen are being produced by the continuous electrolysis of a dilute solution of sodium hydroxide at atmospheric pressure. Liquid water is supplied to the cell at 25°C and gaseous hydrogen and oxygen are to be removed at the same temperature. A current of 3000A at 2.00V is passed through the cell, electrolysing 0.0155 mol water per second.

$$H_2O(l) = H_2(g) + \tfrac{1}{2}O_2(g)$$

At what rate must heating be applied to or from the cell in order to maintain the temperature of the cell at 25°C? $\Delta \bar{H} = 285\ \text{kJ mol}^{-1}$.

Answer
The process is carried out at constant pressure hence

$$\Delta H = q + w_{♀}$$

$w_{♀}$ = electrical work done on the cell

$$= VI$$
$$= 3000 \times 2.00\ \text{J s}^{-1}$$
$$= 6000\ \text{J s}^{-1}$$

$\Delta H = \Delta H$ (due to chemical change)

$$= \Delta \bar{H} \times \text{amount of reaction}$$
$$= +285 \times 10^3\ \text{J mol}^{-1} \times 0.0155\ \text{mol s}^{-1}$$
$$= 4431\ \text{J s}^{-1}$$

now $\quad q = \Delta H - w_{♀}$

$$= (4431 - 6000)\ \text{J s}^{-1}$$
$$= -1569\ \text{J s}^{-1}$$

That is 1.57 kJ s^{-1} (1.57 kW) of cooling.

Example 8.14
Sufficient solid sodium hydroxide at 20°C is dissolved in a solution (originally at 20°C) containing 6 mol NaOH (kg H_2O)$^{-1}$ to bring the concentration up to 12 mol NaOH (kg H_2O)$^{-1}$. Calculate the final temperature if the solution process occurs so rapidly that no heat is transferred and in such a manner that the operator work due to stirring etc. is negligible.

Data

$\Delta \bar{H}_f$ NaOH(*s*), 20°C = −427 kJ mol^{-1}

$\Delta \bar{H}_f$ NaOH(*aq*) 20°C ≈ −470 kJ mol^{-1}

$\Delta \bar{H}_f$ water (*l*) 20°C ≈ −285 kJ mol^{-1}

C_p (12 mol/kg H_2O NaOH) = 4.3 kJ kg^{-1} K^{-1}

Answer
The process consists of both the solution of sodium hydroxide and a change in temperature. The total enthalpy change can be calculated as illustrated in Fig. 8.24. Take an amount of solution containing 1 kg of H_2O.

ΔH (solution of 6 mol of sodium hydroxide at 20°C)

ΔH (to raise the temperature) $\approx C_p \Delta T$

$$= (4.3\ \text{kJ K}^{-1})\,\Delta T$$

$$\Delta H = -260\ \text{kJ} + C_P \Delta T$$

However, for a process carried out at constant pressure,

$$\Delta H = q + w_{♀}$$

In this case $q = 0$ and $w_{\ddagger} \approx 0$ hence

$$\Delta H = 0$$

i.e. $0 = -260\ \text{kJ} + (4.3\text{kJ K}^{-1})\Delta T$

or $\Delta T = 60\text{K}$

hence

$$\text{the final temperature} = 60 + 20°\text{C} = 80°\text{C}$$

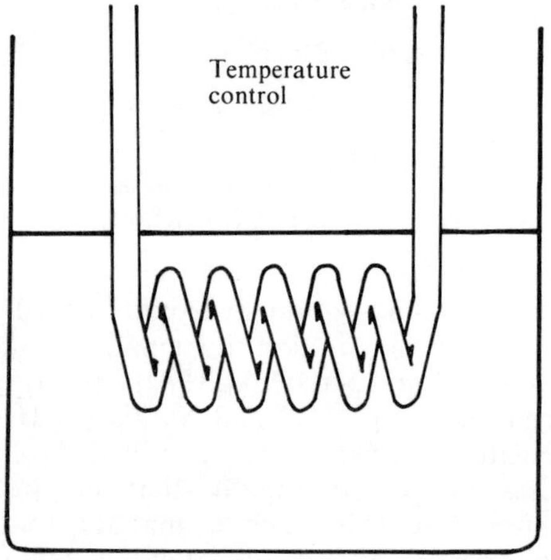

(a) Dissolving NaOH at constant temperature.

(b) Energy supply to raise temperature.

Fig. 8.24 Calculation of enthalpy change on dissolving sodium hydroxide.

8.7 Tabulation of Enthalpies of Materials

In Section 8.6 only changes or differences in enthalpy are considered. This is because absolute values of enthalpy cannot be determined. We can only measure experimentally the value of the enthalpy in one experimental situation relative to its value in some other situation.

8.7.1 Standard enthalpies ($\bar{H}°$)

The most obvious way to proceed is to compare the value of the partial molar enthalpy of a substance with some standard or reference value, exactly as was done for chemical potentials (Section 6.5.3), and values of partial molar volumes (Section 2.6.2). The standard or reference value is called the "standard partial molar enthalpy" or, more commonly, just the "standard enthalpy" of the substance. It is given the symbol $\bar{H}°$. The value chosen depends on the actual substance, the temperature, the phase type and the solvent (if a solute is considered).

8.7.2 Standard enthalpies of reaction ($\Delta\bar{H}°$)

A major use of values of standard enthalpies is in the definition of $\Delta\bar{H}°$ for chemical reactions. An example is the reaction

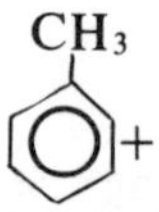

+ HNO_3(in H_2SO_4) =

toluene nitric acid

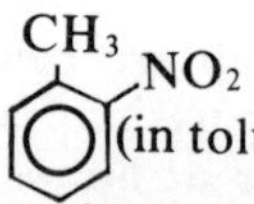

(in toluene) + H_2O(in acid)

o-nitrotoluene

for which

$$\Delta\bar{H}° \equiv \bar{H}°\,(o\text{-nitrotoluene}) + \bar{H}°(\text{water}) - \bar{H}°(\text{nitric acid}) - \bar{H}°(\text{toluene})$$

$\Delta\bar{H}°$ is a very important property of a reaction. It has three major uses.

1. It may be used to estimate values of $\Delta\bar{H}$ for a reaction. In fact few texts differentiate between $\Delta\bar{H}$ and $\Delta\bar{H}°$.

2. It is used to describe the variation of the equilibrium constant with temperature according to the relation (Section 4.8)

$$\frac{d\ln(K_{eq})}{dT} = \frac{\Delta \bar{H}^\circ}{RT^2}$$

3. It is often used as an indication of the order of magnitude and sign of the value of $\Delta\bar{G}^\circ$ for a reaction. Table 6.1 lists a number of values of ΔH° and $\Delta\bar{G}^\circ$ for the formation of substances from their elements. For most examples the signs of the two properties are the same. Note however the values for gaseous $C_{11}H_{24}$ and $C_{11}H_{22}$. This method of estimating values of $\Delta\bar{G}^\circ$ becomes more unreliable as the temperature is increased.

8.7.3 Standard enthalpies of formation ($\Delta\bar{H}_f^\circ$)

Tabulation of standard enthalpies of reaction would require a separate entry for every reaction. Considerable simplification can be made by only tabulating values of $\Delta\bar{H}^\circ$ for a particular class of reaction—the formation of one mole of the substance from its elements in their stable form. For example $\Delta\bar{H}_f^\circ$ (benzoic acid, *s*, 298 K) is determined by the reaction

$$7C(graphite) + O_2(g) + 3H_2(g) = C_6H_5COOH(s)$$

for which

$$\Delta\bar{H}^\circ = -385\ \text{kJ mol}^{-1}$$

so $\Delta\bar{H}_f^\circ\{C_6H_5COOH(s)\} = -385\ \text{kJ mol}^{-1}$

For typical values of $\Delta\bar{H}_f^\circ$ see Table 6.1 on p. 112.

8.7.4 Determination of $\Delta\bar{H}^\circ$ from values of $\Delta\bar{H}_f^\circ$

The use of values of $\Delta\bar{H}_f^\circ$ to determine values of $\Delta\bar{H}^\circ$ for reactions is very simple

$$\Delta\bar{H}^\circ = \Sigma\nu_B\Delta\bar{H}_f^\circ(B)$$

That is we can use values of $\Delta\bar{H}_f^\circ$ in place of values of $\bar{H}^\circ$. Take as an example the chlorination of methane at 298 K.

$$CH_4(g) + 4Cl_2(g) = CCl_4(l) + 4HCl(g)$$

$\Delta\bar{H}^\circ$ for the reaction is defined to be

$$\Delta\bar{H}^\circ = \bar{H}^\circ(CCl_4) + 4\bar{H}^\circ(HCl) - 4\bar{H}^\circ(Cl_2) - \bar{H}^\circ(CH_4)$$

For each substance there exists a formation reaction and corresponding value of $\Delta\bar{H}_f^\circ$. Thus,

(a) $C(s) + 2Cl_2(g) = CCl_4(l)$

$$\Delta\bar{H}_f^\circ(CCl_4) = \bar{H}^\circ(CCl_4) - 2\bar{H}^\circ(Cl_2) - H^\circ(C) = -132\ \text{kJ mol}^{-1}$$

(b) $C(s) + 2H_2(g) = CH_4(g)$

$$\Delta\bar{H}_f^\circ(CH_4) = \bar{H}^\circ(CH_4) - 2\bar{H}^\circ(H_2) - \bar{H}^\circ(C) = -75\ \text{kJ mol}^{-1}$$

(c) $\tfrac{1}{2}H_2(g) + \tfrac{1}{2}Cl_2(g) = HCl(g)$

$$\Delta\bar{H}_f^\circ(HCl) = \bar{H}^\circ(HCl) - \tfrac{1}{2}\bar{H}^\circ(Cl_2) - \tfrac{1}{2}H^\circ(H_2) = -92\ \text{kJ mol}^{-1}$$

From which it can be seen that

$$\Delta H^\circ \equiv \Delta H_f^\circ(CCl_4) + 4\Delta H_f^\circ(HCl) - \Delta H_f^\circ(CH_4)$$

$$= -132\ \text{kJ mol}^{-1} + 4\times(-92)\ \text{kJ mol}^{-1} - (-75)\text{kJ mol}^{-1}$$

$$= -425\ \text{kJ mol}^{-1}$$

Query Why did you leave out the chlorine?
Reply Because the chlorine is an element, its formation reaction is

$$Cl_2(g) = Cl_2(g)$$

Therefore, by definition $\Delta\bar{H}_f^\circ$ (elements in their stable form) $\equiv 0$.

PROBLEMS

8.3. Use values of $\Delta\bar{H}_f^\circ$ from Table 6.1 to determine values of ΔH° (298 K) for the following reactions.

(a) $C(graphite) + 2H_2(g) = CH_4(g)$

(b) $C_2H_2(g) + 2H_2(g) = C_2H_6(g)$

(c) $C_6H_5COOH(s) + 7\tfrac{1}{2}O_2(g) = 7CO_2(g) + 3H_2O(l)$

(d) $\tfrac{1}{2}Br_2(l) + I^-(aq) = \tfrac{1}{2}I_2(s) + Br^-(aq)$

(e) $Al_2O_3(s) + 2OH^-(aq) + 3H_2O(l) = 2Al(OH)_4^-(aq)$

(f) $H_2(g) + Cl_2(g) = 2HCl(aq)$

8.4. Estimate the change in enthalpy when the following processes are carried out at constant temperature and constant pressure.

(values of $\Delta \bar{H} \approx \Delta \bar{H}°$)

(a) 0.2 mol of solid iodine reacts with 0.2 mol of gaseous chlorine, if for the reaction $I_2(s) + Cl_2(g) = 2ICl(g)$

$$\Delta \bar{H}° = 35 \text{ kJ mol}^{-1}$$

(b) 100 cm^3 0.1 M $AgNO_3$ reacts with 100 cm^3 of 0.2 M NH_3, if for the reaction

$$Ag^+(aq) + 2NH_3(aq)$$
$$= Ag(NH_3)_2^+(aq)$$
$$\Delta \bar{H}° = -111 \text{ kJ mol}^{-1}$$

(c) 17 g of ammonia is distilled from aqueous solution if for the process

$$NH_3(g) + \text{water} = NH_3(aq)$$
$$\Delta \bar{H}° = -35.4 \text{ kJ mol}^{-1}$$

(d) Calculate $\Delta \bar{H}°$ for the process

$$HCl(g) = H(g) + Cl(g)$$

if for the processes

$$Cl_2(g) = 2Cl(g) \quad ; \Delta \bar{H}° = 242 \text{ kJ mol}^{-1}$$
$$H_2(g) = 2H(g) \quad ; \Delta \bar{H}° = 432 \text{ kJ mol}^{-1}$$
$$H_2(g) + Cl_2(g) = 2HCl(g); \Delta \bar{H}° = -185 \text{ kJ mol}^{-1}$$

(e) Calculate $\Delta \bar{H}°$ (298 K) for the process

$$C_2H_2(g) + 5/2\, O_2(g)$$
$$= 2CO_2(g) + H_2O(g)$$

(use the data in Table 6.1).

(f) Calculate $\Delta \bar{H}°$ for the reaction

$$U_3O_8(s) + \tfrac{1}{2}O_2(g) = 3UO_3(s)$$

if for the processes

$$U_3O_8(s) + 2C(s) = 3UO_2(s) + 2CO(g);$$
$$\Delta H° = -\ 316 \text{ kJ mol}^{-1}$$

$$UO_3(s) + C(s) = CO(g) + UO_2(s);$$
$$\Delta H° = -2420 \text{ kJ mol}^{-1}$$

$$C(s) + \tfrac{1}{2}O_2(g) = CO(g);$$
$$\Delta H° = -\ 111 \text{ kJ mol}^{-1}$$

Answers (a) 7.0 kJ; (b) −1.11 kJ; (c) +35.4 kJ; (d) +430 kJ mol^{-1}; (e) −1257 kJ mol^{-1}; (f) +6833 kJ mol^{-1}

Change in enthalpy with temperature

8.5. Estimate the difference in enthalpy between

(a) 1 mol of $Al_2O_3(s)$ at 100°C and 500°C if

$$\bar{C}_P^{\circ}(Al_2O_3,s) = 79 \text{ J K}^{-1} \text{ mol}^{-1}$$

(b) 100 g of $BBr_3(l)$ at 20°C and 50°C if

$$\bar{C}_P^{\circ}(BBr_3, l) = 128 \text{ J K}^{-1} \text{ mol}^{-1}$$

(c) 100 g of alloy containing 60 g copper and 40 g of tin at 20°C and 120°C if

$$\bar{C}_P^{\circ}(Sn) = 27 \text{ J K}^{-1} \text{ mol}^{-1}$$
and $$\bar{C}_P^{\circ}(Cu) = 24 \text{ J K}^{-1} \text{ mol}^{-1}$$

(d) A solution containing 0.1 mol of NaCl and 1 kg H_2O at 30°C and 80°C if

$$\bar{C}_P^{\circ}(Na^+, aq) = 46 \text{ J K}^{-1} \text{ mol}^{-1}$$
$$\bar{C}_P^{\circ}(Cl^-, aq) = -136 \text{ J K}^{-1} \text{ mol}^{-1}$$
$$\bar{C}_P^{\circ}(H_2O, l) = 75 \text{ J K}^{-1} \text{ mol}^{-1}$$

Query How can the value of $\bar{C}_P$ for Cl^- be negative?

Reply Such a value means that the heat capacity of a dilute solution of sodium chloride is less than the heat capacity the water would have if it were pure. In a very dilute solution, the partial molar heat capacity of the solvent (water) is made to approach the value for the pure solvent. Hence the sum of the partial molar heat capacities of the ions can be negative. The value of the standard heat capacity, $\bar{C}_P°$, for all ions is taken relative to the value for $H^+(aq)$, which is assigned the value zero

$$\bar{C}_P°(H^+, aq) \equiv 0$$

(e) Carbon dioxide at 300 K and carbon dioxide at 1200 K if

$$C_P°(CO_2) = 32.22 + 0.022\,(T/\text{K}) \text{ J K}^{-1} \text{ mol}^{-1}$$

Answers (a) 31.6 kJ; (b) 1.5 kJ; (c) 3.18 kJ; (d) 208 kJ; (c) 30.6 kJ.

Correlation of enthalpy changes

8.6 For the reaction

$$CH_4(g) + 2O_2(g) = CO_2(g) + 2H_2O(l);$$
$$\Delta \bar{H}^\circ(298\ K) = -889\ kJ\ mol^{-1}$$

(a) How much cooling is needed if a mixture of methane (0.1 mol per second) and air (in 20% excess over that required for combustion) is to react completely and give products at 298 K from reactants at that temperature?

(b) If fuel cells use this reaction and supply 60 kW (60 kJ s^{-1}) of electric power, how much cooling is needed to keep the temperature at 298 K?

(c) What change is made to the answer to (a) if the product water is in the gaseous state?

$$H_2O(l, 298\ K) = H_2O(g, 298\ K);$$
$$\Delta \bar{H}^\circ = 44\ kJ\ mol^{-1}$$

Answers (a) 88.9 kJ s^{-i} = 88.9 kW;
(b) 28.9 kW;
(c) reduction to 80.1 kW.

8.8 Terms which are Often Abused

In this section we wish to take up some terms which seem to us to be useful, but which we think are not treated fairly by their supporters. Perhaps the careful basis we have sought to prepare can be used in making them clear.

8.8.1 Chemical energy

People often speak of chemical energy. Most often when people use the term "chemical energy" they are interpreting the tendency for change. Although they may use values of enthalpy (or internal energy) in their argument, the connection with tendency for change identifies the subject of their reasoning—it is actually chemical potential to which they refer. The phenomenon to which the argument is applied often happens to have values of $\Delta \bar{H}^\circ$ and $\Delta \bar{G}^\circ$ which have the same sign.

8.8.2 Thermal energy

The term "thermal energy", properly applied, describes energy of a substance or system which would alter if the temperature altered. Steam at 500 K is formed from water at 298K with absorption of energy. On lowering the temperature of the surroundings, energy will be transferred to them as heat. Such steam is said to have "thermal energy" and to lose it on cooling. The danger in this expression is that it often leads to thinking of "heat" as though it were "something" possessed by a hot body—not merely transfer of energy. This false thinking, originating in the theory of "caloric", has persisted long after the theory was banished. Probably the main reason for persistence is that this "heat is like a substance" view does not lead to much confusion—until one steps from physics to chemistry or biology. "Thermal energy" is *energy*. Because transfer of energy from a hot body to a cooler one is a spontaneous process it is possible to make use of such transfers in doing operator work. The maximum extent of operator work is governed by two factors—the change in chemical potential and the constancy of the total energy when all interactions are included.

8.8.3 Latent heat

"Latent heat" is an old name for the enthalpy change of a change of phase—from solid to liquid, liquid to gas, etc. The name probably arose from the change in the rate of cooling which occurs when a liquid or a solid phase forms from a gas or liquid phase. There is nothing particularly special about such changes. They arise from phase change and are often of great importance in stabilising temperatures in practical situations.

8.9 Energy and Chemical Existence

So far in this chapter we have taken zeros for the internal energy and for the enthalpy which relate their scales to the properties of the elementary substances in a particular chosen state. We shall now show the connection between this position and that of valency theory. It emerges that we *can* get a more fundamental view of the zero positions for energy scales. It also emerges that the new zero positions for these scales are of no chemical use.

8.9.1 The modern concept of energy

In the modern concepts of physics, the energy and the mass of a system are equivalent properties. The relationship is expressed by Einstein's famous equation,

$$E = mc^2$$

as one of the principles of special relativity.

Here E is the energy of the system,
m is the observed mass of the system,
and c is the velocity of light in free space.

The specific energy of a material is then simply

$$E/m = c^2$$

If we know the molar mass of a chemical substance, $\bar{M}_B$, we can calculate its molar energy

$$\bar{E}_B = \bar{M}_B c^2$$

If we know the difference of molar masses for a chemical reaction we can calculate the energy change associated with the process

$$\Delta\bar{E}(\text{reaction}) = \Delta\bar{M}(\text{reaction})c^2$$

Example 8.15
The molar energy of methane gas ($\bar{M} = 16.0\ \text{g mol}^{-1}$) is

$$\bar{E} = \bar{M}c^2$$
$$= (16 \times 10^{-3}\,\text{kg mol}^{-1}) \times (3 \times 10^8\,\text{m s}^{-1})^2$$
$$= 1.44 \times 10^{15}\ \text{J mol}^{-1}$$

Example 8.16
When solid carbon reacts with gaseous hydrogen to form gaseous methane

$$C(s, 298\text{K}, 10^5\text{Pa}) + 2H_2(g, 298\text{K}, 10^5\text{Pa})$$
$$= CH_4(g, 298\text{K}, 10^5\text{Pa})$$

the change in molar energy is $\Delta\bar{E} = 74\ \text{kJ mol}^{-1}$ and the corresponding change in mass is $\Delta\bar{M}$.

$$\Delta\bar{E} = \Delta\bar{M}\,c^2$$
$$\Delta\bar{M} = \Delta\bar{E}/c^2$$
$$= (74 \times 10^3\text{J mol}^{-1})/(3 \times 10^8\ \text{m s}^{-1})^2$$
$$= 8.2 \times 10^{-13}\ \text{kg mol}^{-1}$$

Such a change of mass cannot be detected.

PROBLEM

8.7. An example of a very high energy reaction is the separation of a proton and an electron

$$H(g) = H^+(g) + e^-(g)$$

If $\Delta\bar{E} = 2.62 \times 10^6\ \text{J mol}^{-1}$

what is the corresponding increase in mass?

The determination of masses is one of the most precise measurements which can be made. Differences of 1 part in 100 million can be detected. But the change of mass in the formation of methane is only about 1 part in 10 million million million. Clearly it is not useful to think of measuring changes of mass when we wish to determine the energy changes associated with ordinary chemical processes. Instead we deal with mass and energy as though they were independent quantities, as we have been accustomed to doing.

8.9.2 Energy storage in molecules

Recognising that substances are comprised of molecules, we may ask how they store the energy, the "internal energy", which is to be associated with them. This question brings us into contact with the theories of valency and their description of the energy of a particular molecule as the sum of the energies involved in its motions and in its interactions with its neighbours.

These molecular motions involve:

translational energy	—	motion from one place to another
rotational energy	—	spinning of molecules about axes in the molecules
internal rotational energy	—	spinning of parts of molecules
vibrational energy	—	movements of parts of a molecule, opposed by the rest, like stretching bonds or wobbling parts of the molecule
electronic energy	—	altering the distribution of electrons
intermolecular energy	—	interaction between molecules
nuclear energy, etc.		

For gaseous carbon dioxide at 300 K, $P = P^\circ$, the following contributions to the internal energy are calculated:

translational energy $(\bar{E}_{trans}) = 3.74\ \text{kJ mol}^{-1}$

rotational energy $(\bar{E}_{rot}) = 2.49\ \text{kJ mol}^{-1}$

vibrational energy $(\bar{E}_{vib}) = 26.16\ \text{kJ mol}^{-1}$

electronic energy $(\bar{E}_{elec}) = -425.86\ \text{kJ mol}^{-1}$

energy of formation,

$$\Delta E_f^\circ\{CO_2(g)\} = -393.47\ \text{kJ mol}^{-1}$$

For translation, rotation and vibration, zero energy has been assigned to a molecule which is "not moving about", "not rotating", and "not vibrating". The electronic energy term has the largest magnitude and is determined as the difference from the energy of the elements, carbon, C(*graphite*), and oxygen, $O_2(g)$, in their reference states at 298 K.

In this method of treatment, the "electronic energy" term is calculated as the difference between the experimentally determined energy of the molecule and the sum of the other terms. These other terms are calculated from models of the behaviour of molecules.

There are gains in this method of analysis of the internal energy. It provides the bases for tests of valency theories and a basis for correlation and prediction of molecular behaviour.

8.9.3 Effect of temperature change on internal energy

An example of the use of the molecular model for energies involves the understanding of the effect of temperature change on the molar internal energy of a substance. The electronic energy does not usually alter much until high temperatures are reached—and for carbon dioxide that means much higher than 1000 K. The translational, rotational and vibrational energies do change significantly with changing temperature. Such a change can be due to changes in the energies (in the quantum levels) or in the distribution of molecules over energy levels.

Theories of molecular energy show that the number of possible energies of any molecule is very much greater than the number which are in use at any time. For this reason it is possible to use a simple statistical treatment, due to Ludwig Boltzmann (Vienna, 1896), to compare the numbers, N_1, N_2, of molecules of one species which have molecular energies equal to ϵ_1 and ϵ_2 at equilibrium. It gives

$$N_1/N_2 = e^{-L(\epsilon_1 - \epsilon_2)/RT}$$

where L is the Avogadro number (6.023×10^{23} molecules mol^{-1}). Translational energies display the effect of this distribution of molecules over energies very clearly. In Fig. 8.25 calculated numbers of nitrogen molecules which have translational energies close to a particular value are shown for three temperatures (100 K, 1000 K and 5000 K). At low temperatures, the energies are low and they lie closer to a single value. At all temperatures a few molecules have energies very much greater than the average, a characteristic which is significant in understanding rates of reaction.

The contributions to the internal energy of carbon dioxide which can be assigned to translation, rotation and vibration are displayed in Fig. 8.26. For this molecule, in the tempera-

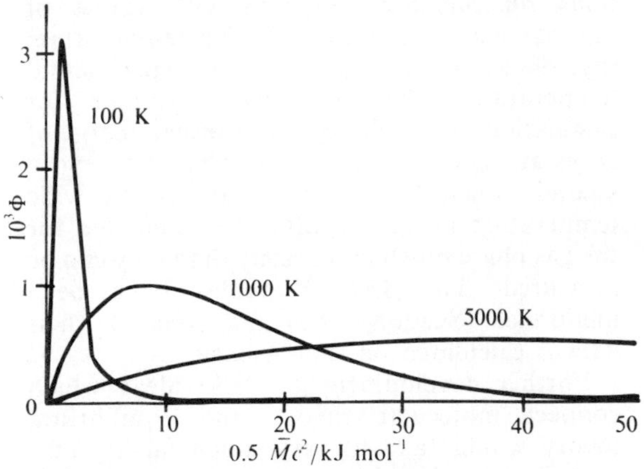

Fig. 8.25 Distribution of translational energy in nitrogen at 100 K, 1000 K, and 5000 K.

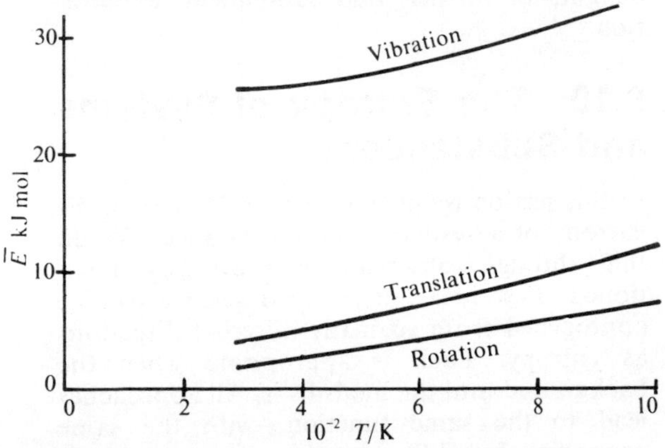

Fig. 8.26 Relative contributions of translation, rotation and vibration to the energy of carbon dioxide between 300 K and 1000 K.

ture range 300 − 1000 K, the changes in these three quantities are all in the same direction and all of about the same amount. Note that direct extrapolation suggests that there is zero translational and zero rotational energy at zero temperature. No straightforward extrapolation of the vibrational energy curve would lead to zero vibrational energy at zero temperature. The difference from zero is the "zero point energy". It is important. For example, it interprets most of the chemical difference between molecules which contain different isotopes (like HF(*g*) and DF(*g*).

Calculations of the changes in internal energy with temperature and of the heat capacities have value in practice. For most chemical reactions the change in the electronic energy terms is much larger than any other change in energy, as it was in our listing for carbon dioxide (Section 8.10.2). Moreover, this term rarely alters much for temperatures less than 1000 K. So one conclusion is that we may usually do calculations *on chemical reactions* with values of internal energy obtained at one temperature, say 298 K, no matter what the experimental temperature. The theoretical formulae for calculation of the changes in internal energy of gases are suited to computer calculation. So a second conclusion is that data for a wide temperature range can often be calculated for the gas phase much more easily than they can be measured. The JANAF tables have been mentioned (Section 4.7.2) and some of their data is calculated on such a basis.

Further development of these ideas which connect molecular theory and equilibrium theory would take us very much farther into both theories than is practicable at present. We have gone a little way in order to point to the delights of further, and subsequent, exploration.

8.10 The Entropy of Systems and Substances

In this section we introduce a new function, the entropy of a system or of a substance. We do this through observations previously mentioned. Few ideas in physical science can be approached from so many different directions as entropy. Each is appropriate when the background and the goals fit it. All approaches lead to the same function, with the same properties, but different approaches emphasise different properties. We shall begin our account with the effects of changing the temperature on equilibria and emphasise that general region of the usefulness of the function.

8.10.1 Approach to the function entropy

Like all definitions of entropy, our approach begins from a group of empirical observations. We recall that the variation of equilibrium constants with temperature shows (Section 4.8) a surprising relationship to the change in enthalpy. There is an observed equation

$$\frac{\mathrm{d}\ln K_{\mathrm{eq}}}{\mathrm{d}T} = \frac{\Delta\bar{H}^\circ}{RT^2} \qquad (1)$$

where $\Delta\bar{H}^\circ$ is the difference in the standard molar enthalpy (Section 8.7.2) of the reactants and products; R is the gas constant (8.314 J K^{-1} mol^{-1}); and T is the thermodynamic temperature. The surprising thing about this empirical relationship is that the equation is obeyed more exactly the more accurate is our knowledge of the quantities in it. Thus, it contains a significant observation about the behaviour of chemical systems.

We also have available a relationship between the equilibrium constant and the difference in standard chemical potentials between the reactants and products in their standard states. The value of $\Delta\bar{G}^\circ$, which is a combination of standard chemical potentials,

$$\Delta\bar{G}^\circ = \sum_i \nu_i \mu_i^\circ \qquad (2)$$

is also

$$\Delta\bar{G}^\circ = -RT\ln K_{\mathrm{eq}} \qquad (3)$$

Combining equations (1, 3) we have

$$-\frac{\mathrm{d}(\Delta\bar{G}^\circ/T)}{\mathrm{d}T} = \frac{\Delta\bar{H}^\circ}{T^2} \qquad (4)$$

The expression on the left can be simplified and the equation rearranged to

$$-\frac{1}{T}\frac{\mathrm{d}\Delta\bar{G}^\circ}{\mathrm{d}T} + \frac{\Delta\bar{G}^\circ}{T^2} = \frac{\Delta\bar{H}^\circ}{T^2} \qquad (5)$$

$$-\frac{\mathrm{d}\Delta\bar{G}^\circ}{\mathrm{d}T} = \frac{\Delta\bar{H}^\circ - \Delta\bar{G}^\circ}{T}$$

The partial molar enthalpies and the chemical potentials can be introduced into equation (5). There is an equation (6), like equation (2).

$$\Delta \bar{H}^\circ = \sum_i \nu_i \bar{H}_i^\circ \qquad (6)$$

If we use both (2) and (6) in equation (5), the equation which results will have similar terms for each substance and these can be equated, so that

$$-\frac{d\mu_i^\circ}{dT} = \frac{\bar{H}_i^\circ - \mu_i^\circ}{T} \qquad (7)$$

For example, for the equation

$$2CO + O_2 = 2CO_2$$

that is $\quad 0 = 2CO_2 - 2CO - O_2$

$$\sum_i \nu_i \mu_i^\circ = 2\mu^\circ_{CO_2} - 2\mu^\circ_{CO} - \mu^\circ_{O_2}$$

$$\sum_i \nu_i H_i^\circ = 2\bar{H}^\circ_{CO_2} - 2\bar{H}^\circ_{CO} - \bar{H}^\circ_{O_2}$$

and the equations (see (7)) become

$$-\frac{d\mu^\circ_{CO_2}}{dT} = \frac{\bar{H}^\circ_{CO_2} - \mu^\circ_{CO_2}}{T}$$

$$-\frac{d\mu^\circ_{CO}}{dT} = \frac{\bar{H}^\circ_{CO_2} - \mu^\circ_{CO}}{T}$$

$$-\frac{d\mu^\circ_{O_2}}{dT} = \frac{\bar{H}^\circ_{O_2} - \mu^\circ_{O_2}}{T}$$

The values of these quantities at 298K are:

CO_2 213.6 J K^{-1} mol^{-1}

CO 197.9 J K^{-1} mol^{-1}

O_2 205.03 J K^{-1} mol^{-1}

The expressions on both sides of equation (7) are very useful and we define a new function, *entropy*, S, for them and for the related quantities for substances which are not in their standard states and also for materials and for systems. In particular, we define the *standard partial molar entropy*, $\bar{S}_i^\circ$, as

$$\bar{S}_i^\circ \equiv -\frac{d\mu_i^\circ}{dT} \equiv \frac{\bar{H}_i^\circ - \mu_i^\circ}{T} \qquad (8)$$

For a substance, i, in any state we define the *partial molar entropy*, $\bar{S}_i$, as

$$\bar{S}_i = \frac{\bar{H}_i - \mu_i}{T} = -\left(\frac{\partial \mu_i}{\partial T}\right)_P \qquad (9)$$

We define the entropy of a material or phase as the sum of the entropies of the substances in it, each entropy being the product of the appropriate partial molar entropy and the amounts of the substance. Thus, for 100m^3 of dry air at 298K and 100 kPa, the total entropy is

$$S(\text{air}) = S_{O_2} + S_{N_2} + S_{Ar} + S_{CO_2} + \ldots$$

$$= \bar{S}_{O_2} \times n_{O_2} + \bar{S}_{N_2} \times n_{N_2} + \ldots$$

Table 8.11 shows that the entropy in this case is 782 kJ K^{-1} mol^{-1}. It follows, since there are equations like (9) for each substance, that the entropy of a material obeys equations (10)

$$\bar{S}_{\text{material}} = -\left(\frac{\partial \bar{G}_{\text{material}}}{\partial T}\right) = \frac{\bar{H}_{\text{material}} - \bar{G}_{\text{material}}}{T} \qquad (10)$$

For a system, we define the entropy as the sum of the entropies of the materials or phases which comprise it. Hence, there are equations (11)

$$S_{\text{system}} = -\left(\frac{\partial G_{\text{system}}}{\partial T}\right) = \frac{H_{\text{system}} - G_{\text{system}}}{T} \qquad (11)$$

Table 8.10
Entropy of 100 m^3 *of Dry Air at* 298K *and* 100 kPa

B	X_B	n_B / mol	$\bar{S}_B$ / JK^{-1}mol^{-1}	S_B / kJ K^{-1}
N_2	0.7809	3151	191.5	603
O_2	0.2095	846	205	173
Ar	0.0093	37.5	155	5.8
CO_2	3×10^{-4}	1.2	214	0.3
Ne	1.8×10^{-5}	0.07	146	0.01
			Total	782

8.10.2 Change of entropy with temperature

The way in which the entropy (of a substance, a material or a system) changes with temperature, without changes of pressure, can be seen by examining the equations which define the entropy. We have

$$\left(\frac{\partial S}{\partial T}\right)_P = \left(\frac{\partial}{\partial T}\frac{H-G}{T}\right)_P$$

$$= -\frac{H-G}{T^2} + \frac{1}{T}\left(\frac{\partial H}{\partial T} - \frac{\partial G}{\partial T}\right)_P \qquad (12)$$

$$= -\frac{S}{T} + \frac{1}{T}C_P + \frac{S}{T}$$

$$= \frac{C_P}{T}$$

or, for a substance in a material

$$\partial \bar{S}/\partial T = \bar{C}_p/T$$

Therefore we can calculate the change in entropy which results from a change in temperature if we have values of the heat capacity, C_p, for that change of temperature

$$\Delta S = \int_{T_1}^{T_2} \frac{C_p}{T}\, dT$$

Example 8.17
The change in entropy on heating gaseous water from 373.16K to 423.16K is

$$\Delta \bar{S} = \int_{373.16}^{423.16} \frac{C_p}{T}\, dT$$

$$= \int_{373.16}^{423.16} \frac{(30.36 + 9.61 \times 10^{-3} T + 11.8 \times 10^{-7} T^2}{T}\, dT$$

$$= 30.36 \ln \frac{423.16}{373.16}$$

$$+ 9.61 \times 10^{-3} (423.16 - 373.16)$$

$$+ 2 \times 11.8 \times 10^{-7} (423.16^2 - 373.16^2)$$

$$= 4.39 \text{ J K}^{-1} \text{ mol}^{-1}$$

8.10.3 Change of entropy with pressure

The change in entropy which results from a change in pressure without change of temperature can be obtained from the equations which define the entropy. We have

$$S = \frac{1}{T}(H - G) \qquad (9-11)$$

so $$\left(\frac{\partial S}{\partial P}\right)_T = \frac{1}{T}\left\{\left(\frac{\partial H}{\partial P}\right)_T - \left(\frac{\partial G}{\partial P}\right)_T\right\}$$

From the argument in Section 6.6.3 it follows that

$$\left(\frac{\partial G}{\partial P}\right)_T = V$$

where V is the volume of the material (or partial molar volume of the substance) to which G (or μ) refers.

From Section 8.6.2,

$$\left(\frac{\partial H}{\partial P}\right)_T = V(1 - \alpha T)$$

Thus $$\left(\frac{\partial S}{\partial P}\right)_T = \frac{1}{T}(V - \alpha V T - V)$$

$$= -\alpha V \qquad (13)$$

Thus, for example, we can calculate the change in partial molar entropy of a substance which results from a change in pressure at constant temperature if we know the way in which the partial molar volume changes with pressure and the value of the coefficient of thermal expansion (28).

Example 8.18

(a) For liquid water at about 298K, α is 4.8×10^{-4} K^{-1}. We have seen (Section 2.7) that the molar volume of liquid water is almost constant and equal to 18.02 cm^3 mol^{-1} between $P°$ and 10 $P°$. Thus the change in molar entropy for this change in pressure is

$$\Delta \bar{S} = -\int_{P°}^{10P°} \alpha \bar{V} dP$$

$$= -\alpha \bar{V}(10P° - P°)$$

$$= -4.8 \times 10^{-4} (\text{K}^{-1})$$

$$\times 18.02 (\text{cm}^3 \text{ mol}^{-1})$$

$$\times 9 \times 101 (\text{kPa})$$

$$= -4.8 \times 10^{-4} (\text{K}^{-1}) \times 18.02$$

$$\times 10^{-6} (\text{m}^3 \text{ mol}^{-1})$$

$$\times 9 \times 101 \times 10^3 (\text{Pa})$$

$$= -7.9 \times 10^{-3} \text{ J K}^{-1} \text{ mol}^{-1}$$

(a very small quantity)

(b) For one mole of gas at 297K, a change in pressure from $P°$ to 10 $P°$ involves an entropy change

$$\Delta \bar{S} = -\int_{P°}^{10P°} \alpha V dP$$

$$= -\int_{P°}^{10P°} \frac{\partial \bar{V}}{\partial T_P} dP$$

$$\approx -\int_{P^\circ}^{10P^\circ} \frac{R}{P}\,\mathrm{d}P$$

$$= -R\ln\frac{10P^\circ}{P^\circ}$$

$$= -19.1\ \mathrm{J\,K^{-1}\,mol^{-1}}$$

8.10.4 Entropy differences between phases

When any two phases are in equilibrium with one another, each substance in one phase is in equilibrium with that substance in the other phase and so that substance has the same value of chemical potential in the two phases (Section 6.6.5). From the definition of the partial molar entropy of the substance in each of the phases α, β

$$\bar{S}(\alpha) = \frac{\bar{H}(\alpha) - \mu(\alpha)}{T}$$

and
$$\bar{S}(\beta) = \frac{\bar{H}(\beta) - \mu(\beta)}{T}$$

the difference between the entropy of a substance in the two phases

$$\bar{S}(\beta) - \bar{S}(\alpha) = \frac{1}{T}(\bar{H}(\beta) - \bar{H}(\alpha)) \qquad (14)$$

Thus if we know the partial molar enthalpies of the substance in both phases we can compute $\Delta\bar{S}$ for the transfer of a substance from one phase to the other. That is, at *equilibrium*

$$\Delta\bar{S}(\alpha\rightarrow\beta) = \Delta\bar{H}(\alpha\rightarrow\beta)/T$$

Query What if the two phases are not at equilibrium?

Reply In that case $\mu(\alpha) \neq \mu(\beta)$ and we are left with the general relation.

$$\Delta\bar{S} = \frac{\Delta\bar{H} - \Delta\bar{G}}{T}$$

Example 8.19
The enthalpy change for formation of gaseous water from liquid at 373K is 40.88 kJ mol^{-1}. The entropy change for this process is therefore

$$\Delta\bar{S} = \frac{\Delta\bar{H}}{T}$$

$$= \frac{40.88\times10^3}{373}\ \mathrm{J\,K^{-1}\,mol^{-1}}$$

$$= 109.5\ \mathrm{J\,K^{-1}\,mol^{-1}}$$

8.10.5 Change of entropy with changes of temperature and phase

The changes in molar entropy of a substance are often required over ranges of temperature which also involve a change of phase. The resulting change in entropy is obtained by simple combination of the methods of Sections 8.11.2 and 8.11.4.

Example 8.20
The entropy change for conversion of *liquid* water at 373.16K to gaseous water at 423.16K (and 100 kPa) is

$$109.5 + 4.39\ \mathrm{J\,K^{-1}\,mol^{-1}}$$

$$= 113.9\ \mathrm{J\,K^{-1}\,mol^{-1}}$$

using results previously obtained.

Query What happens if measurements like these go to zero temperature?

Reply As the temperature becomes small, both C_p and C_p/T become small. In fact, for most changes‡ it appears that the molar entropy change becomes *zero* at zero temperature. If this is the case, it makes sense if the molar entropy of substances is given the value zero at zero temperature.

‡The exceptions to this statement will not concern us.

Solutions to Problems

Chapter 1. Describing Chemical Systems

Problem 1.1

(a) With the naked eye we would classify as uniform materials (homogeneous materials or single phases) most specimens of fresh water, quartz, gasoline, air, table salt, glass and polythene. From the same point of view, composite or heterogeneous character (two or more phases) is apparent in such materials as granite, fruit cake, pepper, plywood and fibreglass. Some materials may be uniform at one level of specification and yet need to be reclassified as composite or heterogeneous when looked at more closely. Thus for many uses muddy water, butter, fresh milk, fog clouds, concrete, blood, beach sand and paint can be regarded as uniform and yet their examination with magnifying instruments can reveal heterogeneity at a finer level of observation.

Fluid composites (emulsions, dispersions, colloids) are likely to break down by sedimentation under gravity or by centrifuging. Solid composites may be crushed and separated by mechanical action, as in the winnowing of wheat from chaff and of rutile (TiO_2) from beach sand and also in the washing of gold from its heterogeneous mixture with silt or sand.

(b) Distilled water, crystallised table salt (sodium chloride), refined cane sugar (sucrose), bicarbonate of soda (sodium hydrogen carbonate) are among the common materials which qualify as pure phases or single substances. As indicated, their purity or the method of purification is often indicated in the common name. Sea water, air, gasoline, jewellers' gold, vanilla essence and tincture of iodine are solution phases or homogeneous mixtures of two or more substances. Their qualities are likely to vary from source to source and between fractions obtained by mechanical separation during passage through a cycle of physical forms. Thus, when a liquid solution is partially evaporated and the vapor is collected and condensed, the residue and the condensate are likely to differ in properties according to the different proportions of the component substances in the two fractions.

(c) Among the elementary substances commonly seen and used are carbon (charcoal), aluminium, iron, chlorine, iodine, magnesium, silicon, lead, copper, tin, zinc, chromium, nickel, silver, gold, mercury, cadmium, oxygen, nitrogen, helium, neon, argon and tungsten.

Familiar compound substances are water (oxide of hydrogen), sodium chloride (table salt), sodium hydrogen carbonate (bicarbonate of soda), ferric oxide (rust), silicon dioxide (silica, quartz, sand), calcium carbonate (marble, limestone), carbon dioxide and carbon monoxide (components of natural and polluted airs).

The earliest simple criterion for classifying a substance as elementary was the absence of decomposition when an enclosed sample was taken to very high temperatures. Antoine Lavoisier demonstrated this principle by counter-example using sunlight focussed by a lens to heat the red calx of mercury and collect the two elementary substances, mercury and oxygen. Michael Faraday used electrolysis to show that caustic soda is a compound substance. Its place in the list of elementary substances was taken by the metal, sodium, obtained as a product of the electrolysis.

(d) In the earth's crust the most abundant of the chemical elements is oxygen. It appears in combination, particularly with silicon (Si), aluminium (Al), iron (Fe), magnesium (Mg), calcium (Ca) and sodium (Na). The elementary substances corresponding to these latter elements are not found in natural terrestrial situations but have been manufactured and put to important uses. A list with all the natural chemical elements and some man-made ones is given in Appendix A.5. Some elements form more than one elementary substance. For example, oxygen forms diatomic oxygen (O_2) and ozone (O_3) and at high temperatures it may be useful to think of monatomic oxygen (O_1) as another substance. Today, electron, X-ray and mass spectrometers can be used to identify the elements present in samples of materials and to measure the number of ratios of their atoms – tasks which in earlier times were completed by inference, only after careful qualitative analysis and quantitative measurement of masses.

Problem 1.2

(a) (i) The percentage composition can be calculated using the relative atomic weights.

symbol	element	AW × coefficient
C	carbon	12.011 × 1
H	hydrogen	1.0079 × 4

relative formula mass = 16.0426
mass percentage of carbon = 74.87
mass percentage of hydrogen = 25.13

(ii) The properties of methane can be understood by assuming that it consists of molecules each with one atom of carbon and four atoms of hydrogen.

(b) $\bar{M}_{CH4}$ = relative formula mass $\times\ 10^{-3}$ kg mol^{-1} = 0.01604 kg mol^{-1}.

(c) Either the density, $\rho = m/V$, or the molar volume, $\bar{V} = n/V$.

Problem 1.3. The density, $\rho = m/V$ for the formalin and the molar mass $\bar{M}_{CH_2O} = m_{CH_2O}/n_{CH_2O}$ for the formaldhyde. These two quantities can be used to convert the mass fraction concentration, $w_{CH_2O} = m_{CH_2O}/m$, into the volume concentration $c_{CH_2O} = n_{CH_2O}/V$.

Problem 1.4.

(a) Chemical change is change in the amounts or distribution of the various chemical substances contained within the boundaries of the system. Chemical change may occur within the system by chemical reaction at one place or throughout the bulk, by transfer from one region to another, or by entry or exit of substances across the boundaries.

Chemical change does not lead to change in the total amounts of any of the chemical elements, nor does it lead to observable change in the total mass of all matter.

(b) $$0 = +1CO_2(g) - 1C(s) - 1O_2(g)$$

$$\triangle m = \triangle m_{CO_2} + \triangle m_{C(s)}, - \triangle m_{O_2(g)}, = 0$$

$$\triangle n_B = +1.5 \text{ mol } CO_2(g), \; -1.5 \text{ mol } C(s), \; -1.5 \text{ mol } O_2(g).$$

Problem 1.5.

(a) We require $\triangle\xi(1)$, $\triangle\xi(2)$ and we know

$$\triangle m_{H_2O} = 36 \text{ g or } \triangle n_{H_2O} = 36 \text{ g}/18 \text{ g mol}^{-1} = 2 \text{ mol}$$

and $$\triangle m_{CO_2} = 33 \text{ g or } \triangle n_{CO_2} = 33 \text{ g}/44 \text{ g mol}^{-1} = 0.75 \text{ mol.}$$

But $$\triangle n_{H_2O} = +2\triangle\xi(1) + 2\triangle\xi(2) = +2 \text{ mol}$$

and $$\triangle n_{CO_2} = +1\triangle\xi(1) = +0.75 \text{ mol.}$$

So $$\triangle\xi(2) = 0.25 \text{ mol.}$$

(b)

$$\triangle n_{CH_4} = -1\triangle\xi(1) - 1\triangle\xi(2) = (-0.75 - 0.25) \text{ mol} = -1.00 \text{ mol}$$

$$\triangle n_{O_2} = -2\triangle\xi(1) - 1.5\triangle\xi(2) = (-1.50 - 0.375) \text{ mol} = -1.875 \text{ mol}$$

$$\triangle n_{CO_2} = +1\triangle\xi(1) + 0\triangle\xi(2) = +0.75 \text{ mol}$$

$$\triangle n_{H_2O} = +2\triangle\xi(1) + 2\triangle\xi(2) = (2 \times 0.75 + 2 \times 0.25) \text{ mol} = 2.00 \text{ mol}$$

$$\triangle n_{CO} = +1\triangle\xi(2) = +0.25 \text{ mol.}$$

Problem 1.6.

(a) One of the several valid sets of 3 equations is

$$C_8H_{18} + 17O_2 = 8CO + 9H_2O \quad \ldots\ldots \quad (1)$$
$$C_8H_{18} + 25O_2 = 8CO_2 + 9H_2O \quad \ldots\ldots \quad (2)$$
$$N_2 + 2O_2 = 2NO_2 \quad \ldots\ldots \quad (3)$$

An alternative to (2) is

$$2CO + O_2 = 2CO_2 \quad \ldots\ldots \quad (2')$$

which is arrived at by the rearrangement

$$(2') = [(2) - (1)]/4.$$

(b) $$\triangle n_{C_8H_{18}} = -1\triangle\xi(1) - 1\triangle\xi(2),$$
$$\triangle n_{O_2} = -17\triangle\xi(1) - 25\triangle\xi(2) - 2\triangle\xi(3),$$
$$\triangle n_{CO} = +8\triangle\xi(1),$$
$$\triangle n_{H_2O} = +9\triangle\xi(1) + 9\triangle\xi(2),$$
$$\triangle n_{N_2} = -1\triangle\xi(3),$$
$$\triangle n_{NO_2} = +2\triangle\xi(3),$$
$$\triangle n_{CO_2} = +8\triangle\xi(2),$$

There are 7 chemical formulae made up of 4 chemical symbols. The number of independent equations is the difference between these numbers.

Problem 1.7.

(a) The accumulation of Cu_2O is
$\Delta n_{Cu_2O} = \nu_{Cu_2O}(1) \times \Delta\xi(1) + \nu_{Cu_2O}(2) \times \Delta\xi(2) = (+2) \times 3 \text{ mol} + (-2) \times 1 \text{ mol} = +3 \text{ mol}.$

(b) We know $\Delta n_{SO_2} = +6 \text{ mol} = +2 \times \Delta\xi(1) + 1 \times \Delta\xi(2)$
and $\Delta n_{Cu} = +12 \text{ mol} = 0 \times \Delta\xi(1) + 6 \times \Delta\xi(2)$

So we can solve for the amounts of the stated reactions
$\Delta\xi(2) = +2 \text{ mol}$ and $\Delta\xi(1) = +2 \text{ mol}.$

Problem 1.8. There are 7 substances (formulae) which may undergo changes in amounts. They are made up from 3 elements (symbols or symbol groupings*). So $4 = 7 - 3$ independent chemical equations are needed.

* In a system such as $CaCO_3$, CaO, CO_2 the groupings CaO and CO_2 can make up $CaCO_3$ – there are 3 elements but only two *separable groupings of elements* and so there is one equation for chemical changes. Transferable groupings of elements occur often in organic chemistry as with CH_3, C_2H_5, etc. and in inorganic chemistry as in nitrates, sulphates, perchlorates, etc.

Chapter 2. Space-filling Properties of Materials

Problem 2.16. We require $Z = P\bar{V}/RT = 1 + BP/RT$

$$\text{At 300 K, } Z = 1 - \frac{2000 \text{ cm}^3 \text{ mol}^{-1} \times 0.1 \times 101.3 \times 10^3 \text{ Pa} \times 1 \text{ Jm}^{-3}/\text{Pa}}{8.314 \text{ JK}^{-1} \text{ mol}^{-1} \times 300 \text{ K} \times 10^6 \text{ cm}^3/\text{m}^3} = 0.9919.$$

At 320 K, 0.9942; 340 K, 0.9957; 360 K, 0.9968.

Chapter 3. Behaviour of Phases

Exercise 3.1. Approx 219 K and 290 K.

Exercise 3.2. $10^2\,(T/\text{K})^{-1}$ approx. 0.7, that is T approx 130 K; $\log P/P^\circ$ approx 1.6, that is P approx $40P^\circ$; 90 K, 55 K.

Exercise 3.3. $T < 4000$ K, $P \approx 200000\,P^\circ$; in the vicinity of -35° C, 5000 P°.

Exercise 3.7. (a) 0.7; (b) 44°C.

Exercise 3.8. The specimen remains liquid and the temperature steadily falls to 30°C. Then "pure" solid *p*-dichlorobenzene begins to appear and continues to do so. The temperature falls less rapidly. The temperature-composition of the liquid follows the line through P to Z. The eutectic solid composite of "pure" solid cyclohexane and "pure" solid *p*-dichlorobenzene begins to separate from the liquid and continues to do so until all liquid has solidified. The temperature remains fixed at −13°C during this period. The temperature of the composite of solid eutetic and "pure" solid *p*-dichlorobenzene then falls steadily below −13°C.

Exercise 3.9. According to Fig. 3.10, at the eutectic temperature, (a) the solid solution of antimony in lead has $x_{Sb} \approx 0.05$, (b) the liquid solution has $x_{Sb} \approx 0.25$.

Exercise 3.10. $k_f = -1.95 \text{ K mol}^{-1}$ (kg water).

Exercise 3.12. 15.5 mol methanol/hr and 36.1 mol ethanol/hr, 2.16 kg mixture hr^{-1}

Exercise 3.13. 112°C at bottom, $\geqslant 82^\circ$C at the top. Very little separation can be achieved. A small amount of near pure water should be recovered at the top. A large amount of the azeotropic mixture, $x_{HNO_3} \approx 0.35$, should be collected at the bottom.

Exercise 3.14. $\leqslant 64°C$ at bottom, $\geqslant 55.2°C$ at top. Approximately equal amounts (moles) of azeotropic mixture (at the top) and near pure methanol (at the bottom) should be recovered.

Problem 3.3. The form of Raoult's Law for one solute B in a solvent A is $p_A \simeq p_A^\circ x_A = p_A^\circ(1 - x_B)$.

When several solute species are present the mole fraction of solvent is calculated by subtracting from unity the sum of mole fractions of *all* solute species.

Problem 3.7. Yes. It is apparent that for each member of this set of four substances, the depression of freezing point is approximately the same at the same value of molality and becomes proportional to the molality at low values of the molality.

$k_f(H_2O) = -1.85$ K(mol per kg − water)$^{-1}$

$RT^2 \bar{M}/\Delta\bar{H} = 8.314 \times JK^{-1} \times (273.15\ K)^2 \times 18.02 \times 10^{-3}\ kg\ mol^{-1} / 6.009 \times 10^3\ Jmol^{-1}$

$= 1.860$ K(mol per kg of water)$^{-1}$

Problem 3.8. At first sight, these data do fit the pattern which emerged in problem 3.7. The depression of freezing point is greater, for a given value of molality. At low molalities with NaCl and KBr, ΔT_f is about twice the expected value. With $MgCl_2$ and K_2SO_4, the depressions are almost three-fold. We can understand these findings by supposing that one mole of NaCl or KCl in aqueous solution provides two mole of ionic solute species. For $MgCl_2$ and K_2SO_4 we suggest three mole of ionic solutes per mole. $MgSO_4$ is more like sucrose and urea. We suggest that the more highly charged ionic species, Mg^{2+} and SO_4^{2-}, tend to form neutral ion pairs, even in water.

Problem 3.9. We require $P_{eq} = p_{H_2O} + p_{sucrose} \simeq p_{H_2O} \simeq x_{H2O(soln)} p^\circ_{H_2O}$.

We know that $p^\circ_{H_2O} = 55.324$ mm Hg at 40°C and that $\pi = 1.5 \times 101.3$ kPa at 18°C.

From $\pi \simeq c_{solutes} RT$, $c_{solutes} = 0.063\ mol\ dm^{-3}$, $x_{solutes} = 0.0011_3$, $p_{H_2O} = 55.261$ mm Hg.

At $\Delta T_f = -3.0$ K, $m_{sugar} \simeq 1.62\ mol\ kg^{-1}$, 3.88 g water remains as solvent for 6.3 mmol (2.15 g) sugar, 94.0 g ice separated out.

Problem 3.10. For urea, $-0.37\ K \approx k_f(water) \times \dfrac{3.0\ g}{60\ g\ mol^{-1}}\ \dfrac{1}{0.250\ kg}$.

For blood, $-0.56\ K \approx -1.85$ K(mol per kg)$^{-1} \times m_{solutes}$. Thus, $m_{solutes} \approx 0.30\ mol\ kg^{-1}$, $c_{solutes} \approx 0.30\ mol\ dm^{-3}$, $\pi \approx 0.30 \times 10^3 mol\ m^{-3} \times 8.314\ JK^{-1}\ mol^{-1} \times 298.15\ K \simeq 750\ kJ\ m^{-3} = 750$ kPa.

Problem 3.11. 123 g mol^{-1}.

Problem 3.12. For the solution at 100°C, 736.2 mm Hg $\approx x_{H_2O} \times 760$ mm Hg

$x_{urea} \approx 0.0313$, $c_{urea} \approx 0.0313/0.018\ dm^3 mol^{-1}$

$\pi_{15°C} \approx 1.74\ mol\ dm^{-3} \times 8.314\ JK^{-1} mol^{-1} \times 288.15\ K$

≈ 4.17 kPa

$T_f \approx 0°C - 1.86$ K(mol per kg water)$^{-1} \times 1.74$ mol per kg water

$\approx -3.2°C$.

Chapter 4. Chemical Reaction and Chemical Equilibrium

Exercise 4.1.

(a) $\lambda^2_{NH_3(g)}$ (b) (i) $\lambda^3_{S_2O_3{}^{2-}} \lambda^4_{Cr_2O_7{}^{2-}} \lambda^{26}_{H_3O^+}$ (ii) $\lambda^8_{Cr^{3+}} \lambda^6_{SO_4{}^{2-}} \lambda^{39}_{H_2O}$

Exercise 4.2.

(a) $1.4 \times 10^{-67} > 1.6 \times 10^{-69}$, the natural direction is forward; (b) $2.94 \times 10^{-2} < 9 \times 10^{-2}$, the natural direction is backward; (c) $14.8 \times 10^{-62} > 2.74 \times 10^{-207}$, the natural direction is forward.

Exercise 4.3.
(a) $10^{-3} < 10^{9} \times 10^{-2}$, forward; (b) $10^{-15} \times 10^{-2} < 10^{-16} \times 1$, advancement;
(c) $10^{-12} \times 10^{-1} \approx 6 \times 10^{-10} \times 1.7 \times 10^{-4} \times 1$, close to equilibrium

Exercise 4.4.
(a) $a_{CH_3OH(g)} < K_{eq}\, a_{CO(g)}\, a^2_{H_2(g)}$ (b) $a_{N_2O_4(g)} > K_{eq}\, a^2_{NO_2(g)}$
(c) $a_{CH_3COO^-(aq)}\, a_{H_3O^+(aq)} = K_{eq}\, a_{CH_3COOH(aq)}\, a_{H_2O}$ (d) $a_{Hg_2^{2+}(aq)}\, a^2_{Cl^-(aq)} = K_{eq}\, a_{Hg_2Cl_2(s)}$
(e) $a_{I_3^-(aq)}\, a^4_{H_2O} = K_{eq}\, a_{H_2O_2(aq)}\, a^3_{I^-(aq)}\, a^2_{H_3O^+(aq)}$

Exercise 4.5.
(a) $a_{O_2(air)} \approx 0.21 \times 20\ \text{kPa}/101.3\ \text{kPa} = 0.041$ (b) $a_{Hg(air)} \approx 10^{-2}\ \text{Pa}/101.3\ \text{kPa} = 0.99 \times 10^{-7}$

Exercise 4.7.
(a) 0.64; (b) 0.03; (c) 4×10^{-4}

Exercise 4.10.
(a) $0.03 \times 0.03 < 1.2 \times 0.9$, forward; (b) $0.32 \times 0.32 < 0.11 \times 1.00$, forward;
(c) $1.00 \times (0.009)^2 > 6.2 \times 10^{-2} \times 0.0075 \times 0.075$, reverse;
(d) (i) $\left[\dfrac{1.4\times10^{-4}\text{mol}}{5\ \text{dm}^3} \Big/ 1\ \text{mol dm}^{-3}\right]^2 > 1.7 \times 10^{-10} \times 1.00$. No.
(ii) $(0.4 \times 2.8 \times 10^{-5})^2 < 1.7 \times 10^{-10} \times 1.00$. Yes.
(e) $a_{ClNO(g)} \simeq 1.00 \times 2.5 \times 10^{-3}\text{mol} \times 8.314\ \text{JK}^{-1}\text{mol}^{-1} \times 503\ \text{K}/500 \times 10^{-6}\text{m}^3 \times 101.3\ \text{kPa}$,
$(0.21)^2 > 0.89(0.12)^2 0.08$, reverse.

Exercise 4.11.
(a) $p_{CO_2} = 0.22 \times 101.3\ \text{kPa}$, $x_{CO_2} = 0.22$ (b) If solubility $= S\ \text{mol dm}^{-3}$, $(1.0\ S)^2 = 6 \times 10^{-5} \times 1.00$
(c) $Na_2CO_3(s) = 2Na^+ + CO_3^{2-} \ldots,$ (1)
$CO_3^{2-} + H_2O = HCO_3^- + OH^-$ (2)
$2H_2O = H_3O^+ + OH^-$ (3)
$K\{3\} = 10^{-14.00}$ at 298 K, $K\{2\} = K\{3\}/4.7 \times 10^{-11} = 2.13 \times 10^{-4}$.
Assume $(c/c^\circ)_{CO_3^{2-}} = 0.1$, $c_{HCO_3^-} = c_{OH^-} = d\ \text{mol dm}^{-3}$, $y_{OH^-} \approx y_{HCO_3^-} \approx y_{H_3O^+}$
Then $(yd)^2 \approx 2.13 \times 10^{-4} \times 0.08 \times 0.1 \times 1.00$
Thus $a_{OH^-} = 1.30 \times 10^{-3}$, $a_{H_3O^+} = 0.766 \times 10^{-11}$, $c_{H_3O^+} = 1.28 \times 10^{-11}\ \text{mol dm}^{-3}$.
(d) $c_{H_3O^+} \simeq 8 \times 10^{-12}\ \text{mol dm}^{-3}$ (e) $p_{NH_3} \approx 0.14 \times 101.3\ \text{kPa}$, $p_{N_2} \approx 43\ \text{kPa}$, $p_{H_2} \approx 79\ \text{kPa}$, $P \approx 136\ \text{kPa}$

Exercise 4.12.
(a) 48.4 (b) 0.20 (c) 179 (d) 6.9×10^{-15} (e) $0.36/(9.8 \times 10^5)$ (f) $3.8 \times 10^{13}/(1.2 \times 10^{33})^2$

Exercise 4.13. (a) 2.18 (b) Decrease T, increase P (by decreasing V, not by adding foreign gas).

Problem 4.1.
(a) $a_{B(g)} \equiv f x_B P/x^\circ P^\circ \equiv f p_B/P^\circ$ (b) $a_{B(in\ A)} \equiv (yc/c^\circ)$
(c) $a_{A(in\ B)} \equiv (fx/x^\circ)_{A(in\ B)}$ (d) $a_{B(in\ A)} \equiv (\gamma m/m^\circ)_{B(in\ A)}$

Problem 4.2. $a_{NO_2(CCl_4)} = 0.98 \times 0.19\ \text{mol dm}^{-3}/1\ \text{mol dm}^{-3} = 0.18$

Problem 4.3. $a_{H_2O(l)} \approx 0.02 \times 100\ \text{kPa}/101.3\ \text{kPa} = 0.02$

Problem 4.4. $a_{N_2O_4(g)} > K\, a^2_{NO_2(g)}$

Problem 4.5. The reaction cannot advance under the given conditions. We find that $a_{products} = 0.20^2 = 0.040$ is greater than $K\, a_{reactants} = 0.35 \times 0.10 \times 1.0 = 0.035$ (This is equivalent to finding that the absolute activity of products is greater than the absolute activity of reactants.)

Problem 4.6. Forward, since $0.001 < 9.0 \times 0.01 \times 1.0^4$

Problem 4.7. Backward, since $1 \times (0.9 \times 10^{-2}\ \text{mol dm}^{-3}/1\ \text{mol dm}^{-3})^2 > 6.2 \times 10^{-2} \times (0.75 \times 0.01\ \text{mol dm}^{-3}/1\ \text{mol dm}^{-3}) \times (1.0 \times 7.5 \times 10^{-2}\ \text{mol dm}^{-3}/1\ \text{mol dm}^{-3})$.

Problem 4.8. At equilibrium $a^3_{I^-(aq)}\, a^4_{H_2O} = K\, a^1_{H_2O_2(aq)}\, a^1_{I_3^-(aq)}\, a^2_{H_3O^+(aq)}$

Problem 4.9. The value of $a^1_{CO_2(g)}\, a^2_{H_2O(g)}\, a^{-1}_{CH_4(g)}\, a^{-2}_{O_2(g)}$ at equilibrium is equal to K(i)

The value of $a^1_{FeNCS^{2+}(aq)}\, a^{-1}_{Fe^{3+}(aq)}\, a^{-1}_{NCS^-(aq)}$, at equilibrium, is equal to K(ii)

The value of $a^1_{CaO(s)}\, a^1_{CO_2(g)}\, a^{-1}_{CaCO_3(s)}$, at equilibrium, is equal to K(iii)

Problem 4.10. $a = fp/p^\circ$. For low P, $f \simeq 1.00$

$$K \approx \frac{(1.00 \times 67.2\,\text{kPa}/101.3\,\text{kPa})^2}{(1.00 + 29.7\,\text{kPa}/101.3\,\text{kPa})\,(1.00 + 3.14\,\text{kPa}/101.3\,\text{kPa})} = 48.4$$

Problem 4.11. We require c_{trans} at equilibrium

Chemical balance: $n_{\text{cis}} + n_{\text{trans}} = 0.1\ \text{mol}, (c_{\text{cis}}/c^\circ) + (c_{\text{trans}}/c^\circ) = 0.1$.

At equilibrium, $(1.0\ c_{\text{trans}}/c^\circ) = 1.5 \times (1.0\ c_{\text{cis}}/c^\circ)$.

Thus, $c_{\text{trans}}/c^\circ = 0.06$, $c_{\text{trans}} = 0.06\ \text{mol dm}^{-3}$.

Problem 4.12. We need to assume $y_{OH^-} = 0.90$, $y_{NH_3} = 1.00$, $pK_w = 14.00$

Then $K\{NH_3 + H_2O = NH_4^+ + OH^-\} = 1.58 \times 10^{-5}$.

Let $(c/c^\circ)_{OH^-} = [OH^-]$, $(c/c^\circ)_{NH_4^+} = [NH_4^+]$, etc.

Then $y_{NH_4^+}\,[NH_4^+]y_{OH^-}[OH^-] = K\, y_{NH_3}\,[NH_3] f_{H_2O} x_{H_2O}$

and $[NH_4^+] = [OH^-] - [H_3O^+]$ (electroneutrality) $\approx [OH^-]$

$[NH_3] = 0.5 - [NH_4^+]$ (nitrogen balance) ≈ 0.5

$0.90\,[OH^-] \times 0.90\,[OH^-] \approx 1.58 \times 10^{-5} \times 1.00 \times 0.5 \times 1.00$, $[OH^-] \approx 3.12 \times 10^{-3} >> [H_3O^+] << 0.5$,

$a_{H_3O^+} \times 0.90 \times 3.12 \times 10^{-3} \approx 1.00 \times 10^{-14} \times (1.00)^2$

$pH \equiv -\log a_{H_3O^+} = 11.45$

Problem 4.13. The system is $C_6H_5COOH(s)/C_6H_5COOH(aq)$, at 20°C, at chemical equilibrium and with pH = 3.0. Known reactions are

$2H_2O = H_3O^+ + OH^-$, $pK_w \approx 14.0$

$C_6H_5COOH(aq) + H_2O = C_6H_5COO^-(aq) + H_3O^+(aq)$, $K = 6.30 \times 10^{-5}$

Let the *solubility* be $S\ \text{mol dm}^{-3}$.

Chemical balance: $[C_6H_5COOH] + [C_6H_5COO^-] = S$

Electroneutrality: $1 \times [H_3O^+] - 1 \times [C_6H_5COO^-] - 1 \times [OH^-] = 0$

Proposed assumptions $[H_3O^+] >> [OH^-]$ (1)

$[C_6H_5COOH] >> [C_6H_5COO^-]$ (2)

All $y \approx 1.0$ (3)

At equilibrium

$a_{C_6H_5COO^-}\, a_{H_3O^+} = 6.30 \times 10^{-5} \times a_{C_6H_5COOH}\, a_{H_2O}$

$10^{-3.0} \times 10^{-3.0} \approx 6.30 \times 10^{-5} \times S \times 1$

$S \approx 0.159 \times 10^{-1} = 0.0159$

→ *Solubility* $\approx 0.016\ \text{mol dm}^{-3}$

$= 0.016\ \text{mol dm}^{-3} \times 122\ \text{g mol}^{-1}$

$= 1.95\ \text{g dm}^{-3}$.

Check assumptions: (1) O.K.; (2) fair (6%), (3) O.K.

Improved solution: $0.017\ \text{mol dm}^{-3}$, $2.07\ \text{g dm}^{-3}$.

Problem 4.14. The system is HAc(0.05 mol), NaAc(0.05 mol), $H_2O(l)$ with total volume $1\ \text{dm}^3$. We require $c_{H_3O^+}$ at equilibrium, knowing that

$K\{HAc + H_2O = H_3O^+ + Ac^-\} = 1.737 \times 10^{-5}$
Let $(yc/c^\circ)_{H_3O^+} = y_{H_3O^+}[H_3O^+]$, $(yc/c^\circ)_{HAc} = y_{HAc}[HAc]$, etc.
At equilibrium,
$y_{H_3O^+}[H_3O^+]\, y_{Ac^-}[Ac^-] = K y_{HAc}[HAc] f_{H_2O}\, x_{H_2O}$
For electroneutrality:
$0 = [Na^+] + [H_3O^+] - [Ac^-] - [OH^-]$ electroneutrality
or $D \equiv [H_3O^+] - [OH^-] = [Na^+] - [Ac^-] \approx 0$
For conservation of Ac: $[HAc] + [Ac^-] = A + B = 0.10$
For conservation of Na: $[Na^+] = B = 0.05$
Thus, $[Ac^-] = B - D \approx B$
$[HAc] = A + D \approx A$
$0.8\,[H_3O^+] \times 0.8 \times 0.05 \approx 1.737 \times 10^{-5} \times 1.0 \times 0.05 \times 1.00$
$c_{H_3O^+} = 2.7 \times 10^{-5}$ mol dm^{-3}
Check approximations: $D \approx [H_3O^+] << B << A$.

Problem 4.18

(a) On addition of more solvent the reaction must go forward to reach a new state of equilibrium. Initially there is one set of equilibrium activities and concentrations

$$\lambda_{products} = \lambda_{reactants}$$
$$a_{HCOO^-(aq)}a_{H_3O^+(aq)} = K_{eq}a_{HCOOH(aq)}a_{H_2O(l)}$$
$$(yc/c^\circ)_{HCOO^-(aq)}(yc/c^\circ)_{H_3O^+(aq)} = K_{eq}(yc/c^\circ)_{HCOOH(aq)}(fx/x^\circ)_{H_2O(l)}.$$

Doubling of volume without reaction would halve the concentrations of both product species while halving the concentration of the reactant, HCOOH, and very slightly increasing the mole fraction of H_2O. Activity coefficients will not be greatly altered. So the activity of the products falls below that of the reactants and the reaction tends to go forward.

(b) In the initial mixture at 2000 K

$$\lambda_{products} = \lambda_{reactants}$$
$$a_{C_6H_6(g)} = K_{eq}(2000\ \text{K})a^3_{C_2H_2(g)}$$

and

$$(fxP/P^\circ)_{C_6H_6(g)} = K_{eq}(2000\ \text{K})(fxP/P^\circ)^3_{C_2H_2(g)}.$$

If pressure P is held, change of temperature will determine the tendency to react by changing K_{eq}. Thus, $d \ln K_{eq}/dT = \Delta\bar{H}^\circ/RT^2$ (Section 4.8) and since $\Delta\bar{H}^\circ$ is here negative, K_{eq} increases as we go to 500 K. Thus the activity of the reactants rises above that of the products and reaction tends to go forward to form more benzene, though the *rate* of approach to equilibrium may become small at lower temperatures.

Problem 4.19. The system is $Ag_2SO_4(s)|Ag^+_2SO_4^{2-}(aq)$.
$K_s\{Ag_2SO_4(s)\} = K\{Ag_2SO_4(s) = 2Ag^+(aq) + SO_4^{2-}(aq)\} = 2 \times 10^{-5}$
Let the *solubility* be S mol dm^{-3}.
Electroneutrality: $1[Ag^+] - 2[SO_4^{2-}] = 0$
Ag balance: $[Ag^+] = 2S$
SO_4 balance: $[SO_4^{2-}] = S$
Assumptions: $y_{Ag^+} \approx 1.0$, $y_{SO_4^{2-}} \approx 1.0$
Approximate solution:
$(1.0 \times 2S)^2 \times (1.0 \times S) \approx 2 \times 10^{-5} \times 1.00$,
$S \approx 1.7 \times 10^{-2}$, *solubility* ≈ 0.017 mol dm^{-3}

Check assumptions: If *ionic strength*, $I = 3 \times 0.017$, and $\log_{10} y_1 \simeq -0.51\sqrt{I}/(1+\sqrt{I})$, $y_1 \approx 0.80$, $y_2 \approx y_1^4 = 0.42$.

First improved solution:

$(0.80 \times 2S)^2 \times (0.42 \times S) \approx 2 \times 10^{-5} \times 1.00$

$S \approx 2.65 \times 10^{-2}$, $I = 3 \times 0.0265$

Problem 4.20. No further reaction will occur.

The system is

$$AgCl(s)|\,Ag^+Cl^-(aq)$$

The chemical process,

$$AgCl(s) = Ag^+(aq) + Cl^-(aq)$$

is at equilibrium. The activity of AgCl(*s*) is not dependent on the amount of the solid present.

Problem 4.21. The system is

$$AgCl(s)|\,Ag^+,\ NO_3^-,\ Na^+,\ Cl^-(aq)$$

with just sufficient Ag^+ to make the reaction

$$AgCl(s) = Ag^+(aq) = Cl^-(aq) \ldots \{1\}$$

occur in reverse.

We require the concentration of Ag^+.

We know $K\{1\} = 2 \times 10^{-10}$ and $c_{Cl^-} \approx 0.01\ \text{mol dm}^{-3}$

(a) We assume $y_{Ag^+} = 1.0 = y_{Cl^-}$.

We have $1.0\,[Ag^+] \times 1.0 \times 0.01 \geqslant 2 \times 10^{-10} \times 1.0$

$c_{Ag^+} \geqslant 2 \times 10^{-8}\ \text{mol dm}^{-3}$.

(b) We assume that $y_{Ag^+} = 0.75 = y_{Cl^-}$.

Then $0.75[Ag^+]\,(0.75 \times 0.01) > 2 \times 10^{-10} \times 1.0$

$c_{Ag^+} > 3.5 \times 10^{-8}\ \text{mol dm}^{-3}$.

Problem 4.22. We require

$$K\{3\} = K\{H_2O(g) = H_2(g) + \tfrac{1}{2}O_2(g)\}$$

Equation {3} is equivalent to {1}−{2}.

Thus $K\{3\} = K\{1\}/K\{2\}$

$= 0.36/9.8 \times 10^5$

$= 3.67 \times 10^{-7}$

Problem 4.23. $$\frac{d \ln K}{d(1/T)} = \frac{-\Delta \bar{H}^\circ}{R},$$

$$\ln \frac{K \text{ at } 400\text{K}}{2 \times 10^4} = \frac{+90.6 + 10^3\ \text{Jmol}^{-1}}{8.314\ \text{JK}^{-1}\text{mol}^{-1}} \left\{ \frac{1}{400\text{K}} - \frac{1}{298\text{K}} \right\}$$

K at $400\text{K} = 2 \times 10^4 \times 8.9 \times 10^{-5} = 1.78$

Problem 4.24. $$2.3026 \log \frac{1.0 \times 10^{-14}}{1.13 \times 10^{-15}} \approx \frac{-\Delta \bar{H}^\circ}{8.314\ \text{JK}^{-1}\text{mol}^{-1}} \left\{ \frac{273\text{K} - 298\text{K}}{298\text{K} \times 273\text{K}} \right\}$$

$$\Delta \bar{H}^\circ = +59.0\ \text{kJ mol}^{-1}.$$

Problem 4.25. The system is CO, H_2, $CH_3OH(g)$.

(a) Halving volume doubles all partial pressures and doubles relative activities.

(b) Addition of He without change of V or T hardly alters p_i's and has negligible effect on relative activities.

(c) Decrease of K. $\dfrac{d \ln K}{dT} = \dfrac{\Delta \bar{H}°}{RT^2}$ is negative.

(d) Forward tendency; no effect; reversal tendency.

Chapter 5. Rates of Reaction

Exercise 5.1.

(a) Total pressure at fixed volume and temperature (gives *rate*/V), total volume at fixed pressure and temperature (gives *rate*), or brown colour intensity (gives dc_{NO_2}/dt, continuous recording)

(b) Titration of residual OH^- (precision), pH meter reading (convenience), electrical conductivity (sensitivity, continuous recording)

(c) Titration of total acid (direct and precise), freezing point or osmotic properties

(d) Visible light absorption (continuous monitoring of c product)

(e) Electrical conductivity (proportional to concentration of dissolved CaF_2).

Exercise 5.3. $dn_{O_2}/dt = -1 \times 1 \text{ mol s}^{-1} - 1 \times 1.5 \text{ mol s}^{-1}$

Problem 5.1.

$$\frac{d\xi}{dt} \equiv \frac{1}{-2}\frac{dn_{C_8H_{18}}}{dt} \equiv \frac{1}{-25}\frac{dn_{O_2}}{dt} \equiv \frac{1}{16}\frac{dn_{CO_2}}{dt} \equiv \frac{1}{18}\frac{dn_{H_2O}}{dt}$$

Problem 5.2.

(a) One correct set is

$$2C_8H_{18} + 25O_2 = 16CO_2 + 18H_2O \quad \{1\}$$
$$2C_8H_{18} + 17O_2 = 16CO + 18H_2O \quad \{2\}$$
$$N_2 + 2O_2 = 2NO_2 \quad \{3\}$$

(b) $d\xi\{1\} = (6.40/16)\text{mol}$, $d\xi\{2\} = (1.44/16)\text{mol}$,
$d\xi\{3\} = (0.01/2)\text{mol}$.

(c) $dn_{H_2O} = 18d\xi\{1\} + 18d\xi\{2\} = 8.82 \text{ mol}$
$dn_{C_8H_{18}} = (-2 \times 0.40 - 2 \times 0.09)\text{mol} = -0.98 \text{ mol}$

(d) *rate*$\{1\} = 0.00100 \text{ mol s}^{-1}$, *rate*$\{2\} = 0.000225 \text{ mol s}^{-1}$
rate$\{3\} = 0.0000125 \text{ mol s}^{-1}$

Problem 5.3.

(a) $rate \equiv \dfrac{d\xi}{dt} \equiv \dfrac{1}{\nu_B}\dfrac{dn_B}{dt}$ The unit is mol s^{-1}.

(b) (i) tonne per year (if interested in total turnover in the utilisation of plant)
(ii) mole per second (if interested in the chemical reaction)
(iii) mole per second (if interested in yield); mole per litre per second (if interested in forecasting and interpreting rates)
(iv) mole or kg per second per kg of catalyst (if interested in catalyst quality)
(v) dm^3 per second if study is conducted at constant pressure; kPa s^{-1} or mmHg per second if study is at constant volume
(vi) mole per second per cubic decimetre

Problem 5.4.

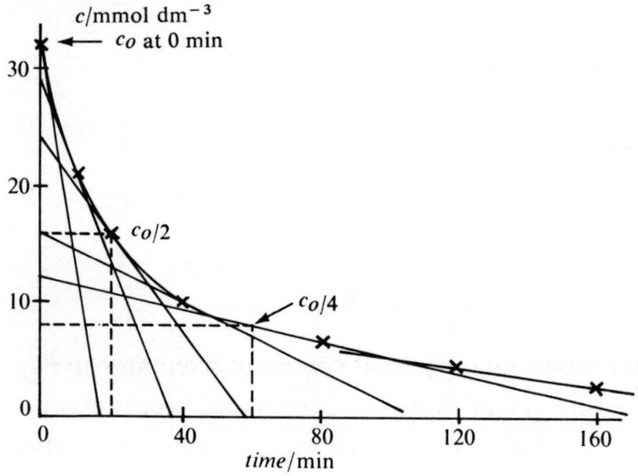

(a), (b)

$\frac{t}{\text{min}}$	$\frac{c}{\text{mmol dm}^{-3}}$	$\frac{rate/V}{\text{mmol min}^{-1}\text{ dm}^{-3}}$
0	33.3	2.03
10	22.2	0.63
20	16.7	0.44
40	11.1	0.16
80	6.67	0.072
120	4.76	0.042
160	3.70	0.025

(a) $t_{1/2} = 20$ min, $t_{3/4} = 60$ min

(c) Graphical treatment to find exponent of concentration (straight line results if rate law is of the simplest type, $rate/V \propto^{nl} c^{a}$)

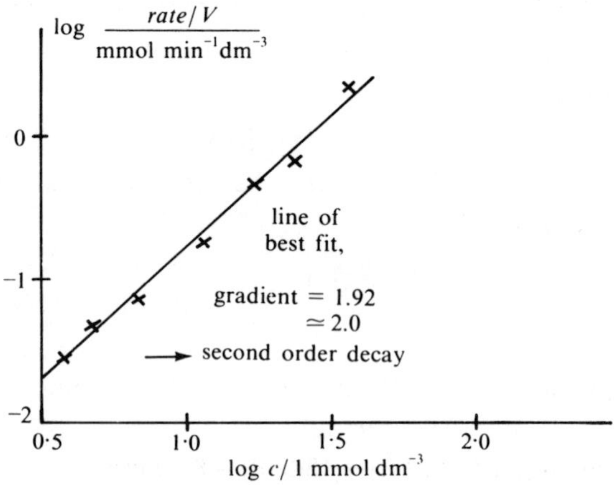

The conclusion that the decay is simple second order can now be tested by reference to the appropriate relationship given in Table 5.3. We substitute $X = c$ and graph reciprocal of concentration versus time. The linear arrangement of points confirms that the decay is second order. The gradient of the line corresponds to the second order decay coefficient. The intercept at zero time corresponds to the reciprocal of initial concentration. This treatment does not suffer from the inaccuracies involved in drawing tangents to obtain estimates of rates of change.

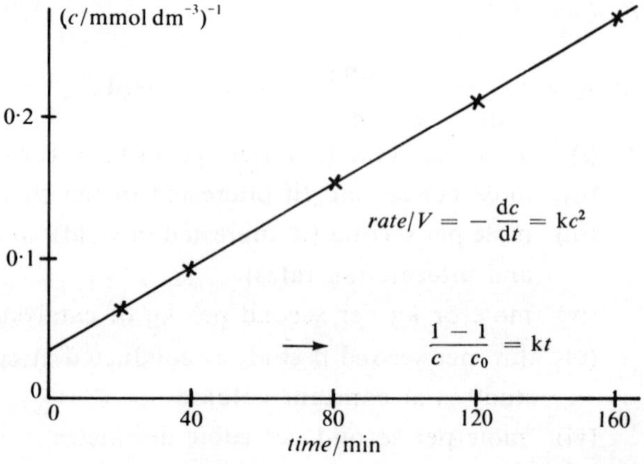

Problem 5.5.

(a) We know that

$$d\xi/dt = -1\,dn_{N_2O_5}/dt = +1\,dn_{N_2O_4}/dt = 2\,dn_{O_2}/dt$$

If the gaseous products enter a fixed volume,

$$dc_{N_2O_4}/dt = 2\,dc_{O_2}/dt$$

However, the concentration of gaseous N_2O_5 will be maintained by continuous evaporation. Thus $dc_{N_2O_5}/dt = 0$.

(b) The total pressure increases at a constant rate. It is clear that the rate of reaction is independent of (that is, zero order with respect to) the concentrations of N_2O_4 and O_2. Since the concentration of N_2O_5 is maintained throughout reaction of the batch, the results offer no means of assessing the dependence of *rate*/*V* on this quantity.

Problem 5.6. This problem can be tackled by the strategy of problem 5.4. However, decay laws of the simplest type (single term, integer exponent) occur quite often for individual batch studies. It is worthwhile to try graphical testing of the simple zero, first and second order rate relations in turn (see p.88). Half lives can be used as a guide to which should be tried first. The constant half life (for N_2O_5) found in this problem is an indication of first order decay. The graph of log $(p_{N_2O_5}/\mathrm{kPa})$ versus *time* turns out to be linear, confirming that the reaction is "first order" with respect to gaseous N_2O_5.

Problem 5.7. H_3O^+ does not appear in the stoichiometric equation,

$$C_{12}H_{22}O_{11} + H_2O = C_6H_{12}O_6 + C_6H_{12}O_6$$

sucrose water glucose fructose

For a given batch we assume that $-dc_{sucrose}/dt = k_{batch}c^{a}_{sucrose}$, with $k_{batch} = kc^{b}_{H_3O^+}$.

Constancy of half life is a symptom of first order decay. We infer that $a = 1.0$. When pH 4 is used instead of pH 5, $c_{H_3O^+}$ is approximately 10 times higher.

With first order decay $k_{batch} = \ln 2/t_{1/2}$. At pH 4, $t_{1/2}$ is 10 times less; so k_{batch} is 10 times greater. Thus $b = 1.0$.

Problem 5.8.

(a) The simplest form of rate law is assumed when designing preliminary experiments. Hydrogen ion does not appear in the stoichiometric equation. Thus, for a given batch,

$Rate/V = (?)\ k_{batch}\ c_{Co(NH_3)_5F^{2+}}$

with $k_{batch} = (?)\ k\ c^{b}_{H_3O^+}$.

Equality of successive half times suggests first order decay ($a = 1$). Doubling of $c_{H_3O^+}$ has doubled k_{batch}. We infer that $b = 1.0$. More comprehensive experiments may now be planned.

(b) $k_{batch} = k\,c^{1}_{H_3O^+}$, with $c_{H_3O^+} = 0.010\ \mathrm{mol\ dm^{-3}}$

$= \ln 2/t_{1/2} = 0.693/(60\ \mathrm{min} \times 60\ \mathrm{s\ min^{-1}})$

$k = 1.92 \times 10^{-2}\ \mathrm{dm^3\ mol^{-1}\ s^{-1}}$.

(c) $$\frac{d\ln k}{dT} \equiv \frac{E_A}{RT^2}, \quad \frac{d\ln k}{d1/T} \equiv -\frac{E_A}{R}$$

$k/\mathrm{dm^3mol^{-1}s^{-1}} = 1.92\times10^{-2}$(at 25°C), 3.84×10^{-2}(at 35°C)

$$\ln\frac{3.84}{1.92} \approx \frac{-E_A}{8.314\ \mathrm{JK^{-1}mol^{-1}}}\left\{\frac{298\mathrm{K} - 308\mathrm{K}}{308\mathrm{K}\times 298\mathrm{K}}\right\}$$

$E_A \approx 52.9\ \mathrm{kJ\ mol^{-1}}$.

Problem 5.9.

$$dP/dt \approx d(n_{gas}RT/V)/dt \approx (RT/V)(d\xi/dt)(dn_{gas}/d\xi)$$
$$\approx (1RT/V)d\xi/dt = 0.063 \text{ mm Hg s}^{-1} = 8.4 \text{ Pa s}^{-1}$$
$$Rate/V \equiv (1/V)d\xi/dt = kc^2_{N_2O}$$
$$= 1.0245 \times 10^{-3} \text{ mol m}^{-3} \text{ s}^{-1} \text{ when } c_{N_2O} = P°/R \times 986\text{K}$$
$$k = 6.7 \times 10^{-6} \text{ m}^3 \text{ mol}^{-1} \text{ s}^{-1} = 6.7 \times 10^{-3} \text{dm}^3 \text{ mol}^{-1}\text{s}^{-1}.$$

Problem 5.10. We require $t_{1/2}$. We know that $c_{H_3O^+} = c_{OH^-} = c$

and $-\dfrac{dc}{dt} = kc^2$, with $k \simeq 1.3 \times 10^{11} \text{ dm}^3 \text{ mol}^{-1} \text{ s}^{-1}$

Thus $\dfrac{1}{c_t} - \dfrac{1}{c_o} = kt$, or $\dfrac{c_o}{c_t} - 1 = c_okt$.

At the half time, $2 - 1 = c_okt_{1/2}$. With $c_o = 10^{-4}\text{mol dm}^{-3}$, $t_{1/2} = 800$ nanoseconds.

Problem 5.11.

(a) $-\dfrac{dc_{FCH_2COOEt}}{dt} = kc_{OH^-}c_{FCH_2COOEt}$ with $k = 14.0 \text{ dm}^3 \text{ mol}^{-1} \text{ s}^{-1}$ and $c_{OH^-} = 1.00 \times 10^{-4} \text{ mol dm}^{-3}$

$$-\frac{dc_{FCH_2COOEt}}{dt} = (14.0 \times 10^{-4} \text{ s}^{-1})c_{FCH_2COOEt}.$$

We require $t_{1/2}$ for a batch hydrolysis under these conditions of first order decay. $\ln 2/k = 8$ min 15s

(b) We require an estimate of k (at 308 K). Assuming that Arrhenius E_A is virtually independent of T,

$$\ln \frac{k \text{ (at 308K)}}{k \text{ (at 298K)}} \approx \frac{-38.4 \times 10^3 \text{Jmol}^{-1}}{8.314 \text{ JK}^{-1}\text{mol}^{-1}} \left\{ \frac{298\text{K} - 308\text{K}}{308\text{K} \times 298\text{K}} \right\}$$

k (at 308K) $= 1.654 \times 14.0 \text{ dm}^3\text{mol}^{-1}\text{s}^{-1}$.

Problem 5.12. From the given form of the overall rate equation we infer that the formula of the transitional complex is $(S_2O_3BrCH_2COO)^{\ddagger 3-}$.

Thus, $k = k_o y_{S_2O_3^{2-}} y_{BrCH_2COO^-} / y_{(S_2O_3BrCH_2COO)^{\ddagger 3-}}$

$$\approx k_o \times 0.81 \times 0.90/0.73 = 1.00\ k_o$$

Note: these are probably poor estimates of the rate coefficients. If $y_1 \approx 0.90$, we would expect $y_2 \approx 0.90^4 = 0.66$ and $y_3 \approx 0.90^9 = 0.39$. Then $k \approx (0.66 \times 0.90/0.39)k° = 1.38\ k°$.

Problem 5.13. The observed rate law matches the stoichiometry of reactants. So it is reasonable to suppose that each molecule of cyclobutene is converted into a molecule of butadiene in an elementary chemical transformation,

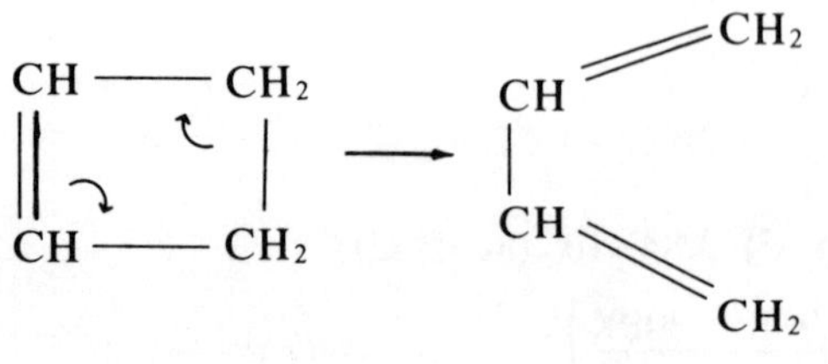

The change in molecular structure is mechanically simple. Concerted rearrangement of the electron pair bonds in a single step is statistically feasible.

Problem 5.14.

(a) *rate* (1)/$V = k_1 c_{CH_3CHO} c_{OH^-}$, *rate* (−1)/$V = k_{-1} c_{CH_2CHO}$,

rate (2)/$V = k_2 c_{CH_3CHO} c_{CH_2^-CHO}$,

rate (3)/$V = k_3 c_{CH_3CH(O^-)CH_2CHO}$

(b) Steps (2) and (3) must be "fast". Step (−1) must be "slow". Then step (1) will be"rate determining".

Problem 5.15.

(a) When a rate law consists of a sum of simple terms we can infer parallel pathways through transitional complexes with different formulae corresponding to the sums of elements appearing in the rate terms. The involvement of a solvent will not be apparent. Thus for the first rate term in the hydrolysis of ethylene epoxide the formula of the transitional complex could be $(C_2H_2O)^{\ddagger}$ or $(C_2H_4O_2)^{\ddagger}$, etc. We could have

(a) 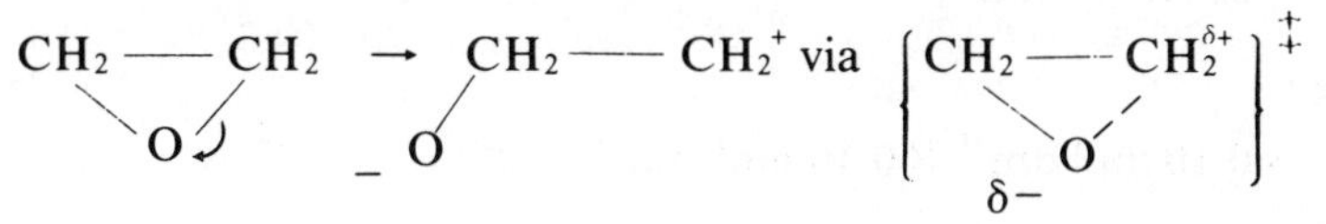

or (b)

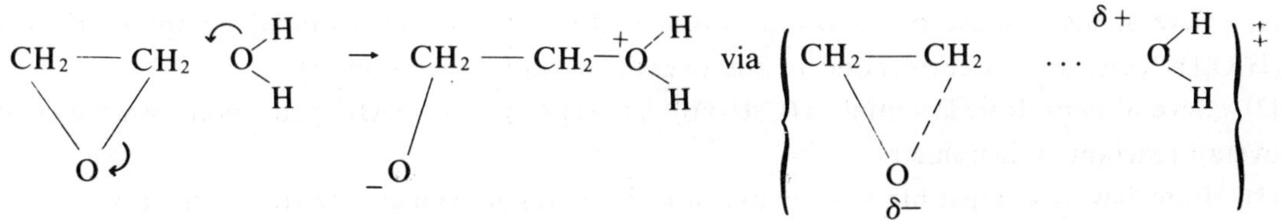

For the second rate term we could have $(C_2H_5O)^{f+}$, $(C_2H_7O_2)^{f+}$, $(C_2H_9O_3)^{f+}$, etc. Protonated ethylene epoxide will be present at equilibrium in a small proportion which increases with concentration of hydrogen ion. We could have

(a′)

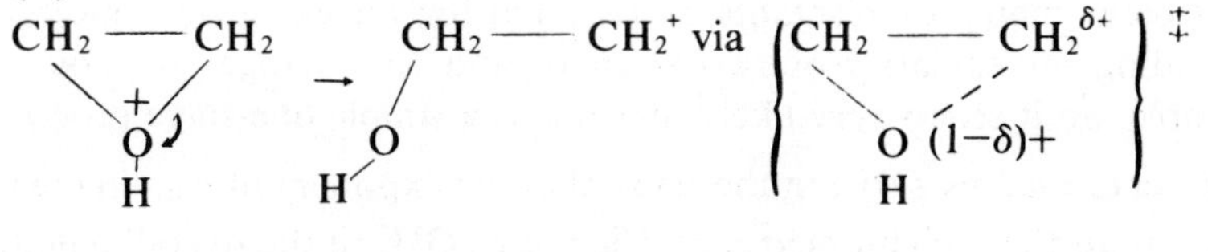

or (b′)

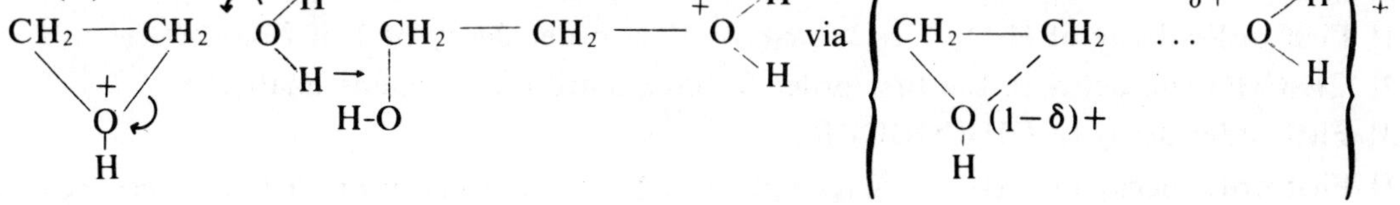

The proton transfer step could be the kinetically significant step

(c′)

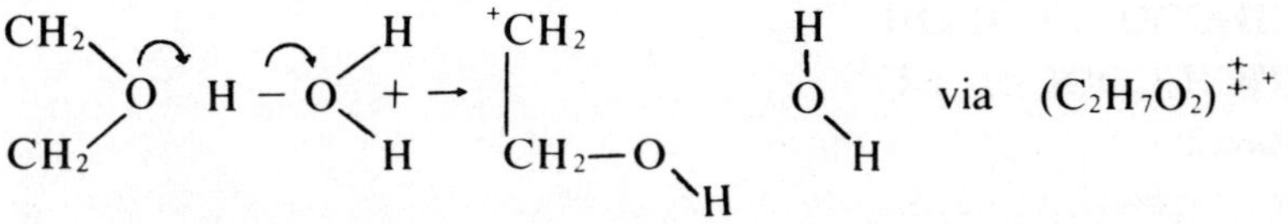

(b) *Rate*/$V = kc_{OH} - c_{CH_3COCH_3}$. Possible formulae of ($\ddagger$) are $(C_3H_7O_2)^{\ddagger -}$, $(C_3H_9O_3)^{\ddagger -}$, etc.
A reasonable structure is shown at right, formed by a bimolecular encounter of predominant species.

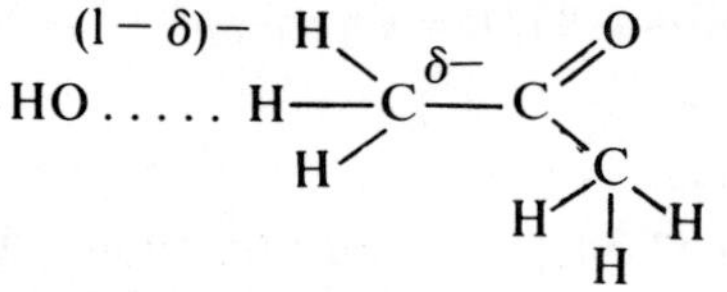

The proposed rate determining step may be written

$$HO^- + CH_3COCH_3 \rightarrow HOH + CH_2^-COCH_3,$$

followed by a fast step

$$I_2 + CH_2^-COCH_3 \rightarrow I^- + ICH_2COCH_3.$$

Problem 5.16.

$$\ln \frac{k\,(60°C)}{k\,(27°C)} \approx \frac{-40\times10^3\,J\,mol^{-1}}{8.314\,J\,K^{-1}mol^{-1}} \left\{ \frac{300K - 333K}{333K \times 300K} \right\}$$

$k(60°C) = 4.90 \times 0.10\ dm^3\ mol^{-1}\ s^{-1}$
rate/$V = kc_Ac_B = 0.49\ dm^3 mol^{-1} s^{-1} \times 0.10\ mol\ dm^{-3} \times 0.10\ mol\ dm^{-3}$
rate = *volume* $\times\ 0.0049\ mol\ dm^{-3}\ s^{-1}$.

Problem 5.17.

(a) {1} Several steps must be involved. Possible formulae of transitional complex $(H_2O_2I)^{\ddagger -}$, $(H_4O_3I)^{\ddagger -}$, etc. do not correspond to the overall reactant stoichiometry.

{2} Several steps. Rate law implies $(C_2H_5O_2)^{\ddagger -}$, $(C_2H_7O_3)^{\ddagger -}$, $(C_2H_3O)^{\ddagger -}$, etc., none of which match overall reactant stoichiometry.

{3} Rate law is compatible with conversion by a single elementary transformation. Possible formulae of ($\ddagger$) are $(C_3H_8O_2)^{\ddagger}$, $(C_3H_{10}O_3)^{\ddagger}$, etc. Overall reactant stoichiometry matches the second of these.

{4} Rate law is compatible with elementary transformation. The complexity of the mechanical rearrangement of the atoms is perhaps too great for the single step conversion to be statistically feasible.

(a) {5} Rate law matches the stoichiometry of reactants as required for an elementary molecular transformation, the corresponding molecularity of two is small and the changes in molecular structure are mechanically simple. So it seems very likely that this is a simple one-stage process.

(b) Reaction {2} may be said to be catalysed by OH^-, in the sense that the exponent of c_{OH^-} in the rate law (one) is greater magnitude than the stoichiometric coefficient of OH^- in the overall equation (zero).

(c) {1} First order decay of H_2O_2 if $c_{I^-} >> c_{H_2O_2}$. First order decay of I^- if $c_{H_2O_2} >> c_{I^-}$

{2} CH_3CHO will decay in the first order fashion, since c_{OH^-} remains unaltered

{3} First order decay of $CH_3COOC_2H_5$.

{4} First order decay of NH_4^+, if $c_{NH_4^+} << {}_{NCO^-}$. First order decay of NCO^-, if $c_{NCO^-} << c_{NH_4^+}$

(d) Typically, reaction rates increase by the order of 5 to 10% per kelvin.

Problem 5.18.

(a) $ICH_2COOC_2H_5 + HO^- = ICH_2COO^- + C_2H_5OH$ {1}

$ICH_2COOC_2H_5 + HO^- = HOCH_2COOC_2H_5 + I^-$ {2}

(b) $dn_{HO^-}/dt = -1$ *rate* {1} $- 1$ *rate*{2}

$= dn_{ICH_2COOC_2H_5}/dt$

(c) $dn_{I^-}/dt = rate\{2\}$

(d) $-\dfrac{dX}{dt} = k c_{OH^-} X = k_{batch} X$

with $k_{batch} = 20\ dm^3 mol^{-1} s^{-1} \times 10^{-5}\ mol\ dm^{-3}$

$t_{1/2} = 58$ min.

(e) $dc_{ClCH_2COO^-}/dt = k\{1\} c_{OH^-} c_{ClCH_2COOC_2H_5}$

$dc_{HOCH_2COOC_2H_5}/dt = k\{2\} c_{OH^-} c_{ClCH_2COOC_2H_5}$

$k\{1\} = 18.0\ dm^3 mol^{-1} s^{-1}$, $k\{2\} = 2.0\ dm^3 mol^{-1} s^{-1}$.

(f) $\ln \dfrac{k \text{ at } T_2}{k \text{ at } T_1} \approx \dfrac{-E_A}{R}\left\{\dfrac{T_1 - T_2}{T_2 \quad T_1}\right\}$

$E_A\{1\} \approx 50\ kJ\ mol^{-1}$, $k\{1\}_{273K} \approx k\{1\}_{298K}/6.3$

$E_A\{2\} \approx 100\ kJ\ mol^{-1}$, $k\{2\}_{273K} \approx k\{2\}_{298K}/40$

At 273K, $k\{1\} \approx 2.8\ dm^3 mol^{-1} s^{-1}$,

$k\{2\} \approx 0.05\ dm^3 mol^{-1} s^{-1}$. Reaction {2} is now less than 2% of the total.

Problem 5.19.

(a) The relationship used here is the linearised form of the integrated first order decay equation. We infer that the concentration of bromine appears in the rate law as a common factor with exponent equal to unity.

(b) We require a mathematical relationship to fit the given values of *rate*/*V* and $c_{pyrazole}$ at fixed c_{Br_2}. By inspection we see that *rate*/*V* and $c_{pyrazole}$ are interdependent. Let us try the simple relationships $rate/V \approx k\, c^1_{pyrazole}$ and $rate/V \approx k\, c^2_{pyrazole}$

supposed exponent	zero	one	two
$\dfrac{10^4 c_{pyrazole}}{mol\ dm^{-3}}$	$\dfrac{10^5 rate/V \div (10^4 c_{pyrazole})^0}{mol\ dm^{-3}\ s^{-1}}$	$\dfrac{10^5 rate/V \div (10^4 c_{pyrazole})^1}{s^{-1}}$	$\dfrac{10^5 rate/V \div (10^4 c_{pyrazole})^2}{dm^3\ mol^{-1}\ s^{-1}}$
1.1	3.0	2.73	2.48
1.5	4.3	2.87	1.91
1.8	5.0	2.78	1.54
2.3	6.2	2.69	1.17

Results obtained assuming that the pyrazole rate exponent is one are essentially constant. Thus we infer that *rate*/*V* is directly proportional to $c_{pyrazole}$. We can write a more detailed form for the rate law,

$$rate/V = k c^1_{Br_2} c^1_{pyrazole}$$

(c)

$$rate/V = k \times 1.0 \times 10^{-5} mol\ dm^{-3} \times 2.3 \times 10^{-4}\ mol\ dm^{-3}$$

$$= 6.2 \times 10^{-5}\ mol\ dm^{-3}\ s^{-1}$$

Thus, $k = 2.7 \times 10^4\ dm^3\ mol^{-1}\ s^{-1}$.

Problem 5.20. $Rate/V \equiv -\mathrm{d}c_{ClO_2}/\mathrm{d}t$.

We could use the method of tangents to determine *rate/V* at various values of *c*, as in page 80, problem 5.4. However, the odds are good enough to warrant doing graphical trials of zero, first and second order decay relationships immediately.

supposed order of decay	zero	first	second
$\dfrac{10^3\ time}{\mathrm{s}}$	$\dfrac{10^{10}c}{\mathrm{mol\ dm^{-3}}}$	$\log\left(\dfrac{10^{10}c}{\mathrm{mol\ dm^{-3}}}\right)$	$\left(\dfrac{10^{10}c}{\mathrm{mol\ dm^{-3}}}\right)^{-1}$
0	1.60	$+0.20_4$	0.62
7	0.90	-0.04_6	1.11
12	0.71	-0.15	1.41
16	0.50	-0.30	2.00
20	0.36	-0.44	2.77
30	0.20	-0.70	5.00
40	0.10	-1.00	10.00

(a) The graph based on first order decay is well fitted by a straight line. We infer that the concentration of ClO_2 is a common factor in the rate law with unit exponent,

$$rate/V = k_{\mathrm{batch}}c^{1}_{ClO_2}$$

(b) $rate/V = 3.5 \times 10^{10}\ \mathrm{dm^3\ mol^{-1}\ s^{-1}}\ c_{Cl}\ c_{ClO_2}$

(c) The formula of the transitional complex is $(Cl_2O_2)^{\ddagger}$. The reaction could involve a single elementary molecular transformation,

$$Cl + ClO_2 \rightarrow ClO + ClO$$

A reasonable structure for the transitional complex might be

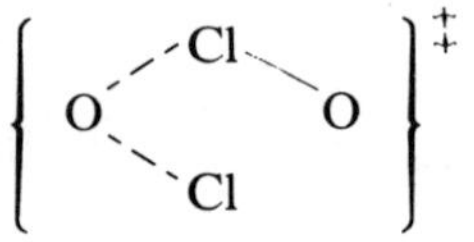

Chapter 6. Chemical Potentials

Exercise 6.1.

(a) hydro-electric power stations, waterwheels...
(b) some types of clocks, counterweighted lifts, sash windows...
(c) jack-hammers, air guns...
(d) steam engines, steam turbines...
(e) mining, quarrying, road cutting construction...
(f) long distance cycling, animal metabolism generally...
(g) lead–acid batteries...
(h) the fuel cell system (which failed in the Apollo 13 moon shot),...

Exercise 6.2.

(a) electrolysis (at high temperature), in which the transfer of electric charge is coupled with the chemical process. . .

(b) evaporation of the water can be driven by thermal action of the sun. . .

(c) burning of the cellulose in the plant material can be used to drive off the iodine. . .

Problem 6.1.

(a) $w_{\text{♀}} \geqslant (237200-237159)\text{J mol}^{-1} \times 55.5\ \text{kmol} \geqslant 2.28\text{MJ}$

(b) (i) μ(*o*-dinitrobenzene) $>$ μ(*p*-dinitrobenzene) Reaction can proceed naturally in the forward direction.

(ii) At equilibrium μ(*p*-dinitrobenzene) $=$ μ(*o*-dinitrobenzene).

(c) The solute in a saturated solution is in equilibrium with the pure solid. Thus μ_{I_2}(*in cyclohexane, sat.*) $=0$.

(d) Deposition of the pure substance from a super-saturated solution can occur naturally. Thus the chemical potential of the solute is greater than the chemical potential of the pure substance.

Exercise 6.3.

(a) $\mu^{\circ}_{HCl(g,298K)} \equiv \Delta G^{\circ}_{f}\{HCl(g,298K)\} = -95\ \text{kJ mol}^{-1}$

(b) (i) $\mu^{\circ}_{CH_3NH_2(g)} = \Delta G^{\circ}_{f} = 32\ \text{kJ mol}^{-1}$(*SI Chemical Data*)

(ii) $\mu^{\circ}_{UF_6(g)} = \Delta G^{\circ}_{f} = -2029\ \text{kJ mol}^{-1}$(*SI Chemical Data*)

(iii) $\mu^{\circ}_{Cu_2O(s)} = \Delta G^{\circ}_{f} = -148\ \text{kJ mol}^{-1}$(*SI Chemical Data*)

(iv) $\mu^{\circ}_{K_2S_2O_3(aq)} = 2\mu^{\circ}_{K^+(aq)^+} + \mu^{\circ}_{S_2O_3{}^{2-}(aq)}$

$= (2\times-67.5+-127.2)\text{kcal mol}^{-1} \times 4.184\ \text{Jcal}^{-1}$

$= -1096\ \text{kJ mol}^{-1}$(*Handbook*)

Problem 6.3.

$-249.2 \times 10^3\ \text{J mol}^{-1} = -8.314\ \text{JK}^{-1}\text{mol}^{-1} \times 1273\text{K} \ln K_{eq}$

$2.303 \log K_{eq} = 23.54,\ K_{eq} = 1.67 \times 10^{10}$.

Problem 6.4.

(a) backwards, at equilibrium.

(b) $3\mu_{Fe} + 2\mu_{O_2} > \mu_{Fe_3O_4}$.

(c) negative

(d) Yes. $\Delta\bar{G} = -131\ \text{kJ mol}^{-1} < 0$

Problem 6.5.

(a) 0

(b) $RT \ln 8 = 5.2\ \text{kJ mol}^{-1}$

(c) $-237\ \text{kJ mol}^{-1}$, 0.98, $\mu^{\circ}\ -\ 44\ \text{J mol}^{-1}$

(d) $zFV = 193\ \text{kJ mol}^{-1}$

(e) $\mu_{I_2}(C_7H_{16}) = \mu_{I_2}(s) = 0,\ a = 0.00679,\ \mu^{\circ} = \mu - RT\ln a = 12.5\ \text{kJ mol}^{-1}$

(f) $>, >, <, <, <, =, >, <$.

Problem 6.6.

(a) $-817\ \text{kJ mol}^{-1}$

(b) 1.4×10^{69}

(c) $-184\ \text{kJ mol}^{-1}$, 1.8×10^{32}

(d) 0.8 GJ

(e) Yes, No

(f) $\Delta\bar{G}^\circ = -678.8$ kJ mol^{-1}, 444.4 kJ mol^{-1}. No, step (ii) will not occur!

Problem 6.7.

(a) $\mu = \mu^\circ + RT\ln a + zF\Phi$, $\mu_2 = \mu_1$, $\mu_2^\circ = \mu_1^\circ$, $z = 1$

$\therefore RT\ln a_2/a_1 = -F(\Phi_2 - \Phi_1)$, 59 mV, 54 mV.

(b) $RT\ln a_2/a_1 = -F(\Phi_2 - \Phi_1)$, $a_2/a_1 \simeq 15$

c_{K^+}(axoplasm) $\simeq 15\ c_{K^+}$(exoplasm).

(c) $\mu_{Cl_2} + \mu_{CN^-} > \mu_{ClCN} + \mu_{Cl^-}$, 0, less, 0,

$\mu = \mu^\circ + RT\ln a$, $a = \exp\{(0 - 7\times 10^3)/RT\}$

$a_{Cl_2}(aq) = 0.06$, $c_{Cl_2}(aq) \simeq a_{Cl_2}(aq) = 0.06$ mol dm^{-3}

(d) $CaF_2(s) = Ca^{2+}(aq) + 2F^-(aq)$, $\Delta\bar{G}^\circ = 49$ kJ mol^{-1}. At equilibrium, $\Delta\bar{G} = 0 = \Delta\bar{G}^\circ$ $+ RT\ln a_{Ca^{2+}}a_{F^-}^2/a_{CaF_2}$.

Let solubility of CaF_2 be S mol dm^{-3}

$a_{Ca^{2+}} \approx S$, $a_{F^-} \approx 2S$, $a_{CaF_2} = 1$

$0 = 49\times 10^3 + 8.314\times 298\ \ln 4S^3$

$\therefore S = 8.7\times 10^{-4}$, solubility of $CaF_2 = 8.7\times 10^{-4}$ mol dm^{-3}.

(e) w_{max}(available) $= -\Delta\bar{G} = -\Delta\bar{G}^\circ = -614$ kJ mol^{-1}.

Chapter 7. Electrochemical Processes and Potentials

Problem 7.1.

(a) $\begin{matrix} H_2(g, pure, P^\circ) \\ +0.35\ HCl\ (aq, c_R) \end{matrix} = \begin{matrix} H_2(g, p_{H_2}) \\ +0.35\ HCl\ (aq, c_l) \end{matrix}$

(b) $$E_{cell} = E^\circ + \frac{RT}{-2F}\ln\frac{a_{H_2(g,pH_2)}a^{0.35}_{HCl(aq,c_L)}}{a_{H_2(g,p^\circ)}a^{0.35}_{HCl(aq,cR)}}$$

(c) $E^\circ_{cell} = 0$

(d) $$E_{cell} = 0 + \frac{RT}{-2F}\ln\left(p_{H_2(g)}/P^\circ\right)$$

(e) If $T = 298.15$ K, $E_{cell} = 88.6$ mV corresponds to 101 Pa.

Problem 7.2.

(a) $Cd(s) + Hg_2SO_4(s) = CdSO_4\frac{8}{3}H_2O(s) + 2Hg(l) + \frac{8}{3}H_2O$

(b) $$E = E^\circ + \frac{RT}{-2F}\ln\frac{a_{CdSO_4\,8/3\,H_2O(s)}a_{Hg(l)}}{a_{Cd(s)}a_{Hg_2SO_4(s)}a^{8/3}_{H_2O}}$$

$$= E^\circ + \frac{4RT}{3F}\ln a_{H_2O},$$ other as being unity.

(c) Provided that the aqueous solutions are at equilibrium (saturated with $CdSO_4$), the activity of water has a definite value for any given temperature. So also has the e.m.f.

Problem 7.3.

(a) $$E = E^\circ + \frac{RT}{-2F}\ln\frac{a_{Cu(s,pure)}a^{0.5}_{CuSO_4(aq,L)}}{a_{Cu(s,raw)}a^{0.5}_{CuSO_4(aq,R)}}$$

(b) $\Delta G^\circ = 0$, thus $E^\circ = 0$

(c) The inert solid particles will simply be uncovered and fall as sludge. The raw copper will have the same chemical activity as pure copper. $E_{cell} = E^\circ = 0$.

(d) $w_{\ddagger} = 2\,F \times 1\text{V} = 2 \times 96.5\ \text{kC mol}^{-1} \times 1\ \text{J C}^{-1}$
$= 193\ \text{kJ mol}^{-1}$.

This work is all "wasted" in order to obtain the required product at a reasonable rate.

Exercise 7.3.

(a) $y_{K^+} = y_{Cl^-} = y_1 = (0.49)^{\frac{1}{2}}$

(b) $y_{Na^+} \approx y_1,\ y_{Mg^{2+}} \approx y_{SO_4^{2-}} \approx y_1{}^4,\ y_{PO_4^{3-}} \approx y_1{}^9$

Exercise 7.4.

$$y_{Na^+}y_{Cl^-} = y_\pm^2(\text{NaCl}) = 0.82^2 = 0.672$$
$$y_{Ba^{2+}}\,y_{Cl^-}^2 = y_\pm^3(\text{BaCl}_2) = 0.55^3 = 0.166$$
$$y_{H^+}^2 y_{SO_4^{2-}} = y_\pm^3(H_2SO_4) = 0.34^3 = 0.0393$$
$$y_{Zn^{2+}}\,y_{SO_4^{2-}} = y_\pm^2(\text{ZnSO}_4) = 0.20^2 = 0.0400$$

Exercise 7.5.

$$a_{NaOH} = a_{Na^+}a_{HO^-} = (\gamma m/m^\circ)_{Na^+}(\gamma m/m^\circ)_{HO^-}$$

with $m_{Na^+} = m_{HO^-} = 17\ \text{mol (kg H}_2\text{O)}^{-1}$,
$m^\circ = 1\ \text{mol (kg H}_2\text{O)}^{-1}$

and $\gamma_{Na^+}\gamma_{HO^-} = \gamma_\pm^2(\text{NaOH}) = 15.82^2$

Thus $a_{NaOH}(aq,\ 17\ \text{mol (kg H}_2\text{O)}^{-1}) = 72.3 \times 10^3$
$a_{H_2SO_4} = 1.604^3 \times (2 \times 17)^2 \times 17 = 81.1 \times 10^3$
$a_{LiCl} = 43.8^2 \times 17 \times 17 = 554 \times 10^3$.

Problem 7.4.

(a) $Pb(s) + PbO_2(s) = PbSO_4(s,L) + PbSO_4(s,R) + 2H_2O$
$+ 1.2\ H_2SO_4(aq,L)$
$+ 0.8\ H_2SO_4(aq,R)$

(b) $E = E^\circ + \dfrac{RT}{-2F}\ln\dfrac{a_{H_2O}}{a_{H^+}^4 + a_{SO_4^{2-}}^2}$ when the composition of sulphuric acid is the same at left, right and in between.

(c) $\Delta\bar{G}^\circ = -zFE^\circ = -2 \times 96.5\ \text{kC mol}^{-1} \times 2.041\ \text{V}$

(d) When the cell has been rested, the solutions at right and left have the same compositions and activities of the ionic species
$2.100\ \text{V} = 2.041\ \text{V} + (RT/-2F)\ln(a_{H_2O}^2/a_{H^+}^4 a_{SO_4^{2-}}^2)$
Thus $a_{H^+}^2 a_{SO_4^{2-}}/a_{H_2O} = 10$. Assuming that
$a_{H_2O} \approx 1.0,\ a_{H^+}^2 a_{SO_4^{2-}} \simeq 10$.

(e) In the recharging process the electrochemical reaction written in (a) is reversed. Sulfuric acid accumulates at the sides of the cell before diffusing to the junction region. So the activities to be used in the Nernst equation may be considerably higher than the activities in the bulk of the solution.

Exercise 7.6.

(a) $Al^{3+} + 3e^- = Al(s)$

(b) $Fe(CN)_6^{3-} + e^- = Fe(CN)_6^{4-}$

(c) $MnO_4^- + 8H^+ + 5e^- = Mn^{2+} + 4H_2O$

(d) $\begin{matrix} CHCOOH \\ \| \\ CHCOOH \end{matrix} + 2H^+ + 2e^- = \begin{matrix} CH_2{-}COOH \\ | \\ CH_2{-}COOH \end{matrix}$

(e) $CO_2 + 5H_2O + 6e^- = CH_3OH + 6OH^-$

Exercise 7.7.

The molar charge of barium ion is +2F and the molar charge of chloride ion is −1F. In the left to right ionic transport process coupled with the cell reaction the molar charge transfer is

$$dQ/d\xi = \nu_{Ba^{2+}}\bar{Q}_{Ba^{2+}} + \nu_{Cl^-}\bar{Q}_{Cl^-}$$
$$= +0.22 \times (+2F) + (-0.56) \times (-1F) = +1.00F.$$

Problem 7.9.

$Zn(s)|ZnSO_4(aq)|ZnSO_4(aq)|ZnSO_4(aq)|Hg_2SO_4(s)|Hg(l)$.

Problem 7.10.

An ionic conductor containing solvated protons, a supply of gaseous hydrogen and an electronic conducting material with a surface which catalyses the electrode process.

(a) The electrode reaction is $2H^+(aq) + 2e^- = H_2(g)$.
The Nernst equation is $E = E° + (RT/-2F)\ln a_{H_2(g)}/a^2_{H^+(aq)}$.
Here $E/V = 0 + (0.0591/-2)\log 1/(10^{-8})^2 = -0.473$

(b) The hydrogen ion activities will become equal, if chemical equilibrium is reached.

Problem 7.11.

$c_{Cu^{2+}}$	1.00	0.91	0.61	0.31	0.11	0.01
$E_{Cu/Cu^{2+}}$	0.34	0.34	0.33	0.32	0.31	0.28 V
$c_{Zn^{2+}}$	0.01	0.10	0.40	0.70	0.90	1.0
$E_{Zn/Zn^{2+}}$	−0.82	−0.79	−0.77	−0.76	−0.76	−0.76 V
E_{cell}	1.16	1.13	1.10	1.08	1.07	1.04 V

Problem 7.12. $Zn(s)|Zn^{2+}(aq)||H^+(aq)|H_2(g)$

(a) $2H^+(aq) + 2e^- = H_2(g)$
$Zn^{2+}(aq) + 2e^- = Zn(s)$

(b) $2H^+(aq) + Zn(s) = Zn^{2+}(aq) + H_2(g)$

(c) $E^\circ_{cell} = E^\circ_R - E^\circ_L = 0.000 - (-0.763)V = 0.763$ V

(d) $\Delta\bar{G}° = -2FE° = -147.3$ kJ mol^{-1}.

(e) L to R

$Zn(s)|Zn^{2+}(aq)||Ag^+(aq)|Ag(s)$

(a) $Ag^+(aq) + e^- = Ag(s)$
$Zn^{2+}(aq) + 2e^- = Zn(s)$

(b) $2Ag^+(aq) + Zn(s) = 2Ag(s) + Zn^{2+}(aq)$

(c) $E^\circ_{cell} = +0.799 - (-0.763)V = 1.562$ V

(d) $\Delta\bar{G}° = -2FE° = -301.5$ kJ mol^{-1}

(e) L to R

$Ag(s)|Ag^+(aq)||H^+(aq)|H_2(g)$

(a) $2H^+(aq) + 2e^- = H_2(g)$
$Ag^+ + e^- = Ag(s)$

(b) $2H^+ + 2Ag(s) = H_2(g) + 2Ag^+(aq)$

(c) $E^\circ_{cell} = 0.000 - 0.799\ V = -0.799\ V$

(d) $\Delta\bar{G}^\circ = -2FE^\circ = +154.2\ kJ\ mol^{-1}$

(e) R to L.

Problem 7.13.

Electrode reaction: $Fe^{3+}(aq) + e^- = Fe^{2+}(aq)$

Nernst equation: $E = E^\circ + (RT/-1F)\ln(a_{Fe^{2+}}/a_{Fe^{3+}})$

$= +0.771V - 0.0591V \times \log(0.9 \times 10^{-4}/0.9 \times 2.0)$

$= 0.771 + 0.254 = 1.025V.$

Problem 7.14.

The hydrogen electrode is arbitrarily assigned zero value of standard electrode potential. All voltages are measured relative to this reference value.

Problem 7.15.

(a) $Ag(s)\,|\,AgCl(s)\,|\,Cl^-(aq)$

$Na(in\ Hg)\,|\,Na^+(aq)$

(b) $AgCl(s) + e^- = Ag(s) + Cl^-(aq)$

$Na^+(aq) + e^- = Na(in\ Hg)$

(c) $AgCl(s) + Na(in\ Hg) = Ag(s) + Cl^-(aq) + Na^+(aq)$

(d) $E = E^\circ + (RT/-F)\ln a_{Na^+(aq)}a_{Cl^-(aq)}/a_{Na(in\ Hg)}$

(e) $\mu_{AgCl(s)} - 1F\ \ _R = \mu_{Ag(s)} + \mu_{Cl^-(aq)}$,

$\mu_{Na^+(aq)} - 1F\ \ _L = \mu_{Na(in\ Hg)}$.

Chapter 8. Energies of Chemical Materials

Exercise 8.1. Because numerous approaches may be used in analysing these situations we do not wish to write out a particular "ideal" answer. The reader would benefit from discussing his own analysis with a colleague or tutor.

Exercise 8.2. As above.

Exercise 8.3.

U (acid layer) $= \bar{U}_{HNO_3} \times n_{HNO_3} + \bar{U}_{H_2SO_4} \times n_{H_2SO_4} + \ldots$

U (gas phase) $= \bar{U}_{N_2} \times n_{N_2} + \bar{U}_{O_2} \times n_{O_2} + \bar{U}_{CO_2} \times n_{CO_2} + \ldots$

Problem 8.1.

(a) The system is closed. There is no transfer of energy to the environment by material. The change in volume of the system is negligible. There is no direct transfer of energy to the environment by working. The useful work is 80% of the maximum, $-\Delta G$. That is, energy transfer by work done on external devices is $0.80 \times 259\ kJ\ mol^{-1} = 207\ kJ\ mol^{-1}$. The energy transferred to the environment by heating must be $258 - 207 = 51\ kJ\ mol^{-1}$.

(b) When there is no power production (useful work) the energy transferred to the environment by heating must be $-\Delta E = +258\ kJ\ mol^{-1}$.

Exercise 8.6. See page 150 for equations

(a) $\Delta\bar{U}\{2\} = \bar{U}_{p\text{-nitrotoluene}(l)} - \bar{U}_{toluene(l)}$

$+ \bar{U}_{H_2O(in\ H_2SO_4)} - \bar{U}_{HNO_3(in\ H_2SO_4)}$

(b) (i) $d\xi\{1\}/d\xi\{2\} = 65/35$

(ii) $$\frac{-dU}{dn_{roluene}} = 0.65\bar{U}_{o-nitrotoluene} + 0.35\bar{U}_{p-nitrotoluene} + \bar{U}_{H2O} - \bar{U}_{toluene} - \bar{U}_{HNO3}$$

Problem 8.2.

$$0 = CuSO_4(s) + 5H_2O(g) - CuSO_45H_2O(s)$$

$$\Delta\bar{H} = -770 + 5\times(-242) - (-2278)\ \text{kJ mol}^{-1} = +298\ \text{kJ mol}^{-1}$$

$$\Delta E = \Delta(H - PV) = w_{♀} - P\Delta V + q$$

$$\Delta H = w_{♀} + q, \text{ when } P = P_{\text{envir.}}$$

$$\Delta H = q, \text{ when } w_{♀} = 0. \quad \text{Thus } q = 298\ \text{kJ mol}^{-1} \times 2\ \text{mol.}$$

Exercise 8.8.

$$\Delta\bar{H}\{H_2O(s) = H_2O(l)\} = \bar{H}\{H_2O(l)\} - \bar{H}\{H_2O(s)\}$$

$$\Delta\bar{H}\{I_2(s) = I_2(g)\} = \bar{H}\{I_2(g)\} - \bar{H}\{I_2(s)\}$$

$$\Delta\bar{H}\{NaCl(s) = NaCl(aq)\} = \bar{H}\{Na^+(aq)\} + \bar{H}\{Cl^-(aq)\} - \bar{H}\{NaCl\,(s)\}$$

Exercise 8.9. See page 150 for equations

(a) $\Delta\bar{H}\{2\} = \bar{H}_{\text{p-nitrotoluene(l)}} - \bar{H}_{\text{toluene(l)}}$

$\qquad + \bar{H}_{H_2O(\text{in } H_2SO_4)} - \bar{H}_{HNO_3(\text{in } H_2SO_4)}$

(b) (i) $d\xi\{1\}/d\xi\{2\} = 65/35$

(ii) $$-\frac{dH}{dn_{\text{toluene}}} = 0.65\,\bar{H}_{\text{o-nitrotoluene}} + 0.35\,\bar{H}_{\text{p-nitrotoluene}} + \bar{H}_{H_2O} - \bar{H}_{\text{toluene}} - \bar{H}_{HNO_3}$$

Problem 8.3.

(a) $\Delta\bar{H}(a) = \Delta\bar{H}_f\{CH_4(g)\} \times 1 + \Delta\bar{H}_f^\circ\{C(\text{graphite})\} \times (-1) + \Delta\bar{H}_f^\circ\{H_2(g)\} \times (-2)$

$\qquad = -75 \times 1 + 0 \times (-1) + 0 \times (-2) = -75\ \text{kJ mol}^{-1}$

(b) $\Delta\bar{H}(b) = -85 \times 1 + 227 \times (-1) + 0 \times (-2) = -312\ \text{kJ mol}^{-1}$

(c) $\Delta\bar{H}(c) = -3227.9\ \text{kJ mol}^{-1}$

(d) $\Delta\bar{H}(d) = -65\ \text{kJ mol}^{-1}$

(e) $\Delta\bar{H}(e) = +11\ \text{kJ mol}^{-1}$

(f) $\Delta\bar{H}(f) = \Delta\bar{H}_f^\circ\{H^+(aq)\} \times 2 + \Delta\bar{H}_f^\circ\{Cl^-(aq)\} \times 2 + \Delta\bar{H}_f^\circ\{H_2(g)\} \times (-1) + \Delta H_f^\circ\{Cl_2(g)\} \times (-1$

$\qquad = 0 \times 2 + (-167) \times 2 + \text{Ox}(-1) \times \text{Ox}(-1) = -334\ \text{kJ mol}^{-1}$

Problem 8.7.

$$\Delta\bar{E} = \Delta\bar{M}\,c^2, \Delta\bar{M} = 0.29 \times 10^{-10}\text{kg mol}^{-1}.$$

Appendix:

Quantities and symbols used frequently in Physical Chemistry

A.1 Basic Physical Quantities

	symbol	*unit*	*symbol for unit*
number of particles	N	one	
cartesian coordinates	x,y,z	metre	m
time	t	second	s
mass	m	kilogram	kg
electric charge	Q	coulomb	C
temperature (thermodynamic)	T	kelvin	K
(Celsius temperature $t/°C \equiv T/K - 273.15$)			
amount of the chemical species, B (mole number of B)	n_B	mole	mol

A.2 Related Physical Quantities for Bodies of Matter

	symbol	*definition*	*unit*
length	l		m
height	h		m
displacement (along path)	ds	$\lvert dx,dy,dz \rvert$	m
area			
– of cross section	A	$A_{xy} \equiv l_x l_y$	m^2
– of surface	S		

volume	V		m^3
velocity	v	$v_x = dx/dt$	ms^{-1}
acceleration	a	$a_x = dv_x/dt$	ms^{-2}
force	F	$F_x = ma_x$	newton, $N \equiv kg\,m^{-2}$
pressure	P	$P = F/A$	pascal, $Pa \equiv N\,m^{-2}$
work	w	$F.ds$	joule, $J \equiv N\,m$

A.3 Physical Constants

velocity of light $c = 2.997\,924\,580 \times 10^8\ ms^{-1}$

Planck constant $h = 6.626\,176 \times 10^{-34}$ Js

Avogadro constant $L = 6.022\,045 \times 10^{23}\ mol^{-1}$

Faraday constant $F = 96.484\,56 \times 10^3\ C\ mol^{-1}$

Molar gas constant $R = 8.314\,41\ JK^{-1}\ mol^{-1}$

Relative atomic masses ("atomic weights") — see tables of "atomic weights" based on ^{12}C

Gravitational acceleration $g = 9.81\ N\ kg^{-1}$ near earth's surface

A.4 General Chemical Symbols and Equations

formula for a chemical substance or molecular species	B (solvent may be A)
phase types — general	(β) (or (α))
gaseous	(g)
liquid	(l)
solid	(s)
aqueous	(aq)
amount of B	n_B (mole number)
change in amount of B	$\Delta n_B = (n_B)_{II} - (n_B)_I$
stoichiometric coefficient of B	ν_B $\Delta n_A/\nu_A = \Delta n_B/\nu_B$
equation for chemical reaction	$0 = \Sigma_B \nu_B B$
change in extent of reaction	$\Delta\xi = \Delta n_B/\nu_B$
total mole number	$n = \Sigma_B n_B$
mole fraction of B	$x_B = n_B/n$
mass fraction of B	$w_B = m_B/m$
molality of B in solution	$m_B = n_B/m_{solvent}$
concentration of B	$c_B = n_B/V$
partial pressure of B in gaseous mixture	$p_B = x_B P$

specific volume	$v = V/m$
density of mass	$\rho = m/V$
molar volume of a material	$V = V/n$
molar volume of pure B	$\bar{V}_B = V/n_B$ given T,P
molar volume of B in solution with A	$\bar{V}_B = (dV/dn_B)$ given n_A, T, P
molar volume of B in dilute solution in A	$\bar{V}_{B \text{ in } A} = (dV/dn_B)$ given n_A, T, P, as n_B increases from zero

molar volume of gaseous material

$$P\bar{V} = RT + B/\bar{V} + C/\bar{V}^2 + .., \quad \text{given } T$$
or
$$P\bar{V} = RT + B^*P + C^*P^2 + .., \quad \text{given } T$$
or
$$P\bar{V} = ZRT \, (Z = \text{compression factor})$$

equation for transfer of B from phase (α) to phase (β)	$B(\alpha) = B(\beta)$
boiling point	T_b at given P
freezing point	T_f at given P
equilibrium vapor pressure	P_{eq} at given T
critical point	T_c, P_c fixed for pure substances
triple point	T_t, P_t fixed for pure substances
gradient of equilibrium line for phase transfer	$(dP/dT)_{eq} = \triangle\bar{H}/(T\triangle\bar{V})$
freezing point depression	$\triangle T_f \simeq -k_f x_B \simeq -k_f' m_B$
boiling point elevation	$\triangle T_b \simeq k_b x_B \simeq k_b' m_B$
vapor pressure of solvent	$p_A \simeq p_A^\circ x_{A(l)}$, Raoult's law
vapor pressure of solute	$p_B \simeq K x_{B(soln)}$, Henry's law
osmotic pressure of solution	$\pi \simeq RTc_B$
distribution of solute B between solvents A and A'	$c_{B(A')}/c_{B(A)} \simeq K$
number of substances	$\mathscr{B}$
number of independent reactions	$\mathscr{R}$
number of independent components	$\mathscr{C} = \mathscr{B} - \mathscr{R}$
number of phases	$\mathscr{P}$
number of degrees of freedom	$\mathscr{F} = \mathscr{C} + 2 - \mathscr{P}$ (phase rule)
absolute activity of B	λ_B
tendency for reaction	$\Pi\lambda_{reactants} - \Pi\lambda_{products}$
relative activity of B	a_B $\lambda_B = \lambda_B^\circ a_B$, given (β, T, c°)
activity coefficient of B	y_B $a_B = (yc/c^\circ)_B$ $c^\circ = 1$ mol dm^{-3}
activity coefficient of B	γ_B $a_B = (\gamma m/m^\circ)_B$ $m^\circ = 1$ mol B (kg$-$A)$^{-1}$

activity coefficient of B	f_B	$a_B = (fx/x^\circ)_B$, $x^\circ = 1$,
	or	$a_{B(g)} = \left\| \frac{fx}{x^\circ} \frac{P}{P^\circ} \right\|_{B(g)}$
		$x^\circ = 1$, $P^\circ = 101.3 \times 10^3$ Pa
equilibrium constant of reaction	K_{eq}	$\equiv 1/\{\Pi_B \lambda_B^{\circ\ \nu_B}\}$ = equilibrium value of $\Pi_B a_B^{\nu_B}$
rate of reaction or transfer		$d\xi/dt = (1/\nu_B)(dn_B/dt)$
rate of reaction divided by volume		$rate/V = (1/V) d\xi/dt$
flux of B	J_B	= rate of transfer of B divided by area
rate coefficient	k	$rate/V \simeq k c_A^{a} c_B^{b} c_X^{x} \ldots$
formula of transition species	$(\ddagger)$	$(A_aB_bX_x)^{\ddagger}$
rate constant	k°	$rate/V = \dfrac{k^\circ a_A^{a} a_B^{b} a_X^{x} (c^\circ)^{a+b+x}}{y_{\ddagger}}$
net rate/V of elementary reaction $aA + bB \rightleftharpoons cC + dD$		$rate/V = k_f^\circ a_A^{a} a_B^{b} (c^\circ)^{a+b}/y_{\ddagger} - k_r^\circ a_C^{c} a_D^{d} (c^\circ)^{c+d}/y_{\ddagger}$
reference transition frequency		$RT/Lh \approx 10^{13}\ s^{-1}$ at 298 K
absolute rate equation for elementary step, or $aA + aB \rightarrow$ products		$rate = \dfrac{RT}{Lh} \left\{ \dfrac{c^\circ V}{\lambda^\circ y} \right\}_{\ddagger} \lambda_A^{a} \lambda_B^{b}$ $rate/V = \dfrac{RT}{Lh} K_{\ddagger} a_A^{a} a_B^{b} (c^\circ)/y_{\ddagger}$
potential for transfer of B from (α) to (β)	$\mu_{B(\beta)} - \mu_{B(\alpha)}$	$\dfrac{w_{\text{♀}}(minimum)}{dn_B(\alpha \rightarrow \beta)}$
potential of neutral species B in (β)	$\mu_{B(\beta)}$	$\mu_{B(\beta)}^\circ + RT \ln a_{B(\beta)} + \bar{M}_B (gh)_\beta$
standard potential of B in β	$\mu_{B(\beta)}^\circ$	$RT \ln \lambda^\circ_{B(\beta)}$, given T, x°, P°
reference potentials	zero	elements and H^+ in reference states at given T
potential for chemical reaction	$\Delta \bar{G}$	$\Sigma_B \nu_B \mu_B$
Gibbs energy of a phase (β)	G_β	$\Sigma_B \mu_B n_B$
Gibbs energy of a system	G	$\Sigma_\beta G_\beta$
equation for transfer of electrons through wiring	$e^-(L) = e^-(R)$	L = "left", R = "right"

electric current through wiring	$I_{L\to R}$	$= \mathrm{d}Q_{L\to R}/\mathrm{d}t$ $= -1F\mathrm{d}n_{e^-(L\to R)}/\mathrm{d}t$
terminal voltage	$\Delta\mathscr{V} = \mathscr{V}_R - \mathscr{V}_L$	$= \dfrac{w_{♀}\text{(minimum)}}{\mathrm{d}Q_{L\to R}}$
potential of electrons in phase (M)	$\mu_{e^-(M)} = -1F$	
electromotive force	E_{cell}	$= \Delta\mathscr{V}$ at $I = 0$
equation for electrochemical process	$0 = \Sigma_B \nu_B B$	including charged species
electron number or faraday number	z	for given equation for electrochemical process
potential for electrochemical process	$\Delta\bar{G}$	$= -zFE_{cell} = \Sigma\nu_B(\mu^\circ + RT\ln a)_B$
equation for transfer of charged species, B^{z+}	$B^{z+}(L) = B^{z+}(R)$	left and right hand ionic conducting phases
electric potential in a phase (β)	$\Phi_{(\beta)}$	
electrochemical potential of B^{z+} in phase (β)	$\mu_{B^{z+}(\beta)}$	$(\mu^\circ + RT\ln a)_{B^{z+}(\beta)} + zF\Phi_\beta$
equation for electrode reaction	$ze^-(M) = \Sigma_B\nu_B B$	involving charged species
electrode potential	$E_{electrode}$	$\Sigma_B\nu_B(\mu^\circ + RT\ln a)_B/(-zF)$
equation for transfer of 1F across junction	$\Sigma_B\nu_B\{B^{z+}(L) = B^{z+}(R)\}$	$\Sigma\,\nu_B z_B F = 1F$
junction potential	$E_{junction}$ or $\Delta\Phi_J$ at $I = 0$	$\Sigma\quad \nu_B\{(\mu^\circ + RT\ln a)_{B^{z+}(R)}$ $-(\mu^\circ + RT\ln a)_{B^{z+}(L)}\}$
cell potential (electromotive force)	$E_{cell} = E_{electrode,R} - E_{electrode,L} + E_{junction}$	

A.5 Table of Atomic Weights, 1973

(scaled to relative atomic mass, $A_r(^{12}C) = 12$)

The values of $A_r(E)$ given here apply to elements as they exist in materials of terrestrial origin and to certain artificial elements.

Element	*Atomic weight*	*Element*	*Atomic weight*
Actinium	—	Mercury	200.5_9
Aluminium	26.98154	Molybdenum	95.9_4
Americium	—	Neodymium	144.2_4
Antimony	121.7_5	Neon	20.17_9
Argon	39.94_8	Neptunium	237.0482
Arsenic	74.9216	Nickel	58.70
Astatine	—	Niobium	92.9064
Barium	137.3_4	Nitrogen	14.0067
Berkelium	—	Nobelium	—
Beryllium	9.01218	Osmium	190.2
Bismuth	208.9804	Oxygen	15.999_4
Boron	10.81	Palladium	106.4
Bromine	79.904	Phosphorus	30.97376
Cadmium	112.40	Platinum	195.0_9
Caesium	132.9054	Plutonium	—
Calcium	40.08	Polonium	—
Californium	—	Potassium	39.09_8
Carbon	12.011	Praseodymium	140.9077
Cerium	140.12	Promethium	—
Chlorine	35.453	Protactinium	231.0359
Chromium	51.996	Radium	226.0254
Cobalt	58.9332	Radon	—
Copper	63.54_6	Rhenium	186.207
Curium	—	Rhodium	102.9055
Dysprosium	162.5_0	Rubidium	85.467_8
Einsteinium	—	Ruthenium	101.0_7
Erbium	167.2_6	Samarium	150.4
Europium	151.96	Scandium	44.9559
Fermium	—	Selenium	78.9_6
Fluorine	18.99840	Silicon	28.08_6
Francium	—	Silver	107.868
Gadolinium	157.2_5	Sodium	22.98977
Gallium	69.72	Strontium	87.62
Germanium	72.5_9	Sulphur	32.06
Gold	196.9665	Tantalum	180.947_9
Hafnium	178.4_9	Technetium	—
Helium	4.00260	Tellurium	127.6_0
Holmium	164.9340	Terbium	158.9254
Hydrogen	1.0079	Thallium	204.3_7
Indium	114.82	Thorium	232.0381
Iodine	126.9045	Thulium	168.9342
Iridium	192.2_2	Tin	118.6_9
Iron	55.84_7	Titanium	47.9_0
Krypton	83.80	Tungsten	183.8_5
Lanthanum	138.905_5	Uranium	238.029
Lawrencium	—	Vanadium	50.941_4
Lead	207.2	Xenon	131.30
Lithium	6.94_1	Ytterbium	173.0_4
Lutetium	174.97	Yttrium	88.9059
Magnesium	24.305	Zinc	65.38
Manganese	54.9380	Zirconium	91.22
Mendelevium	—		

Index